De Gruyter Graduate

Baumslag, Fine, Kreuzer, Rosenberger • A Course in Mathematical Cryptography

Gilbert Baumslag, Benjamin Fine,
Martin Kreuzer, Gerhard Rosenberger

A Course in Mathematical Cryptography

—

DE GRUYTER

Mathematics Subject Classification 2010
94A62, 11T71, 68P25, 13P10, 20F10, 20F65, 20F36

Authors

Prof. Dr. Gilbert Baumslag †

Prof. Dr. Benjamin Fine
Fairfield University
Department of Mathematics
1073 North Benson Road
Fairfield, CT 06430
USA
fine@fairfield.edu

Prof. Dr. Martin Kreuzer
Universität Passau
Fakultät für Informatik und Mathematik
Innstr. 33
94032 Passau
Germany
martin.kreuzer@uni-passau.de

Prof. Dr. Gerhard Rosenberger
Universität Hamburg
Fachbereich Mathematik
Bundesstr. 55
20146 Hamburg
Germany
gerhard.rosenberger@math.uni-hamburg.de

ISBN 978-3-11-037276-2
e-ISBN (PDF) 978-3-11-037277-9
e-ISBN (EPUB) 978-3-11-038616-5

Library of Congress Cataloging-in-Publication Data
A CIP catalog record for this book has been applied for at the Library of Congress.

Bibliographic information published by the Deutsche Nationalbibliothek
The Deutsche Nationalbibliothek lists this publication in the Deutsche Nationalbibliografie;
detailed bibliographic data are available on the Internet at http://dnb.dnb.de.

www.degruyter.com

Preface

Historically, secret codes and ciphers were placed in the realm of espionage and diplomacy. Although some people considered the mathematics of devising and breaking codes, it remained for a long time a discipline on the fringes of mathematics. Several things changed this view. First, sophisticated mathematical techniques were developed during the Second World War to aid in the cryptanalysis of the Enigma code and other war time ciphers. Then the widespread usage of computers and the advent of the internet led to the need of sending financial and other sensitive information over public channels. This sparked an intensive development of mathematical cryptography, both symmetric and public key.

Traditionally, **cryptography** refers to the science and/or art of devising and implementing secret codes and ciphers, while **cryptanalysis** is the science and/or art of breaking them. The whole disciple, cryptography plus cryptanalysis, is usually called **cryptology**. Sometimes however, cryptography is used in place of cryptology.

At present most people are familiar with the phrase *"you are now entering an encrypted page ... "*. Cryptography has become essential as bank transactions, credit card information, contracts, and sensitive medical information are sent through insecure channels. Clearly, this must be done in encrypted form. This leads us to the concept of public key cryptography which deals with the problem of hiding secret data on open channels when every potential attacker knows the encryption method.

True public key methods were born with the rise of high speed computing and with the discovery by Diffie and Hellman in 1976 (actually known earlier to British Intelligence) of a true one-way function. The use of the computer was essential because arithmetic with very large numbers was necessary. A second public-key method, the RSA method, was developed by Rivest, Shamir, and Adleman a year later. These beginnings turned cryptography into a major discipline for both study and research in mathematics and in computer science. Today many, if not most, universities offer courses in cryptography. These courses are really of two different types. The first is geared to the mathematical aspects of the subject while the second deals with the computer science and engineering aspects, that is how to implement a mathematical cryptosystem or cryptanalysis on a computer or other physical devices.

This book is concerned with the mathematical, especially algebraic, aspects of cryptography. It grew out of many courses presented by the authors over the past twenty years at various universities: City University of New York for Gilbert Baumslag, Fairfield University, City University of New York, and the University of Dortmund for Ben Fine, University of Regensburg, University of Dortmund, and University of Passau for Martin Kreuzer, and University of Dortmund, University of Passau, and the University of Hamburg for Gerhard Rosenberger. These courses covered a wide range of topics within cryptography and were presented at a variety of levels. The authors have had numerous Ph.D. and Masters level students in the discipline which has become an extremely popular area. In the following 15 chapters we cover a wide range of topics in mathematical cryptography. They are primarily geared towards graduate students and advanced undergraduates in mathematics and computer science, but may also be of interest to researchers in the area. We feel that this book could be a suitable text for first and second year Master level courses which may even include elliptic curve methods, group based techniques, as well as cryptography using Gröbner bases.

The book can be roughly divided into four parts. The first part, consisting of Chapters 1 to 4, covers the general ideas of cryptography and cryptanalysis. Starting from the most basic constructions, we define all the relevant material: cryptosystems, key spaces, as well as encryption and decryption maps. We also discuss the distinction between symmetric and asymmetric encryption and describe the basic outline of public key cryptography. Moreover, we introduce statistical cryptanalysis and point out some ideas from complexity theory. In Chapter 4 we present additional cryptographic protocols such as authentication, digital signatures and zero-knowledge proofs.

Part two, comprising Chapters 5 through 8, deals with number theoretic cryptographic methods. Chapters 5 and 6 introduce the fundamental ideas needed from number theory. Then, in Chapter 7, we apply these to develop common number theoretic public key cryptosystems and protocols: Diffie-Hellman, RSA, ElGamal, and Rabin. In Chapter 8 we provide a detailed explanation of elliptic curves and elliptic curve cryptography.

Chapters 9 to 12 form the third part which presents group-based cryptography. It deals with cryptosystems and crytographic protocols whose platforms are non-abelian groups. Combinatorial group theory plays the central role here, and we cover the necessary material on finitely presented groups in Chapter 9. In Chapter 10 we introduce free group cryptography and the Wagner-Magyarik method which was the seminal idea in the discipline. We also discuss in that chapter the use of matrix groups and a secure back-up password protocol that utilizes combinatorial group theory. In the first half of Chapter 11 we introduce the best-known public-key group based cryptosystems: the Ko-Lee protocol and the Anshel-Anshel-Goldfeld protocol. Braid groups were suggested as potential platforms for these protocols. So, the second half of Chapter 11 is concerned with the general theory of braid groups and braid-group cryptography. In Chapter 12 we collect some further suggestions for using finitely presented groups in a cryptographic setting.

The final part, consisting of Chapters 13, 14, and 15 deals with Gröbner basis and lattice-based methods. Chapter 13 starts by introducing Gröbner bases of commutative polynomial ideals and Buchberger's Algorithm to compute them. They are used both in commutative Gröbner basis cryptosystems and in some techniques of algebraic cryptanalysis explained in the second part of this chapter. In Chapter 14 we extend Gröbner bases to non-commutative polynomial rings and even two-sided free modules over these. Thereby quite general Gröbner basis cryptosystems can be constructed which include most of the methods presented in the earlier chapters. In the last chapter, Chapter 15, we give a short overview of lattices and lattice-based cryptography.

Special thanks are due to many of our students who, over the years, have looked at and made suggestions on the evolving versions of this book. We are especially grateful to Anja Moldenhauer for providing excellent and detailed help with both typesetting and editing parts of this book. Further, we cordially thank our wives and families whose massive amounts of patience and support were absolutely indispensable for the creation of a voluminous monograph such as this one.

While this book was being completed, tragically our colleague, co-author and friend, Gilbert Baumslag, died. Gilbert was one of the foremost researchers in the world in infinite group theory and developed many computational techniques for finitely presented groups. Over the past decade he had been active in transferring this knowledge to group-based cryptography. It is our hope that this book honors his memory and his contribution.

Fairfield, Passau, and Hamburg,
in January 2015,

Benjamin Fine,
Martin Kreuzer,
and Gerhard Rosenberger

Contents

1 Basic Ideas of Cryptography

1.1 Mathematical Cryptography

The subject of this book is **mathematical cryptography**. By this we mean the mathematics involved in cryptographic protocols. We will define, and make precise, all these terms as we proceed. As the field has expanded, using both commutative and non-commutative algebraic objects as cryptographic platforms, we felt that a book describing and explaining all these mathematical methods would be of considerable value.

Cryptography or **cryptology** is loosely the science of encrypting and decrypting secret codes, and the related task of breaking or uncovering secret codes. The science of cryptography touches on many other disciplines, both within mathematics and computer science and in engineering. In mathematics, cryptology uses, and touches on, algebra, number theory, graph and lattice theory, algebraic geometry and probability and statistics. Analysis of cryptographic security leads to using theoretical computer science especially complexity theory. The actual implementation of cryptosystems, and the hard work of carrying out security analysis for specific cryptosystems falls into engineering and practical computer science and computing. In this book we will look primarily at the first part, the mathematics of cryptographic protocols. We will not look at all at hardware implementation. In Section 1.2 we will present many of the terms and mathematical formulations in cryptology. Section 1.2 will be further expanded in Chapter 2.

Up until fairly recently, cryptography was mainly concerned with message confidentiality – that is sending secret messages so that interceptors or eavesdroppers cannot decipher them. The discipline was primarily used in military and espionage situations and as recently as the 1956 Encyclopedia Brittanica article on Cryptography said that there seemed to be only limited use in business and commerce. Two things changed all that. The first was the increased capability and use of computers and computing. This both allowed more complex encryption systems that could not be done by hand but could be on a computer, and required more complex encryption, since cryptanalysis, that is, code breaking, was enhanced by the computer. The second thing that skyrocketed the use of cryptographic methods was the discovery of workable one way functions that then allowed for public key cryptosystems. This allowed the transmission of sensitive data over public airwaves even though any potential attacker could view this data and further the attacker knew the encryption technique. In Section 1.3 we give a very brief history of cryptography while in Section 1.5 and then again in Chapter 2 we describe the basic ideas differentiating classical or symmetric key cryptography from public key cryptography.

An important aspect of cryptographic protocols are their security, that is the ability of the encryption to withstand attacks from unwanted adversaries. Since modern cryptography is done on a computer, cryptographic security must bring in ideas from computer science and complexity theory. We will present some of these ideas from a mathematical viewpoint in Chapter 3. The book [MSU1] by Myasnikov, Shpilrain and Ushakov provides a much more extensive discussion of complexity theory and its relationship and use relative to cryptography.

Traditionally, the main mathematical tools involved in cryptographic protocols were number theoretic. To encrypt an alphabet with N letters the letters were considered as modular integers $0, 1, 2, ..., N - 1$ in the modular ring $\mathbb{Z}_N$. Number theoretic functions on $\mathbb{Z}_N$ were then used. The main public key cryptographic methods, Diffie-Hellman and RSA, are based on supposedly hard number theoretic problems, the discrete logarithm problem and the factorization problem respectively. We touch on these ideas in Section 1.4 and then much more fully in Chapter 6. We will discuss the main traditional public key methods in Chapter 7. In an attempt to build cryptosystems with smaller necessary key spaces, algebraic geometry was introduced to cryptography. The concepts of elliptic curves and their corresponding elliptic curve groups were combined with the Diffie-Hellman concepts to build elliptic curve cryptography. We discuss elliptic curve methods in Chapter 8.

The traditional cryptographic methods, both symmetric key and public key, such as the RSA algorithm, Diffie-Hellman, and elliptic curve methods, are number theory based. Hence, from a theoretical point of view, they depend on the structure of abelian groups. Although there have been no successful attacks on the standard protocols, there is a feeling that the strength of computing machinery has made these techniques less secure. The big cloud in this direction are quantum algorithms and the possibility that a workable quantum computer can be built. Quantum algorithms can break the present versions of both Diffie-Hellman and RSA. We will briefly touch on quantum algorithms in Chapter 3.

As a result of this, there has been an active line of research to develop and analyze new cryptosystems and key exchange protocols based on non-commutative cryptographic platforms. This line of investigation has been given the broad title of **non-commutative algebraic cryptography**.

Up to this point the main sources for non-commutative cryptographic platforms have been non-abelian groups. In cryptosystems based on these objects, algebraic properties of the platforms are used prominently in both devising cryptosystems and in cryptanalysis. In particular the difficulty, in a complexity sense, of certain algorithmic problems in finitely presented groups, such as the conjugator search problem, has been crucial in encryption and decryption. We give an introduction to these group theoretic ideas in Chapters 9 and 10. Chapter 11 deals with the main public key methods using non-abelian groups, the Ko-Lee method and the Anshel-Anshel-Goldfeld method.

The main sources for non-abelian groups are combinatorial group theory and linear group theory. Braid group cryptography, where encryption is done within the classical braid groups, is one prominent example. The one-way functions in braid group systems are based on the difficulty of solving group theoretic decision problems such as the conjugacy problem and conjugator search problem. Although braid group cryptography had initially a lot of success, various potential attacks have been identified. Besides discussing braid groups and braid group cryptography, Chapter 11 introduces what is necessary for a general non-abelian cryptographic protocol. The study and cryptanalysis of potential platform groups has had a strong positive effect on both group theory and complexity theory. Motivated in large part by cryptography, there has been tremendous interest in **asymptotic group theory** and **generic properties**. We discuss these in several places in this book.

Chapter 12 describes further uses of non-abelian groups in cryptography. One idea that we will describe is to replace the social security number by a finitely presented group and then encode a persons' information securely within that group.

Gröbner bases have also been employed both to develop non-commutative cryptosystems and to cryptanalyze other systems that are polynomial based. Chapters 13 and 14 describe these Gröbner basis techniques.

The use of lattices as cryptographic platforms has also generated a generated a great deal of interest, primarily because these methods seem to be resistant to attacks by quantum algorithms. We provide an overview of lattice based cryptography in Chapter 15.

1.2 Cryptography, Cryptanalysis and Cryptosystems

In this section we start to introduce the basic terminology and mathematics used in cryptography. In later sections we will make these ideas more precise.

Cryptography refers to the science and/or art of sending and receiving coded messages. Coding and hidden ciphering is an old endeavor used by governments and militaries and between private individuals from ancient times. Recently it has become even more prominent because of the necessity of sending secure and private information, such as credit card numbers, over essentially open communication systems.

Traditionally **cryptography** is the science of devising and implementing secret codes or **cryptosystems**. **Cryptanalysis** is the science of breaking cryptosystems while **cryptology** refers to the whole field of cryptography plus cryptanalysis. In most modern literature cryptography is used synonymously with cryptology. Theoretically cryptography uses mathematics, computer science and engineering.

A **cryptosystem** or **code** is an algorithm to change a plain message, called the **plaintext message**, into a coded message, called the **ciphertext message**. In general both the **plaintext message** (uncoded message) and the **ciphertext message** (coded message) are written in some N-letter alphabet which is usually the same for both plaintext and code. The method of coding, or the encoding algorithm, is then a transformation of the N-letters. The most common way to perform this transformation is to consider the N letters as N integers modulo N and then perform a number theoretical function on them. Therefore most encoding algorithms use modular arithmetic and hence cryptography is closely tied to number theory. In this section we give a brief overview of cryptography and some number theoretic algorithms used in encryption. The subject is very broad, and as mentioned above, very current, due to the need for publicly viewed but coded messages. There are many references to the subject. The book by Koblitz [Ko1] gives an outstanding introduction to the interaction between number theory and cryptography. It also includes many references to other sources. The books by Buchmann [Buc] and Stinson [Sti2] describe the whole area.

Modern cryptography is usually separated into **classical cryptography**, also called **symmetric key cryptography**, and **public key cryptography**. In the former, both the encoding and decoding algorithms are supposedly known only to the sender and receiver, usually referred to as Bob and Alice. In the latter, the encryption method is public knowledge but only the receiver knows how to decode. We make this more precise in Chapter 6 when we discuss public key methods in detail. Here we present first the basic terminology used in classical cryptography.

The message that one wants to send is written in **plaintext** and then converted into code. The coded message is written in **ciphertext**. The plaintext message and ciphertext message are written in some **alphabets** that are usually the same. The process of putting the plaintext message into code is called **enciphering** or **encryption** while the reverse process is called **deciphering** or **decryption**. Encryption algorithms break the plaintext and ciphertext message into **message units**. These are single letters or pairs of letters or more generally k-vectors of letters. The transformations are done on these message units and the encryption algorithm is a mapping from the set of plaintext message units to the set of ciphertext message units. Putting this into a mathematical formulation we let

$$\mathcal{P} = \text{set of all plaintext message units and}$$
$$\mathcal{C} = \text{set of all ciphertext message units.}$$

The encryption algorithm is then the application of an injective map

$$f : \mathcal{P} \to \mathcal{C}.$$

The map f is the **encryption map**. The left inverse map

$$g : \mathcal{C} \longrightarrow \mathcal{P}$$

is the **decryption** or **deciphering map**. The collection $\{\mathcal{P}, \mathcal{C}, f, g\}$, consisting of a set of plaintext message units, a set of ciphertext message units, an encryption map and its left inverse is called a **basic cryptosystem**. Later in this chapter we will present a broader definition of a cryptosystem which also includes an index set called the **key space**. For the present we will use cryptosystem to mean basic cryptosystem.

Breaking a code is called **cryptanalysis**. An attempt to break a code is called an **attack**. Most cryptanalysis depends on a statistical frequency analysis of the plaintext language used (see exercises). Cryptanalysis depends also on a knowledge of the form of the code, that is, the type of cryptosystem used.

We now present some simple examples of cryptosystems and cryptanalysis.

Example 1.2.1. The simplest type of encryption algorithm is a **permutation cipher**. Here the letters of the plaintext alphabet are permuted and the plaintext message is sent in the permuted letters. Mathematically if the alphabet has N letters and σ is a permutation on $1, ..., N$, the letter i in each message unit is replaced by $\sigma(i)$. For example suppose the plaintext language is English and the plaintext word is BOB and the permutation algorithm is

$$
\begin{array}{ccccccccccccc}
a & b & c & d & e & f & g & h & i & j & k & l & m \\
b & c & d & f & g & h & j & k & l & n & o & p & r
\end{array}
$$

$$
\begin{array}{cccccccccccc}
n & o & p & q & r & s & t & u & v & w & x & y & z \\
s & t & v & w & x & a & e & i & z & m & q & y & u
\end{array}
$$

then $BOB \rightarrow CTC$.

Example 1.2.2. A very straightforward example of a permutation encryption algorithm is a **shift algorithm**. Here we consider the plaintext alphabet as the integers $0, 1, ..., N - 1 \pmod N$. We choose a fixed integer k and the encryption algorithm is

$$f(m) \equiv (m + k) \pmod N.$$

This is often known as a **Caesar code** after Julius Caesar who supposedly invented it. It was used by the Union Army during the American Civil War. For example if both the plaintext and ciphertext alphabets were English and each message unit was a single letter then $N = 26$. Suppose $k = 5$ and we wish to send the message $ATTACK$. If $a = 0$ then $ATTACK$ is the numerical sequence $0, 19, 19, 0, 2, 10$. The encoded message would then be $FYYFHP$.

Any permutation encryption algorithm which goes letter to letter is very simple to attack using a statistical analysis. If enough messages are intercepted and the plaintext language is guessed then a frequency analysis of the letters will suffice to crack the code. For example in the English language the three most commonly occurring letters are E, T and A with a frequency of occurrence of approximately 13 %, 9 % and 8 % respectively. By examining the frequency of occurrences of letters in the ciphertext, the letters corresponding to E, T and A can be uncovered. In the next chapter we present an example of a statistical attack in English.

Polyalphabetic ciphers are an attempt to thwart statistical attacks. In these coding methods, different letters may be encrypted with different alphabets. One variation of the basic Caesar code is the following where message units are k-vectors. It is actually a type of polyalphabetic cipher called a **Vigenére code**.

Example 1.2.3. In this code, message units are considered as k-vectors of integers modulo N from an N letter alphabet. Let $B = (b_1, ..., b_k)$ be a fixed k-vector in $\mathbb{Z}_n^k$. The Vigenére code then takes a message unit

$$(a_1, ..., a_k) \rightarrow (a_1 + b_1, ..., a_k + b_k) \pmod{N}.$$

The vector $(b_1, ..., b_k)$ in the example above is called a **key**. Another version of a polyalphabetic cipher in given in the next example which further uses the idea of a **key** needed to unlock the code.

Example 1.2.4. Suppose we have an N-letter alphabet. We then form an $N \times N$ matrix P where each row and column is a distinct permutation of the plaintext alphabet. Hence P is a permutation matrix on the integers $0, ..., N - 1$. Bob and Alice decide on a **keyword**. The keyword is placed above the plaintext message and the intersection of the keyword letter and plaintext letter below it will determine which cipher alphabet to use. Let us make this precise with a 9 letter alphabet $A, B, C, D, E, O, S, T, U$. Here for simplicity we assume that each row is just a shift of the previous row, but any permutation can be used.

Key Letters

		A	B	C	D	E	O	S	T	U
a	A	a	b	c	d	e	o	s	t	u
l	B	b	c	d	e	o	s	t	u	a
p	C	c	d	e	o	s	t	u	a	b
h	D	d	e	o	s	t	u	a	b	c
a	E	e	o	s	t	u	a	b	c	d
b	O	o	s	t	u	a	b	c	d	e
e	S	s	t	u	a	b	c	d	e	o
t	T	t	u	a	b	c	d	e	o	s
s	U	u	a	b	c	d	e	o	s	t

Suppose the plaintext message is STAB DOC and Bob and Alice have chosen the keyword BET. We place the keyword repeatedly over the message

$$B \quad E \quad T \quad B \quad E \quad T \quad B$$
$$S \quad T \quad A \quad B \quad D \quad O \quad C$$

To encode we look at B which lies over S. The intersection of the B key letter and the S alphabet is a t so we encrypt the S with T. The next key letter is E which lies over T. The intersection of the E key letter with the T alphabet is c. Continuing in this manner and ignoring the space we get the encryption

$$\text{STAB DOC} \rightarrow \text{TCTCTDD}$$

Example 1.2.5. A final example, which is not number theory based, is the so-called **Beale Cipher**. This has a very interesting history which is related in the popular book

Archimedes Revenge by P. Hoffman (see [Hof]). Here letters are encrypted by numbering the first letters of each word in some document like the Declaration of Independence or the Bible. There will then be several choices for each letter and a Beale cipher is quite difficult to attack.

1.3 A Very Brief History of Cryptography

Until relatively recent times cryptography was mainly concerned with message confidentiality, that is, sending secret messages so that interceptors or eavesdroppers cannot decipher them. The discipline was primarily used in military and espionage situations. As explained in the introduction this changed with the vast amount of confidential data that had to be transmitted over public airways so the field has expanded to may different types of cryptographic techniques such as digital signatures and message authentications.

Cryptography and encryption does have a long and celebrated history. In the Bible, in the book of Jeremiah, they use what is called an **Atabash Code**. In this the letters of the alphabet, Hebrew in the Bible, but can be used with any alphabet, are permuted first to last. That is, in the Latin alphabet Z would go to A and so on.

The Kabbalists and the Kabbala believe that the Bible - written in Hebrew where each letter also stands for a number- is a code from heaven. They have devised elaborate ways to decode it. This idea has seeped into popular culture where the book "The Bible Code" became a bestseller.

In his military campaigns Julius Caesar would send out coded messages. His method, which we looked at in the last section, is now known as a Caesar code. It is a shift cipher. That is each letter is shifted a certain amount to the right. A shift cipher is a special case of an affine cipher that will be elaborated upon in the next section. The Caesar code was resurrected and used during the American Civil War.

Coded messages produced by most of the historical methods reveal statistical information about the plaintext. This could be used in most cases to break the codes. The discovery of frequency analysis was done by the Arab mathematican Al-Kindi in the ninth century and the basic classical substitution ciphers became more or less easily breakable.

The use of ciphers and code books flourished during the political intrigues in Europe during the middle ages and the use of codes and code books became prevalent in diplomacy. About 1470 Leon Alberti developed a method to attempt to thwart statistical analysis. His innovation was to use a polyalphabetic cipher where different parts of the message are encrypted with different alphabets. His method was mistakenly attributed to the French cryptographer Blaise de Vigenére, who was a century later than Alberti. The type of polylaphabetic code as described in Example 1.2.4 is now referred to as a Vigenére code. Further work was done on polyaphabetic codes by Johannes Trithemius and G. B. Belasso in the sixteenth century. Belasso seems to be the first to describe, in a book, the mechanics of polyalphabetic codes. Vigenére was a mathematician and cryptographer who worked and wrote about on many different types of ciphers. Famous mathematicians, Charles Dodgson (Lewis Carrol of *Alice in Wonderland* fame) and Charles Babbage, also wrote on the Vigenére code and Babbage actually broke a version of it.

A different way to thwart statistical attacks is to use blanks, that is, meaningless letters within the message. Mary, Queen of Scots, used a random permutation cipher with neutrals in it where a neutral was a random meaningless symbol. Unfortunately for her, her messages were decoded, and she was beheaded.

There have been various physical devices and aids used to create codes. Prior to the widespread use of the computer, the most famous cryptographic aid was the **Enigma machine** developed and used by the German military during the Second World War. This was a rotor machine using a polyalphabetic cipher. An early version was broken by Polish cryptographers early in the war, so a larger system was built that was considered unbreakable. British cryptographers led by Alan Turing broke this and British knowledge of German secrets had a great effect on the latter part of the war.

The development of digital computers allowed for the development of much more complicated cryptosystems. Further this allowed for the encryption using anything that can be placed in binary formats whereas historical cryptosystems could only be rendered using language texts. This has revolutionized cryptography. The use of computers also required the development of more complex encryption methods since computing capability enhanced the ability to cryptanalyze, that is break cryptosystems.

In 1976 Diffie and Hellman developed the first usable public key exchange protocol. This allowed for the transmission of secret data over open airways. A year later, Rivest, Adelman and Shamir, developed the RSA algorithm a second public key protocol. There are now many and we will discuss them later. In 1997 it became known that public key cryptography had been developed earlier by James Ellis working for British Intelligence and that both the Diffie-Hellman and RSA protocols had been developed earlier by Malcom Williamson and Clifford Cocks respectively.

Both Diffie-Hellman and RSA require very large key spaces (see Chapter 2). In an attempt to use the same ideas but reduce the key space size it was suggested that Diffie-Hellman be applied to other abelian groups. To do this, algebraic geometry was introduced into cryptography. In 1985 Neil Koblitz and independently Victor Miller suggested the use of elliptic curves over finite fields and their corresponding groups as possible cryptographic platforms. These methods have been quite successful and result, in many cases, in faster encryption and smaller key spaces than standard RSA methods.

As cryptographic applications became more prevalent it was felt that there should be an encryption standard. In 1976, the Data Encryption Standard (DES) was adopted as the standard for symmetric key encryption. DES is what is called a **block cipher** and has the structure of a so-called **Feistel network**. In Chapter 2 we discuss Feistel networks and DES in detail.

The original DES is now insecure for many applications and there have been successful attacks on it, the first being in 1999. Subsequently a competition was initiated to find a secure replacement. The winner was developed by Rijmen and Daemen and their method is called the Rijndael cipher. This method is also a block cipher that proceeds with several rounds of encrypting blocks and then mixing blocks. In 2001 the National Institute of Standards and Technology of the United States adopted a standardization of the Rijndael cipher, now called **AES** for Advanced Encryption Standard, as the industry standard for a symmetric key encryption. Although not universally used it is the most widely used. We will discuss AES further in Chapter 2.

Finally within the past twenty years the ideas of quantum algorithms and the possibility of a workable quantum computer led researchers to start to consider public key methods using non-abelian groups. In 1999 Ko, Lee *et al.* (see [KLCHKP]) and Anshel-Anshel-Goldfeld (see [AAG1]) introduced public key exchange methods involving the braid groups as platforms. There has been an active line of research over the past fifteen years in this area.

This has been a whirlwind look at cryptography's past. The complete history of cryptography is both extensive and fascinating. The books by David Kahn [Kha] and Simon Singh [Sin] provide a wealth of information.

1.4 Encryption and Number Theory

Most of traditional cryptographic methods are number theory based. At the simplest level these techniques use modular arithmetic, however a wide array of complicated number theory is used in cryptography. In Chapters 4 and 5 we will give a full description of modular arithmetic and its use in cryptography. In this introductory section we describe some elementary number theoretically derived cryptosystems. This is to provide a flavor of how number theory enters cryptography.

In applying a cryptosystem to an N letter alphabet we consider the letters as integers modulo N. The encryption algorithms then apply number theoretic functions and use modular arithmetic on these integers. One example of this was the shift or Caesar cipher described in Example 1.3.2. In this encryption method, a fixed integer k is chosen and the encryption map is given by

$$f : m \rightarrow m + k \, (\mathrm{mod}\, N).$$

The shift algorithm is a special case of an **affine cipher**. Recall that an **affine map** on a ring R is a function $f(x) = ax + b$ with $a, b, x \in R$. We apply such a map to the ring of integers modulo n, that is, $R = \mathbb{Z}_n$, as the encryption map. Specifically suppose we have an N letter alphabet and consider the letters as the integers $0, 1, ..., N-1 \, (\mathrm{mod}\, N)$, that is in the ring $\mathbb{Z}_N$. We choose integers $a, b \in \mathbb{Z}_N$ with the greatest common divisor $(a, N) = 1$ and $b \neq 0$. a, b are called the **keys** of the cryptosystem. The encryption map is then given by

$$f : m \rightarrow am + b \, (\mathrm{mod}\, N).$$

Example 1.4.1. Using an affine cipher with the English language and keys $a = 3, b = 5$ encode the message EAT AT JOE'S. Ignore spaces and punctuation.

The numerical sequence for the message ignoring the spaces and punctuation is

$$4, 0, 19, 0, 19, 9, 14, 4, 18.$$

Applying the map $f(m) \equiv 3m + 5 \, (\mathrm{mod}\, 26)$, we get

$$17, 5, 62, 5, 62, 32, 47, 17, 59 \rightarrow 17, 5, 10, 5, 10, 6, 21, 17, 7.$$

Now rewriting these as letters we get

$$\text{EAT AT JOE'S} \rightarrow \text{RFKFKGVRH}.$$

Since $(a, N) = 1$ the integer a has a multiplicative inverse modulo N. The decryption map for an affine cipher with keys a, b is then

$$g = f^{-1} : m \rightarrow a^{-1}(m - b) \, (\mathrm{mod}\, N).$$

Since an affine cipher, as given above, goes letter to letter it is easy to attack using a statistical frequency approach. Further if an attacker can determine two letters and knows that it is an affine cipher the keys can be determined and the code broken. For example suppose the affine cipher is

$$f(n) \equiv an + b \, (\mathrm{mod}\, N)$$

and an attacker has learned that $f(n_1) = m_1, f(n_2) = m_2$. This gives the attacker a two-by-two system of linear equations

$$an_1 + b = m_1$$
$$an_2 + b = m_2.$$

This is easily solvable for a, b, if $(n_1 - n_2, N) = 1$, breaking the code.

To provide better security in an affine cipher it is preferable to use k-vectors of letters as message units. In this case the form of an affine cipher becomes

$$f : v \rightarrow Av + B$$

where v and B are k-vectors from $\mathbb{Z}_N^k$ and A is an invertible $k \times k$ matrix with entries from the ring $\mathbb{Z}_N$. Here the computations are done modulo N. Since v is a k-vector and A is a $k \times k$ matrix the matrix product Av produces another k-vector from $\mathbb{Z}_N^k$. Adding the k-vector B again produces a k-vector so the ciphertext message unit is also a k-vector. The keys for this affine cryptosystem are the enciphering matrix A and the shift vector

B. The matrix A is chosen to be invertible over $\mathbb{Z}_N$. This condition is equivalent to the determinant of A being a unit in the ring $\mathbb{Z}_N$. The decryption map is given by

$$v \to A^{-1}(v - B).$$

Here A^{-1} is the matrix inverse over $\mathbb{Z}_N$ of A and v is a k-vector. The **enciphering matrix** A and the shift vector B are now the keys of the cryptosystem.

A statistical frequency attack on such a cryptosystem requires knowledge, within a given language, of the statistical frequency of k-strings of letters. This is more difficult to determine than the statistical frequency of single letters. As for a letter to letter affine cipher, if $k + 1$ message units, where k is the message block length, are discovered, then the code can be broken. As in the example for a cipher on single letters in this an attacker obtains a $(k + 1) \times (k + 1)$ system of linear equations which can be solved (see exercises).

Example 1.4.2. Using an affine cipher with message units of length 2 in the English language and keys

$$A = \begin{pmatrix} 5 & 1 \\ 8 & 7 \end{pmatrix}, \quad B = (5, 3)$$

encode the message EAT AT JOE'S. Again ignore spaces and punctuation.

Message units of length 2, that is 2-vectors of letters are called **digraphs**. We first must place the plaintext message in terms of these message units. The numerical sequence for the message EAT AT JOE's ignoring the spaces and punctuation is as before

$$4, 0, 19, 0, 19, 9, 14, 4, 18.$$

Therefore the message units are

$$(4, 0), (19, 0), (19, 9), (14, 4), (18, 18)$$

repeating the last letter to end the message.

The enciphering matrix A has determinant 1 which is a unit modulo 26 and hence is invertible so it is a valid key.

Now we must apply the map $f(v) \equiv Av + B \pmod{26}$ to each digraph. For example

$$A \begin{pmatrix} 4 \\ 0 \end{pmatrix} + B = \begin{pmatrix} 5 & 1 \\ 8 & 7 \end{pmatrix} \begin{pmatrix} 4 \\ 0 \end{pmatrix} + \begin{pmatrix} 5 \\ 3 \end{pmatrix} = \begin{pmatrix} 20 \\ 32 \end{pmatrix} + \begin{pmatrix} 5 \\ 3 \end{pmatrix} = \begin{pmatrix} 25 \\ 9 \end{pmatrix}$$

where everything is done modulo 26.

Doing this to the other message units we obtain

$$(25, 9), (22, 25), (5, 10), (1, 13), (9, 13).$$

Now rewriting these as digraphs of letters we get

$$(Z, J), (W, Z), (F, K), (B, N), (J, N).$$

Therefore the coded message is

$$\text{EAT AT JOE'S} \to \text{ZJWZFKBNJN}.$$

Example 1.4.3. Suppose we receive the message ZJWZFKBNJN and we wish to decode it. We know that an affine cipher with message units of length 2 in the English language and keys

$$A = \begin{pmatrix} 5 & 1 \\ 8 & 7 \end{pmatrix}, \quad B = (5,3)$$

are being used.

The decryption map is given by

$$v \rightarrow A^{-1}(v - B)$$

so we must find the inverse matrix for A. For a 2×2 invertible matrix $\begin{pmatrix} a & b \\ c & d \end{pmatrix}$ we have

$$\begin{pmatrix} a & b \\ c & d \end{pmatrix}^{-1} = \frac{1}{ad - bc} \begin{pmatrix} d & -b \\ -c & a \end{pmatrix}.$$

Therefore in this case, recalling that multiplication is modulo 26,

$$A = \begin{pmatrix} 5 & 1 \\ 8 & 7 \end{pmatrix} \implies A^{-1} = \begin{pmatrix} 7 & -1 \\ -8 & 5 \end{pmatrix}.$$

The message ZJWZFKBNJN in terms of message units is

$$(25, 9), (22, 25), (5, 10), (1, 13), (9, 13).$$

We apply the decryption map to each digraph. For example

$$A^{-1}(v - B) = \begin{pmatrix} 7 & -1 \\ -8 & 5 \end{pmatrix} \left(\begin{pmatrix} 25 \\ 9 \end{pmatrix} - \begin{pmatrix} 5 \\ 3 \end{pmatrix} \right) = \begin{pmatrix} 7 & -1 \\ -8 & 5 \end{pmatrix} \begin{pmatrix} 20 \\ 6 \end{pmatrix} = \begin{pmatrix} 4 \\ 0 \end{pmatrix}.$$

Doing this to each digraph we obtain

$$(4, 0), (19, 0), (19, 9), (14, 4), (18, 18)$$

and rewriting in terms of letters

$$(E, A), (T, A), (T, J), (O, E), (S, S).$$

This gives us

$$\text{ZJWZFKBNJN} \rightarrow \text{EATATJOESS}.$$

1.5 Public Key Cryptography

Presently there are many instances where secure information must be sent over open communication lines. These include, for example, banking and financial transactions, purchasing items via credit cards over the internet and similar things. This led to the development of **public key cryptography**. Roughly, in classical cryptography only the sender and receiver know the encoding and decoding methods. Further it is a feature of such cryptosystems, such as the ones that we've looked at, that if the encrypting method is known then the decrypting can be carried out. In public key cryptography the encryption method is public knowledge but only the receiver knows how to decode. More precisely in a classical cryptosystem once the encrypting algorithm is known the decryption algorithm can be implemented in approximately the same order of magnitude of time. In a public key cryptosystem, developed first by Diffie and Hellman, the decryption algorithm is much more difficult to implement. This difficulty depends on the type of computing machinery used, and as computers get better, new and more secure public key cryptosystems become necessary.

The basic idea in a public key cryptosystem is to have a **one-way function** or **trapdoor function**. That is a function which is easy to implement but very hard to invert. Hence it becomes simple to encrypt a message but very hard, unless you know the inverse, to decrypt. The trapdoor is the simple knowledge that one needs to invert.

The standard model for public key systems is the following. We suppose that Alice's **public key** is an injective map f_A from the set of plaintext message units to the set of ciphertext message units. This map f_A is made public. Alice's **private key** is g_A, the left inverse of f_A and is held by her in secret. Since g_A is the left inverse of f_A it follows that for any message m we have $g_A(f_A(m)) = m$. Similarly Bob has a public key f_B that is public knowledge and a private key g_B that he holds secretly.

Alice wants to send a message to Bob. The encrypting map f_A for Alice is public knowledge as well as the encrypting map f_B for Bob. On the other hand, the decryption algorithms g_A and g_B are secret and known only to Alice and Bob respectively. Let m be the message Alice wants to send to Bob. She sends $f_B(m)$. The map f_B is public knowledge and it is easy to encrypt. To decrypt Bob now applies g_B to what he has received. Recall that presumably only he knows his private key g_B. Doing this he obtains

$$g_B\left(f_B(m)\right) = m$$

and hence recovers the message.

If Bob receives a message how can he be certain that it actually comes from Alice. Within this standard model it is relatively easy to build in **verification** or **authentication**. As above let m be the message Alice wants to send to Bob. Now instead of sending $f_B(m)$ she sends

$$f_B\left(g_A(m)\right).$$

Now to decode Bob applies first g_B, which is his private key that only he knows. This gives him

$$g_B\left(f_B\left(g_A(m)\right)\right) = g_A(m).$$

He then looks up f_A which is publicly available and applies

$$f_A\left(g_A(m)\right) = m$$

to obtain the message. This supplies **verification** and **authentication** in the following manner. Suppose m is Alice's verification; signature, social security number etc. If Bob receives $f_B(m)$ it could be sent by anyone since f_B is public. On the other hand, since only Alice supposedly knows g_A getting a reasonable message from $f_A(g_B f_B g_A(m))$ would verify that it is from Alice. Applying g_B alone should result in nonsense.

Getting a reasonable one-way function can be a formidable task. Presently, the most widely used public key systems are based on difficult to invert number theoretic functions. Diffie-Hellman in 1976 developed the original public key idea using the **discrete logarithm problem** usually abbreviated to **discrete log problem** or **DLP**.

In modular arithmetic it is easy to raise an element to a power but difficult to determine, given an element, if it is a power of another element. Specifically if G is a finite group, such as the cyclic multiplicative group of $\mathbb{Z}_p$ where p is a prime, and $h = g^k$ for some k then the **discrete logarithm** or **discrete log** of h to the base g is any integer t with $h = g^t$. In the Diffie-Hellman system there is a public key exchange between the communicating parties based on the discrete log problem. This public key will indicate a private encryption system.

In 1977 Rivest, Adelman and Shamir developed the **RSA algorithm** which is presently one of the most widely used public key cryptosystems. It is based on the difficulty of factoring large integers and in particular on the fact that it is easier to test for primality than to factor. As in the Diffie-Hellman protocol there is a public key exchange using the difficulty of factoring followed by an encryption.

We will discuss both the Diffie-Hellman protocol and the RSA protocol in detail in Chapter 7.

1.6 Cryptosystems and the Key Space

In the beginning of this chapter we defined a **basic cryptosystem** as a tuple $\{\mathcal{P}, \mathcal{C}, f, g\}$ where $\mathcal{P}$ is the set of plaintext message units, $\mathcal{C}$ is the set of ciphertext message units, f is an injective map from $\mathcal{P}$ to $\mathcal{C}$, and g is its left inverse. The map f is the **encryption map** or **encryption algorithm**, while its left inverse g is the **decryption map** or **decryption algorithm**. Now we place this in a more general context. We call this wider model a (general) cryptosystem. Roughly, a (general) cryptosystem is an indexed collection of basic cryptosystems, indexed by a set $\mathcal{K}$ called the **key space**. Formally, it is defined as follows.

Definition 1.6.1. A **cryptosystem** is a tuple $(\mathcal{P}, \mathcal{C}, \mathcal{K}, \mathcal{E}, \mathcal{D})$ where

$\mathcal{P}$ = set of plaintext message units, called the **plaintext space**,

$\mathcal{C}$ = set of ciphertext message units, called the **ciphertext space**,

$\mathcal{K}$ = an index set, called the **key space**.

The elements $k \in \mathcal{K}$ are called **keys**.

$$\mathcal{E} = \text{a set of injective maps } f_k : \mathcal{P} \to \mathcal{C}$$

indexed by the key space. This is called the set of **encryption maps**. Hence, for each $k \in \mathcal{K}$, there is an injective map $f_k : \mathcal{P} \to \mathcal{C}$.

$$\mathcal{D} = \text{a set of maps } g_{k'} : \mathcal{C} \to \mathcal{P}$$

also indexed by the key space. This is called the set of **decryption maps**.

The central property of a cryptosystem is that, for each $k \in K$, there exists a corresponding key $k' \in \mathcal{K}$ and a decryption map $g_{k'} : \mathcal{C} \to \mathcal{P}$ such that $g_{k'}$ is the left inverse of f_k. In other words, for every plaintext message unit m, we have $g_{k'}(f_k(m)) = m$.

In our previous language this means that for each $k \in \mathcal{K}$ we have a basic cryptosystem $\{\mathcal{P}, \mathcal{C}, f_k, g_{k'}\}$. To place further emphasis on the key space, we say that for every encryption map $f_k : \mathcal{P} \to \mathcal{C}$, the index k is the corresponding **encryption key**, and for every decryption map $g_{k'} : \mathcal{C} \to \mathcal{P}$, the corresponding index k' is called the **decryption key**.

Using this model, we can easily distinguish symmetric from asymmetric cryptosystems. In a symmetric key cryptosystem, if the encryption key k is given, it is easy to find the corresponding decryption key k'. In fact, most of the time we have $k' = k$. In an asymmetric or public key cryptosystem, even if the encryption key k is known, it is infeasible to find or compute the decryption key k'. This is of course equivalent to how it was phrased in the short description in Section 1.5: if the encryption map is known, it is infeasible to find the inverse. In most public key cryptosystems the encryption key is made public while the decryption key is secret and known only to the party receiving the communication.

Notice that the key space must be fairly extensive to prevent a potential attacker from simply doing an **exhaustive search** to find the decryption key. Before we discuss specific cryptosystems, we introduce general cryptographic protocols.

1.7 Cryptographic Protocols

Besides secure confidential message transmission there are many other tasks that are important in cryptography, both public key and symmetric key. Although it is not entirely precise, we say that a **cryptographic task** is where one or more people must communicate with some degree of secrecy. The set of algorithms and procedures needed to accomplish a cryptographic task is called a **cryptographic protocol**. A cryptosystem is just one type of cryptographic protocol.

Definition 1.7.1. Suppose that several parties want to manage a cryptographical task. Then they must communicate with each and cooperate. Hence each party must follow certain rules and implement certain agreed upon algorithms. The set of all such methods and rules to perform a cryptographical task is called a **cryptographic protocol**.

We now list several cryptographic tasks. These will be discussed in more detail later in the book together with cryptographic protocols to implement them.

(1) **Authentication:** This is the process of determining that a message, supposedly from a given person, both does come from that person and has not been tampered with. Included in authentication are the concepts of **hash functions** and **digital signatures**. Hash functions will be discussed in Chapter 2. Another important usage is **password identification**. We will discuss a secure backup password system using non-commutative methods later in the book.

(2) **Key Exchange and Key Transport:** In a key exchange two people usually called Bob and Alice exchange a secret shared key to be used in some symmetric encryption. In a key transport one party transports to another a secret key that is to be used.

(3) **Secret Sharing:** This is a method where some secret is to be shared by k people but not available to any proper subset of them. There is a beautiful simple solution to the general problem given by Shamir. We will describe this also in Chapter 2.

(4) **Zero Knowledge Proof** This is an argument that convinces someone that you have solved a problem, for example a combinatorial problem, without giving away the solution. This is tied to authentication.

We will look in detail at protocols for each of these cryptographic tasks in Chapter 4.

1.8 Exercises

1.1 Is every encryption map with a finite set of ciphertexts surjective?

1.2 If we encrypt using a shift cipher explain why discovering one encrypted letter will break the system?

1.3 All integers can be expressed in binary form, that is only using digits of 0, 1. Explain why in a number theoretic cipher it is only really necessary to encrypt a bit, that is, encrypt a 0 or a 1?

1.4 Suppose we have an encryption system and we have a method that correctly guesses each encrypted letter with a probability of 0.5. If we have a message 20 letters long, what is the probability of randomly guessing the plaintext? (Randomly guessing indicates that guessing two letters in a row are independent and hence the probabilities multiply.)

1.5 Encrypt the message *NO MORE WAR* using an affine cipher with single letter keys $a = 7, b = 5$.

1.6 An encryption is done letter by letter using an affine cipher with single letter keys a, b. Using the English alphabet so that $N = 26$ we find that 6 is encrypted by 23 and 15 is encrypted by 24. What are the keys a, b?

1.7 Encrypt the message *NO MORE WAR* using an affine cipher on two vectors of letters and an encrypting key

$$A = \begin{pmatrix} 5 & 2 \\ 1 & 1 \end{pmatrix}, \quad B = (3, 7).$$

1.8 What is the decryption algorithm for the affine cipher given in the last problem?

1.9 How many different affine enciphering transformations are there on single letters with an N letter alphabet?

1.10 If we use an affine cipher on $N, N \geq 2$, single letters with $n \rightarrow an + b$ with b not congruent to 0 modulo N and $(a - 1, N) = 1$, show that there is always a unique fixed letter. (This can be used in cryptanalysis).

1.11 Let $N \in \mathbb{N}$ with $N \geq 2$ and $n \mapsto an + b$ with $(a, N) = 1$ be an affine cipher on an N letter alphabet. Show that if any two letters are guessed $n_1 \rightarrow m_1, n_2 \rightarrow m_2$ with $(n_1 - n_2, N) = 1$ then the code can be broken.

1.12 Let $\mathcal{A}$ be an N-letter alphabet. As before consider $\mathcal{A}$ as $\mathbb{Z}_N$ and treat them as the modular ring. Consider plaintext messages as elements of $\mathbb{Z}_N$ and consider the encryption map $f(n) \equiv n^2 \pmod{N}$. Is this a good encryption map? Why or why not?

2 Symmetric Key Cryptosystems

2.1 Mixed Encryption

Modern cryptography is usually separated into **symmetric key cryptography** and **public key cryptography**. In symmetric key cryptography, both the encryption and decryption keys are supposedly known only to the sender and receiver, usually referred to as Bob and Alice. In public key cryptography, the encryption map is public knowledge but only the receiver knows how to decode. In this chapter we look at the standard methods for efficient symmetric key cryptography, while in Chapter 7 we discuss public key methods in detail.

In general, symmetric key encryption methods are much more efficient in terms of time and storage requirements then public key methods. Therefore, in real practice what is used is **mixed encryption**. By this we mean that the message m is encrypted by some symmetric key method that is dependent on a key k called the **session key**. It is assumed that this encrypted message is **semantically secure**, that is secure against cipher text only attacks (see Chapter 3). The session key k is transmitted or shared by some public key protocol. Notice that there are several different keys in this protocol. The user's public and private keys which are then used to share or transport the session key. The session key is then used to encrypt and decrypt the message. There are key spaces for the public and private keys used in the key transfer and a key space used for the session key. These key spaces may or may not be the same. A mixed encryption system is often called a **digital envelope**.

In outline we have the following for real world mixed encryption.

(1) The communicating parties have agreed upon a symmetric key method that will be used for message encryption and decryption. It is assumed that the encryption and decryption algorithms in this method are known, once a given key is known. Each transmission session will use a different key, called the **session key**, so that once the session key is known the communicating parties can encrypt and decrypt messages based on it. It is further assumed that the encryption protocol is semantically secure.

(2) The users agree upon a public key method and choose necessary parameters and public and private keys.

(3) The communicating parties then use the public key protocol to share or transport the session key k.

(4) Once the session key is known to the communicating parties, the message m can be encrypted and decrypted using the agreed upon symmetric key method.

(5) The security of the mixed encryption protocol is then dependent on both the security of the private key method as well as the security of the symmetric key method.

2.2 Block Ciphers

Real world applications of cryptography usually proceed by agreeing on a secret key via a public channel, and then encrypting and decrypting by some symmetric key method which uses the common secret key. Frequently it is desired that this symmetric key encryption be very efficient, so that it can be done rapidly on a computer. The symmetric key encryption methods in most common use are either **block ciphers** or **stream ciphers**. Block ciphers are defined as follows.

Definition 2.2.1. Let A be an alphabet, and let n and m be positive integers. A **block cipher** with **input block length** n and **output block length** m over the alphabet A is a cryptosystem $(\mathcal{P}, \mathcal{C}, \mathcal{K}, \mathcal{E}, \mathcal{D})$ where
(1) $\mathcal{P} = A^n$, i.e., plaintext message units are n-tuples of letters from A,
(2) $\mathcal{C} = A^m$, i.e., ciphertext message units are m-tuples of letters from A,
(3) $\mathcal{K}$ is an arbitrary (finite) set,
(4) $\mathcal{E}$ is the set of encryption maps $f_k : A^n \to A^m$, and
(5) $\mathcal{D}$ is the set of decrpytion maps $g_k : A^m \to A^n$.

Notice that, for simplicity, we are assuming that an encryption key and the corresponding descryption key agree. This is no restriction, since the decryption key can be calculated efficiently from the encryption key in symmetric cryptography.

In many actual implementations, the encryption and decryption maps are bijective, and input and output block lengths are equal. The usual alphabet is $A = \{0, 1\}$, of course. Thus the encryption maps $f_k : \{0, 1\}^n \to \{0, 1\}^n$ are permutations. Since there exist $(2^n)!$ such permutations, and since the size of the key space is generally much smaller than this huge number, only a small part of the set of possible encryption maps is realized in practice.

For practical purposes it is essential that block ciphers are very efficient to encrypt and decrypt once one has the key. Since the input block length is n, messages of longer length are encrypted by dividing them into blocks of length n. In this way, using a block cipher, we can encrypt a message of arbitrary length. The total message length must be then divisible by n. To accomplish this, some form of **padding** is used, that is adding meaningless letters to the end of the message. There are different padding schemes and we refer to [Buc] for a discussion of these.

Very simple block ciphers are obtained by using a fixed permutation or a fixed substitution. They are called **permutation ciphers** (or **transpositions**) and **substitution ciphers**, respectively. Such cryptosystems were used historically, but as we explained in Chapter 1, they are not secure. Therefore, to increase security, modern block ciphers are created by iterating the basic operations of substitution and permutation.

Iterated Block Ciphers

The difficulty of creating a secure block cipher which is easy to implement and allows efficient encryption and decryption is usually met by repeatedly applying simple operations such as substitutions and permutations. The resulting ciphers are called **iterated block ciphers** or **product ciphers**. An input block of plaintext is subjected to repeated applications of a fixed set of mappings. Each iteration in this scheme is called a **round**. Each round may use a separate encryption maps and a separate key. For the construction of iterated block ciphers, several methods have been developed.

A **substitution-permutation network** is a type of iterated block cipher that applies rounds consisting of substitutions and permutations. For efficiency reasons the substitutions are performed on small parts of the input (or intermediate) tuples. A substitution box or **S-box** substitutes a small block of bits with another block of bits. The permutation steps, however, affect all the bits. They take the outputs of all the S-boxes of one round, permute the combined bit tuple, and feed them into the next round or into a key addition step. It is desirable that the permutation distributes the outputs of any S-box to as many S-box inputs as possible. Further at each round, the round key is combined with some group operation such as XOR. The operation XOR is simply addition modulo 2 on the bits $\{0, 1\}$. Decryption is done by simply reversing the encryption process. Here the inverses of the S-boxes and permutations are applied by using the round keys in reverse order. The ideas used in a substitution-permutation network become important when we discuss the modern standard block cipher protocol AES.

Another important type of iterated block cipher is a **Feistel network**. It became the basis for the first widely accepted encryption standard. In Section 2.4 we introduce this topic in detail.

There are two main methods for cryptanalysis of block ciphers; **linear cryptanalysis** and **differential cryptanalysis**. We will discuss these in Chapter 3.

Block Cipher Modes of Operation

There are several different variations of block ciphers, each addressing certain security problems of the basic design. We will discuss security and secrecy in the next chapter and then delve more deeply into these variations. These variations are called **block cipher modes of operation**. Here we briefly discuss several of them. More complete discussions can be found in [Buc].

The most basic model for a block cipher is called the **electronic codebook mode** or **ECB mode**. This directly follows the procedure described above where a message of arbitrary size is decomposed into blocks of length n. In this mode, the plaintext message must be padded so that its length is divisible by n. There are security problems with the ECB mode similar to the problems we described when the block length is one. The main disadvantage is that identical plaintext blocks are encrypted each time into identical blocks. It follows that the ECB method does not hide any patterns in the data well at all. This is especially true for selected plaintext-ciphertext attacks. The ECB mode, although quite simple in its construction and operation, is generally not recommended for encryption unless some form of randomness is included. Hence there are variations to address these security weaknesses. The general idea in these variations is to introduce some randomness into the block cipher. These include an additional input called an **initialization vector** to create a **probabilistic encryption**. The initialization vector, which is randomly chosen, is used to form the first ciphertext block, which is then used as a new initialization vector. Hence encryption at each step depends not only on the encryption algorithm, but on the previous encryption.

A very common variation is the **cipher block chaining mode** or **CCM mode**. Here the encryption of a block depends both on an encryption key and the previous blocks encrypted. The initialization vector is chosen randomly or pseudo-randomly which is then XOR'ed to the first plaintext block. This method encrypts the same plaintext block with possibly a different ciphertext block depending on the initialization vector and where the block appears in the overall message. This method avoids many of the security problems of the standard ECB mode.

Another variation is the **cipher feedback mode**, abbreviated CFB, which tries to copy a stream cipher. In this method the initialization vector is first encrypted and then XOR'ed to the first plaintext block.

A further variation is called the **output feedback mode**, abbreviated OFB. This method repeatedly encrypts the initialization vector to create a keystream. Again this is done in order to obtain the properties of a stream cipher (see Section 2.3).

2.3 Stream Ciphers

To decrypt a message using a block cipher, the receiver must wait until the message is completed. Often it is necessary to decode immediately or synchronously between the sender and receiver as the message arrives. This is the basic rationale for **stream ciphers**. Roughly a stream cipher generates a string of keys that then encrypts each letter individually in a message. We make this more precise.

Definition 2.3.1. A **stream cipher** is a symmetric key cipher where plaintext characters are combined with a pseudorandom key generator called a **key stream**. In a stream cipher the plaintext characters are encrypted one at a time and the encryption of successive characters varies during the encryption.

Hence in a stream cipher, each character in a plain text is encrypted with a different key. Given a starting key $k \in \mathcal{K}$, where $\mathcal{K}$ is the key space, there is a pseudorandom function $g_k : \mathcal{K} \to \mathcal{K}$ which generates a sequence of keys $(s_1, s_2, \ldots, s_n, \ldots)$. The function g_k is called a **keystream generator** while the sequence $(s_1, s_2, \ldots)$ is the **keystream**. Suppose that $\mathcal{A}$ is the plaintext alphabet and $a = a_1 a_2 a_3 \cdots a_k$ is a plaintext message with each $a_i \in \mathcal{A}$. The key k is chosen and it generates the keystream $s_1, s_2, \ldots, s_n, \ldots$. For each s_i there is an encryption map f_{s_i} and a corresponding decryption map g_{e_i}. Let F denote the overall encryption map. Then we would have

$$F(a_1 a_2 \cdots a_k) = f_{s_1}(a_1) f_{s_2}(a_2) \cdots f_{s_k}(a_k).$$

In practice the plaintext characters are usually single bits $0, 1$. Stream ciphers typically encrypt and decrypt at higher speeds than block ciphers and can usually be decrypted synchronously with the encryption. Stream ciphers are an attempt to approximate the security of so-called **one-time pads**. We will discuss this in the next chapter.

Very important in working with stream ciphers, and with many cryptographic systems in general, is the **XOR operation**. This is simply the operation on bits $\{0, 1\}$ that is addition modulo 2. Hence if $\oplus$ represents XOR we have

$$0 \oplus 0 = 0$$
$$0 \oplus 1 = 1$$
$$1 \oplus 0 = 1$$
$$1 \oplus 1 = 0.$$

In the most common form of a synchronous stream cipher the plaintext and the key stream are given in bits $0, 1$ and the keystream is combined with the plaintext via the **XOR operation**. For example suppose that the plaintext message is $a = 11001$ and the keystream is $011100\ldots$ then the encrypted message is

$$1 \oplus 0, 1 \oplus 1, 0 \oplus 1, 0 \oplus 1, 1 \oplus 0 = 10111.$$

Here $\oplus$ denotes addition modulo 2 so that $0 \oplus 1 = 1$ and $1 \oplus 1 = 0$.

Notice that in this type of stream cipher, both the sender and receiver must be exactly in synchronization. If a digit is somehow lost, the message cannot be recovered. There are several other types of stream ciphers that attempt to alleviate this drawback. One is called a **self synchronizing stream cipher** where the encryption of a bit is based not only on the key stream but on previous encrypted bits.

Generation of Stream Ciphers

Stream ciphers require sequences of pseudo-random digits. These are sequences that behave as if they are random. Here we will discuss two procedures to generate pseudo-random sequences and hence stream cipher key generation. One method is software technical and one is hardware technical. First we need the concept of a **linear congruence generator**. Here we use some ideas that we will clarify in Chapter 4 when we review all the necessary number theory. For a given natural number n we denote by $\mathbb{Z}_n$ the ring of integers modulo n (see Chapter 4). Elements of $\mathbb{Z}_n$ are then residue classes of integers modulo n. If a is an integer we will denote the corresponding reside class in $\mathbb{Z}_n$ by $\bar{a}$.

Definition 2.3.2. Let $n \in \mathbb{N}$ and $\bar{a}, \bar{b} \in \mathbb{Z}_n$. A bijective map $f : \mathbb{Z}_n \to \mathbb{Z}_n$ given by $x \mapsto \bar{a}x + \bar{b}$ is called a **linear congruence generator**.

Notice that if $a \in \mathbb{Z}$ and $\bar{a}$ is the corresponding residue class in $\mathbb{Z}_n$ then the linear congruence generator $x \mapsto \bar{a}x + \bar{b}$ is bijective if and only if $(a, n) = 1$ (see Chapter 5).

If we choose a large modulus n, linear congruence generators are used to generate pseudo-random integers. In using a linear congruence generator

$$f : x \mapsto \bar{a}x + \bar{b}$$

the integers a, b should be chosen so that the function f has no fixed point in $\mathbb{Z}_n$. Then $\bar{b} \neq \bar{0}$ for otherwise $\bar{0}$ is a fixed point. Hence let $\bar{b} \neq \bar{0}$. If $\bar{a} = 1$ then f has no fixed point but then the function is just a linear shift which is insecure. Therefore let $\bar{a} \neq 1$. Then f has a fixed point in $\mathbb{Z}_n$ if $(a - 1, n) = 1$ because then there exists a $d \in \mathbb{Z}$ with $\overline{(d(a-1))} = \bar{1}$ and then $x = -\overline{db}$ is a fixed point in $\mathbb{Z}_n$. Therefore altogether for a linear congruence generator we should choose a, b so that $(a, n) = 1$, $(a - 1, n) > 1$, $\bar{a} \neq \bar{1}$, $\bar{b} \neq \bar{0}$.

Using the idea of a linear congruence generator we now give a procedure for **software generation of a stream cipher**.
(1) Choose a seed $s \in \mathbb{Z}$ by key agreement or as a random number
(2) Let $n \in \mathbb{N}, a, b \in \mathbb{Z}$ and $f : \mathbb{Z}_n \to \mathbb{Z}_n, x \to \bar{a}x + \bar{b}$ a linear congruence generator. Determine the sequence

$$x_0 = \bar{s} \in \mathbb{Z}_n, \quad x_1 \equiv f(x_0) \,(\text{mod } n), \quad x_2 \equiv f(x_1) \,(\text{mod } n), \ldots$$

(3) Transform the sequence of plaintext units into a sequence of residue classes $m_0, m_1, \ldots$ in $\mathbb{Z}_n$.
(4) Encrypt the m_i into

$$c_i = m_i + x_i \in \mathbb{Z}_n.$$

The secret key is $\bar{s} \in \mathbb{Z}_n$.

Before moving onto the hardware procedure we make some comments on the cryptanalysis of this software generated stream cipher.
(1) The integer n should be very large and the residue classes should occur with the same probability. Further the function f should not have a fixed point.
   To accomplish this we must choose f and $\bar{s} \in \mathbb{Z}_n$ so that the **period length** $x_0, x_1, \ldots$ is as large as possible. Best would be the maximal length n.
(2) If we know sufficiently many plain text units which follow each other and we know the linear congruence generator used then we may calculate $\bar{s}$.

Theorem 2.3.3 (Maximal period length for $n = 2^m$). *Let $n \in \mathbb{N}$ with $n = 2^m$, $m \geq 1$ and let $a, b \in \mathbb{Z}$ such that $f : \mathbb{Z}_n \to \mathbb{Z}_n, x \mapsto \overline{a}x + \overline{b}$ is a linear congruence generator. Further let $s \in \{0, 1, \ldots, n-1\}$ be given and $x_0 = \overline{s}, x_1 = f(x_1), \ldots$. Then the sequence $x_0, x_1, \ldots$ is periodic with the maximal period length $n = 2^m$ if and only if the following hold:*
(1) *a is odd.*
(2) *If $m \geq 2$ then $a \equiv 1 \pmod 4$.*
(3) *b is odd and $\overline{b} \neq \overline{0}$.*

Proof. We show that (1),(2) and (3) hold if the period length is maximal. First we must have $(a, n) = 1$ since f is a linear congruence generator. Further f has no fixed point because the period length is maximal.

We show that $a \equiv 3 \pmod 4$ is not possible if $m \geq 2$. Suppose that $a \equiv 3 \pmod 4$ and $m \geq 2$. Then $a + 1 \equiv 0 \pmod 4$ and it follows that

$$(1 + a + a^2 + \cdots + a^{2i-1}) = (1 + a^2 + a^4 + \cdots + a^{2i-2})(1 + a) \equiv 0 \pmod 4. \quad (\star)$$

We now consider

$$x_{i+1} - x_i = f(x_i) - f(x_{i-1}) = (\overline{a}x_i + \overline{b}) - (\overline{a}x_{i-1} + \overline{b}) = \overline{a}(x_i - x_{i-1})$$

for $i \geq 1$. It then follows recursively that

$$x_k - x_0 = (x_k - x_{k-1}) + (x_{k-1} - x_{k-2}) + \cdots + (x_1 - x_0)$$
$$= \overline{a}^{k-1}(x_1 - x_0) + \overline{a}^{k-2}(x_1 - x_0) + \cdots + (x_1 - x_0)$$
$$= (x_1 - x_0)(\overline{1} + \overline{a} + \overline{a}^2 + \cdots + \overline{a}^{k-1}) \text{ for } k \geq 1.$$

Therefore

$$x_{2i} \equiv x_0 \pmod 4$$
$$x_{2i+1} \equiv x_1 \pmod 4$$

from the relation $(\star)$ above. Hence half of the elements in the sequence have the same residue class as x_0 modulo 4 and the other half the same as x_1 modulo 4 which gives a contradiction to the maximality of the period length. Therefore $a \equiv 1 \pmod 4$ if $m \geq 2$.

To show (3) notice that in a sequence with maximal period length the residue class $\overline{0}$ must occur. Hence without loss of generality we may assume that $x_0 = \overline{0}$. Then $x_1 = \overline{b}$ and recursively we have

$$x_k = (\overline{1} + \overline{a} + \cdots + \overline{a}^{k-1})\overline{b}$$

for $k \geq 1$ since $x_0 = \overline{0}$ and $x_1 = \overline{b}$. All elements in the sequence are multiples of $\overline{b}$. There is an $x_i = \overline{1}$ and therefore $\overline{b}$ is invertible in $\mathbb{Z}_n$ with $n = 2^m$ and hence b is odd.

Now assume that (1),(2) and (3) are satisfied. The theorem follows directly if $n = 2$ since then if $x_0 = \overline{0}$ then $x_1 = \overline{1}$ and if $x_0 = \overline{1}$ then $x_1 = \overline{0}$. Now suppose that $m \geq 2$ so that $n \geq 4$. We show that we obtain the maximal period length $n = 2^m$ for $x_0 = \overline{0}$ which proves the theorem.

Let $x_0 = \overline{0}$. Then as before we obtain recursively

$$x_k = (\overline{1} + \overline{a} + \cdots + \overline{a}^{k-1})\overline{b}$$

for $k \geq 1$. Since b is odd we have

$$x_k = \overline{0} \quad \text{if and only if } (\overline{1} + \overline{a} + \cdots + \overline{a}^{k-1}) = \overline{0} \in \mathbb{Z}_n.$$

We write $k = 2^r t$ with $r \geq 0$ and t odd. Then

$$\overline{0} = (\overline{1} + \overline{a} + \cdots + \overline{a}^{k-1}) = (\overline{1} + \overline{a} + \cdots + \overline{a}^{2^r-1})(\overline{1} + \overline{a}^{2^r} + (\overline{a}^{2^r})^2 + \cdots + (\overline{a}^{2^r})^{t-1}).$$

The second factor is congruent to 1 modulo 2 and hence

$$2^m | (1 + a + \cdots + a^{k-1}) \quad \text{if and only if } 2^m | (1 + a + \cdots + a^{2^r-1}).$$

The integer $1 + a + \cdots + a^{2^r-1}$ is divisible by 2^r since it is a sum of 2^r odd numbers but not divisible by 2^{r+1}. It follows that $r \geq m$ if and only of $2^m | k$.

Therefore $x_k = \overline{0}$ occurs for $k \geq 1$ for the first time when $k = n = 2^m$. □

We now describe a hardware procedure for generating a stream cipher. This is called the **linear shift feedback register** that we will abbreviate as LSFR.

The idea is as follows. We tie together n memory cells in a row. Each memory cell c_i contains a bit a_i. A feedback function R integrates the memory cells.

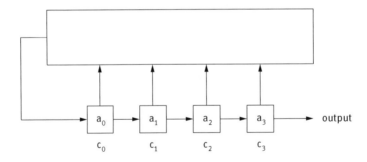

Fig. 2.1: Linear Shift Feedback Register.

(1) By each pulse the content a_i of c_i, $0 \leq i < n - 1$ will be passed to c_{i+1}.
(2) By each pulse the content a_n of c_n is the output.
(3) By each pulse the sequence $a_0, \ldots, a_{n-1}$ will be integrated by means of R and the result will be written modulo 2 in the cell c_0.

Example 2.3.4. We use a LSFR with four cells c_0, c_1, c_2, c_3 and a feedback function

$$R(a_0, a_1, a_2, a_3) = a_1 + a_3 \in \mathbb{Z}_2.$$

We use the initialization $(1, 0, 0, 0)$. We then have:

Pulse		0	1	2	3	4	5	6
a_0		1	0	1	0	0	0	1
a_1		0	1	0	1	0	0	0
a_2		0	0	1	0	1	0	0
a_3		0	0	0	1	0	1	0
—		—	—	—	—	—	—	—
Output		—	0	0	0	1	0	1

and the final output is 0.

The mathematical interpretation of this example is as follows. The sequence (a_0, a_1, a_2, a_3) corresponds to the element

$$8a_0 + 4a_1 + 2a_2 + a_3 \in \mathbb{Z}_{16}.$$

The reset process is then

$$8(a_1 + a_3) + 4a_0 + 2a_1 + a_2 \in \mathbb{Z}_{16}.$$

This gives a function $\mathbb{Z}_{16} \to \mathbb{Z}_{16}$.

We mention that linear feedback functions are relatively insecure. To break a LSFR of length m, or equivalently to calculate the feedback function R, we need in general m bits of plain text and 2^{m-1} bits of the pseudo random sequence of bits.

2.4 Feistel Networks, DES and AES

As cryptographic applications became more prevalent, it was felt that there should be a standard cryptosystem. In 1976 DES or Data Encryption Standard was adopted as the standard for symmetric key encryption. DES is a block cipher which uses what is called a **Feistel network**. A Feistel network has the advantage that the encryption and decryption algorithms are very similar, even identical in some cases, requiring only a reversal of the key schedule. Therefore the size of the code or circuitry required to implement such a cipher is minimized. A Feistel network is an iterated cipher with an internal function called a round function. Feistel networks were developed originally at IBM by the physicist Horst Feistel.

A general **Feistel network** is a block cipher that combines several **rounds** of encryption. The alphabet is binary bits 0, 1. We assume a block cipher of block length m and the number of rounds is fixed at r. The resulting Feistel network has block length $2m$. Given an initial key k we generate a keystream of length r, $s_1, \ldots, s_r$. These are called the **round keys**. As before let f_{s_i} be the encryption map for the key stream element s_i and f_k the encryption map for the initial key k. Let a be a plaintext message of length $2m$. Divide a into a left half L_0 and a right half R_0 each of length m. Inductively we form the sequence

$$(L_i, R_i) = (R_{i-1}, L_{i-1}) \oplus f_{s_i}(R_{i-1}) \quad \text{for } 1 \le i \le r$$

where as before $\oplus$ is XOR. The encryption then of $a = L_0 R_0$ is

$$f_k(L_0, R_0) = (R_r, L_r).$$

To decrypt we inductively form

$$(R_{i-1}, L_{i-1}) = (L_i, R_i \oplus f_{s_i}(L_i)) \quad \text{with } 1 \le i \le r.$$

The original DES used a 64 bit block size and a 56 bit key. This had been shown to be insecure for many applications and there have been successful attacks on it, the first being in 1999. Subsequently a competition was initiated to find a secure replacement. The winner was developed by Joan Daemen and Vincent Rijmen (see [DR]) and their method is called the **Rijndael cipher**. This method is a block cipher that proceeds with several rounds of encrypting blocks and then mixing blocks. In 2001 the National Institute of Standards and Technology adopted a standardization of the Rijndael cipher, now called **AES** for Advanced Encryption Standard, as the industry standard for a symmetric key encryption.

Although not universally used, it is the most widely used symmetric key encryption method. The mathematics in AES is done over the finite field $\mathbb{F}_q$ for $q = 2^8$. We briefly describe the method. It is called a **substitution-permutation cipher** and works rapidly in both software and hardware. It does not use a Feistel network. AES has a fixed blocksize of 128 bits and a key size any multiple of 32 bits with 128 as a minimum. It operates on a 4×4 array of bytes termed the **state** and the calculations, as mentioned above are done in the finite field $\mathbb{F}_q$, where $q = 2^8$. It has four basic steps: key expansion, initial round, rounds, final round.

(1) In **key expansion** a set of round keys is derived from the initial key.
(2) In the **initial round** each byte of the initial state is combined with the first round key using bitwise XOR.
(3) Each **round** then consists of four steps.
    (a) In the first step, called **subbytes**, each byte is replaced by another according to a lookup table.
    (b) In the second step, called **shiftrows**, the rows of the state are shifted cyclically a certain number of steps.
    (c) The third step, called **mixcolumns**, mixes the columns of each state combining the bytes.
    (d) The final step, called **addroundkey**, adds the roundkey with the state via bitwise XOR.
(4) In the **final round**, the subbytes, shiftrows and addroundkey are done but no column mix.

Since its adoption there have been no successful mathematical attacks on AES. For more details see [Buc].

2.5 One-Way Functions and Trapdoors

Public key cryptography depends on **one-way functions**. These are maps $f : X \mapsto Y$, such that it is easy to compute $f(x)$ for any $x \in X$ but hard to compute the preimage $f^{-1}(y)$ for most randomly selected $y \in Y$. Hard to compute can be measured in terms of complexity.

An encryption algorithm is a **public key encryption map** or **trapdoor function** if the algorithm is a one-way function from the set of plaintext messages to the set of ciphertext messages. That is, it is easy to compute ciphertext messages from plaintext messages given a key K but infeasible to decrypt without the key. Mathematically, if f is a trapdoor function if there exists some secret k, such that given $f(x)$ and k it is easy to compute x but computationally infeasible to compute x if k is not known. The key k that makes the function easy to decrypt is called the **trapdoor**.

Trapdoor functions arose in cryptography after the development of the Diffie-Hellman and RSA public key crytographic protocols. Diffie and Hellman introduced the term trapdoor function. We will discuss these in Chapter 7.

In a public key cryptosystem there are generally two types of one-way functions. The first is the encryption algorithm. Finding the inverse of this is called the **cracking problem**. The other is the function upon which the encryption algorithm or trapdoor function is based.

2.6 Exercises

2.1 We use the standard allocation

$$A = 01, \quad B = 02, \ldots, \quad Z = 26.$$

Calculate the plaintext number M for the plaintext message

Louisa is born on Christmas Day

2.2 Let $F : \mathbb{Z}_{24} \to \mathbb{Z}_{24}$ by $x \mapsto \overline{5}x + \overline{3}$. Calculate the period length for $x_0 = \overline{0}$.

2.3 Let $\mathcal{A}$ be an alphabet and $\mathcal{A}^*$ be the set of words in $\mathcal{A}$. Show that $\mathcal{A}^*$ together with the concatenation of words is a free monoid.

Recall that a monoid is a structure with one associative operation and an identity.

2.4 Recall that a block cipher encrypts n-letter blocks by n-letter blocks. Hence if $\mathcal{A}$ is the plaintext alphabet then a block cipher is a map $f : \mathcal{A}^n \to \mathcal{A}^n$. Show that a block cipher must be a permutation of $\mathcal{A}^n$.

2.5 Given a 4-bit LFSR with feedback function

$$R(a_0, a_1, a_2, a_3) = c_4 a_0 + c_3 a_1 + c_2 a_2 + c_1 a_3 \in \mathbb{Z}_2.$$

Starting with $(1, 0, 0, 0)$, we get the output string $(0, 0, 0, 1, 1, 1, 1, 0)$. Calculate the values of c_1, c_2, c_3, c_4 that break this LFSR.

2.6 Let $\mathcal{A} = \{0, 1\}$, $n = 4$ and π is a permutation on $\{1, 2, 3, 4\}$. Let

$$e_\pi : \{0, 1\}^4 \rightarrow \{0, 1\}^4$$

be defined by

$$b_1 b_2 b_3 b_4 \mapsto b_{\pi(1)} b_{\pi(2)} b_{\pi(3)} b_{\pi(4)}.$$

The plaintext is $m = 1011000101001010$. Calculate the ciphertext.
If π is the permutation $\left(\begin{smallmatrix} 1 & 2 & 3 & 4 \\ 2 & 3 & 4 & 1 \end{smallmatrix} \right)$, give the explicit output for the ciphertext for m.

2.7 Given a linear congruence generator $f : \mathbb{Z}_n \rightarrow \mathbb{Z}_n$ with $n = 2^{32}$ and the first three values

$$x_0 = 1394449523, \quad x_1 = 3456347474 \quad \text{and} \quad x_2 = 1689033221.$$

(a) Calculate the next two values x_3 and x_4.
(b) Calculate x_{10^n} for $n = 1, 2, \ldots, 10$.

2.8 Given an LFSR with n memory cells in series.
(a) Find an $n \times n$ matrix A such that the state of the shift feedback register after i pulses is given by

$$t_i = t_0 A^i = t_{i-1}$$

where we have the initialization $t_0 = (1, 0, \ldots, 0)$.
(b) Prove that the maximal period length of the shift feedback register is $2^n - 1$.

3 Cryptanalysis and Complexity

3.1 Cryptanalysis and Cryptanalytic Attacks

As explained in Chapter 1, cryptology has two parts: **cryptography** and **cryptanalysis**. Cryptanalysis is the part of cryptology that involves the methods used to break a code or to attempt to break a code. By breaking a code or cryptographic protocol, we mean the ability, under appropriate assumptions, to either determine the secret key or determine the plaintext from the ciphertext without finding the key.

An attempt to break a code is called an **attack**. The security of a cryptographic protocol depends on its ability to withstand attacks. The person mounting an attack is called an **attacker** or an **eavesdropper**. It is common practice to call the communicating parties Bob and Alice and a potential eavesdropper Eve.

An attack usually attempts to take advantage of some weakness in the encryption algorithm that allows one to decode. Cryptanalysis depends also on a knowledge of the form of the code, that is, the type of encryption algorithm employed. Most classical cryptanalysis depends on a statistical frequency analysis of the ciphertext versus the plaintext. An example of a statistical frequency attack is presented in the next section.

First we describe various types of cryptoanalytic attacks. The security of a cryptosystem or general cryptographic protocol depends on its ability to withstand these attacks. In addition, the attack and its chance of success depend upon the security assumptions made. By this we mean the assumptions about what information is available to the eavesdropper or attacker.

(1) **Ciphertext Only Attack:** In a ciphertext only attack, an attacker has access to the ciphertext but not the corresponding plaintext. This is the weakest type of attack. A not very efficient method, but one that can always be attempted, is an **exhaustive search**. The assumption is that the attacker knows the encryption method but needs the secret key. To find the secret key the attacker makes an exhaustive search of the key space. Hence for cryptographic security, the keyspaces in encryption protocols must be large enough so that it is computationally infeasible to do an exhaustive search.

In a ciphertext only attack, if it is known that the encryption is some sort of substitution encryption, then the main method is a statistical attack. This will be examined in the next section.

(2) **Known Plaintext Attack:** In a known plaintext attack an attacker has access to some selections of plaintext and the corresponding ciphertext. A known plaintext attack was actually part of the technique used in breaking the Enigma code.

(3) **Chosen Plaintext Attack:** In a chosen plaintext attack the attacker has access to a given plaintext and the corresponding ciphertext of his choice. Thus he can determine possible decryptions.

(4) **Chosen Ciphertext Attack:** In a chosen ciphertext attack the attacker can decrypt parts of the ciphertext but does not know the whole decryption key.

All of these methods of attack depend upon looking at some portion of an encoded message. There are other attacks that do not depend on looking at encrypted messages at all. These are called **side channel attacks**. For example, by looking at how much computer time or energy is expended in sending a coded message, an attacker might determine the relative size of the encryption key. In these notes we will not touch on side channel attacks.

In assessing cryptographic security we must also consider whether we have a possible **passive attacker** or a possible **active attacker**. A passive attacker can only look at encrypted transmissions but cannot interfere or change them at all. On the other hand, an active attacker has the ability to perhaps modify encrypted transmissions. A very powerful technique along these lines is called the **man in the middle attack**. Suppose Bob wants to communicate with Alice. The attacker gets in the middle and intercepts Bob's message. He then alters it in such a way that he can see how it is to be encrypted. He then sends it on to Alice. For example, suppose that Bob wants to get into a bank account and has a password. The attacker gets in the middle and intercepts the password. The attacker then sends the password on to the bank so that Bob does not realize that his password has been stolen. Later, the attacker uses the stolen password to access Bob's account.

3.2 Statistical Methods

The most common type of attack, and the weakest type of attack, is a ciphertext only attack. Here the attacker has access only to the ciphertext and not any of the plaintext that it came from. The attacker assumes that the encryption is some sort of substitution encryption. Under this assumption, the main method of cryptanalysis is a statistical frequency attack.

A **statistical frequency attack** on a given cryptosystem requires knowledge, within a given language, of the statistical frequency of k-strings of letters where k is the size of the message unit. In general if $k > 1$ this is more difficult to determine than the statistical frequency of single letters. As for a letter to letter affine cipher, if $k + 1$ message units, where k is the message block length, are discovered, then the code can be broken (see Chapter 1).

Example 3.2.1. The following table gives the approximate statistical frequency of occurrence of letters in the English language. The passage below is encrypted with a simple permutation cipher without punctuation. we will use a frequency analysis to decode it.

letter	frequency	letter	frequency	letter	frequency
A	.082	B	.015	C	.028
D	.043	E	.127	F	.022
G	.020	H	.061	I	070
J	.002	K	.008	L	.040
M	.024	N	.067	O	.075
P	.019	Q	.001	R	.060
S	.063	T	.091	U	.028
V	.010	W	.023	X	.001
Y	.020	Z	.001		

GTJYRIKTHYLJZAFZRLHDHTN
HDNHAGZVDGVHAHFGZVDGVZR
TVMOVFUVHEOMJZDUHRLVUHRZKA
GZUYHRVTAGZVDGVHDNVDKZDVVTZNK
GTJYRIKTHYLZARLVAGZVDGVIS
NVEEMIYZDKAVGTVRGINVAGTJYRHD
HMJAZAZARLVAGZVDGVISFTVHCZDK
ISFTVHCZDKGTJYRIAJARVUARLV
BLIMVAPFWVGRZAZAGHMMVNGTJYRIMIKJ

We first make a frequency analysis letter by letter. We obtain the following data:

$$a = 19 \quad b = 1 \quad c = 2 \quad d = 15 \quad e = 3 \quad f = 6 \quad g = 20 \quad h = 18 \quad i = 11$$
$$j = 10 \quad k = 9 \quad l = 8 \quad m = 8 \quad n = 7 \quad o = 2 \quad p = 1 \quad q = 0 \quad r = 16$$
$$s = 3 \quad t = 14 \quad u = 5 \quad v = 31 \quad w = 1 \quad x = 0 \quad y = 9 \quad z = 21$$

This would indicate that $V = e$ and $\{H, R, A, Z\} = \{a, t, o, i\}$. Playing then with the message and trying various permutations of letters with frequency values close to the total frequency values as above we finally arrive at a decoding.

$$a = h \quad b = f \quad c = g \quad d = n \quad e = v \quad f = s \quad g = k \quad h = l \quad i = z$$
$$j = w \quad k = c \quad l = m \quad m = u \quad n = d \quad o = i \quad p = y \quad q = x \quad r = t$$
$$s = a \quad t = r \quad u = p \quad v = e \quad w = b \quad x = q \quad y = j \quad z = o$$

so that the message is.

CRYPTOGRAPHY IS BOTH AN ART AND A SCIENCE. AS A SCIENCE IT RELIES HEAVILY ON MATHEMATICS, COMPUTER SCIENCE AND ENGINEERING. CRYPTOGRAPHY IS THE SCIENCE OF DEVELOPING SECRET CODES. CRYPTANALYSIS IS THE SCIENCE OF BREAKING CRYPTOSYSTEMS. THE WHOLE SUBJECT IS CALLED CRYPTOLOGY.

3.3 Cryptographic Security

The security of a cryptographic protocol is its ability to withstand attacks under given computing and theoretical security assumptions. In reality there are two types of cryptographic security; **theoretical security** and **practical security**. Practical security is also called **computational security** and relies on the computational infeasibility of breaking a cryptographic protocol given security assumptions on the attack. By security assumptions we mean several things.

(1) What type of cryptographic protocol are we using, and, what is the practical difficulty of the "hard" problem that the protocol is based upon?

(2) What is the goal of the attacker – for example to break the code completely or recover the secret key or just disrupt transmission?

(3) What computational power does the attacker have?

On the other hand, theoretical security or **information-theoretic security** is based on the impossibility of breaking the protocol, even assuming infinite computing power. While theoretical security is stronger than computational security, in most cases it is less practical. Generally, except for the one-time pad that we will discuss in Section 3.4, there are no cryptosystems with perfect security. Hence when we speak of cryptographic security we are speaking of computational security.

3.3.1 Security Proofs

There is no known computational problem that has been proved to be computationally difficult for a reasonable model of computation. Therefore, cryptographic security is generally reduced to tying the feasibility of breaking the protocol to the solvability of a known "hard" problem such as the halting problem for a Turing machine. These hard problems can either be proved to be intractable, or shown to be difficult, in a complexity sense. We discuss a bit of complexity theory in Section 3.7.

A **security proof**, or **security reduction**, for a cryptographic protocol is a proof that shows that breaking the protocol is equivalent to, or dependent upon, solving some known difficult problem. Hence if $\mathcal{P}$ is a difficult problem, and $\mathcal{E}$ denotes some protocol, then a security proof for $\mathcal{E}$ is a proof that says that breaking the encryption method $\mathcal{E}$ implies a solution to the problem $\mathcal{P}$. Therefore cryptographic security proofs are relative.

The most common public key protocols, Diffie-Hellman and RSA, do not have any true security proofs, since the difficulty of the problems that they depend upon is not really known. The Diffie-Hellman protocol, that we will introduce in Chapter 5, is dependent upon the discrete log problem that we briefly discussed in Chapter 2. There are no known efficient algorithms to solve this problem but its theoretical difficulty is unknown. What is known is that the protocol has not be broken in the forty years since its introduction. Similarly the RSA protocol is based on the difficulty of factoring large integers. Again there is no security proof but no efficient algorithm has been developed to break the whole system.

In Chapter 5 we will examine various algorithms to attempt to solve both the discrete log problem and the factorization problem. These algorithms then become potential attacks on the Diffie-Hellman and RSA protocols whose security depends on the difficulty of solving these problems.

3.4 Perfect Security and the One-Time Pad

Claude Shannon, in a famous paper in 1949 (see [Sha3]), placed a mathematical model on cryptographic security in terms of probability. In his mathematical model, Shannon introduced the concept of placing a probability distribution on the set of plaintext message units, on the set of keys, and on the set of ciphertext message units. From these probability distributions, he defined both **perfect security** or **perfect secrecy** and what it means in a mathematical sense to break a code.

In general an encryption protocol has perfect security if an observer learns nothing about a plaintext message (expect possibly its maximal length) by observing the corresponding cipher text. To place this in a mathematical context, we let $\mathcal{P}$ denote the set of plaintext messages, $\mathcal{C}$ the set of ciphertext messages, and $\mathcal{K}$ the set of keys. We assume that there are probability distributions on all three sets. We refer to [Fre] for any information on probability theory.

Definition 3.4.1. Let $\mathcal{P}$ be the set of plaintext messages, $\mathcal{C}$ the set of ciphertext messages and $\mathcal{K}$ the set of keys for a cryptographic protocol $\mathcal{E}$. Then $\mathcal{E}$ has **perfect security** if for any given plaintext message P and corresponding ciphertext message C we have $Prob(P|C) = Prob(P)$, that is, the conditional probability of determining the plaintext message P, given knowledge of the ciphertext message C, is exactly the same as the absolute probability of determining the plaintext P.

Shannon then proved the following strong theorem, that describes when a cryptographic protocol is perfectly secure.

Theorem 3.4.2. *Let $\mathcal{P}$ be the set of plaintext messages, $\mathcal{C}$ the set of ciphertext messages and $\mathcal{K}$ the set of keys for a cryptographic protocol $\mathcal{E}$. We assume that all three sets have probability distributions on them. Suppose that the sets $\mathcal{P}, \mathcal{C}$ and $\mathcal{K}$ are all infinite and $Prob(P) > 0$ for any plaintext P. Then $\mathcal{E}$ has perfect security if and only if the key space is uniformly distributed and if for any plaintext P and any ciphertext C there is exactly one key k that encrypts P by C.*

From this definition, Shannon proved that the one-time pad, that we describe shortly, has perfect security.

Perfect security is difficult to obtain. Closely related is **semantic security**. This is the appropriate notion of security relative to a passive attacker. A cryptosystem is **semantically secure** relative to given security assumptions if it is infeasible for a passive attacker to determine relevant information about a plaintext given an observation of a ciphertext.

The one type of cryptographic protocol that has been proved to be perfectly secure is the **one-time pad** or **Vernam one-time pad**. This was developed for telegraphic security in the nineteenth century but was patented by Gilbert Vernam in 1917. In the paper of Shannon, mentioned above, it was proved that this method has perfect security. In a one-time pad the plaintext are sequences of bits of a given length and the ciphertext is the plaintext XOR'ed with a single time use key given also by a sequence of bits of the same length as the plaintext. Formally:

Definition 3.4.3. Suppose the sets $\mathcal{P}$ of plaintext messages, $\mathcal{C}$ of ciphertext messages and $\mathcal{K}$ of keys are all given by elements of $\{0, 1\}^n$. That is, plaintext messages, ciphertext messages and keys are all random bit strings of fixed length n. For a given key $k \in \mathcal{K}$ the encryption function is given by $F_k(p) = p \oplus k$ for $p \in \mathcal{P}$. We assume that the distribution on all three sets is the uniform distribution and a key k is only used once. The resulting encryption protocol is called a **one-time pad** or **Vernam one-time pad**.

Shannon proved that the one-time pad, under the assumptions provided in the definition, is perfectly secure, as long as the keys are randomly chosen and used only once.

Theorem 3.4.4 (Shannon). *A one-time key pad has perfect secrecy if the keys are randomly chosen from the uniform distribution of keys and a key is used only once.*

Shannon also showed that if a key is used more than once the perfect security is lost. A one-time pad is also not secure against an active attacker who can change, to his advantage, the length of the message.

Although the one-time pad is theoretically secure there are many problems with its practical use.
(1) The one-time pad requires perfectly secure and random keys. Hence,to be able to to use it we need a secure method to generate random bits. Theoretically, if pseudo-random numbers are used the perfect security falls by the wayside.
(2) For the one-time pad, there must be secure generation and secure transmission of the keys which must be as long as the plaintext.
(3) A key can be used only once.

For these reasons the one-time pad, while important theoretically, is not used to a great extent in encryption. However, stream ciphers, as described in the previous chapter, are a method to attempt to mimic the important properties of the one-time pad. For this reason stream ciphers are often called **Vernam ciphers**.

3.4.1 Vigènere Encryption and Polyalphabetic Ciphers

Another method to copy the security properties of the one-time pad is the use of **polyalphabetic ciphers**. By this we mean that a plaintext letter can be encrypted differently in a different place. The first version of a polyalphabetic coding system was given by Alberti in about 1466. This was mentioned in Chapter 1. Such cryptosystems now are called **Vigènere cryptosystems** after Vigènere who included the method in his book in 1586. The famous Enigma machine was a type of Vigènere cipher.

In the simplest variation of a Vigènere cryptosystem we have the following. We begin with an N-letter alphabet. This is the plaintext alphabet. We then form an $N \times N$ matrix P where each row and column is a distinct permutation of the plaintext alphabet. Hence P is a permutation matrix on the integers $0, \ldots, N - 1$. The communicating parties decide on a **keyword**. This is private between the communicators. To encrypt the keyword is placed above the plaintext message over and over until each plaintext letter lies below a letter in the keyword. The ciphertext letter corresponding to a given plaintext letter is determined by the alphabet at the intersection of the keyword letter and plaintext letter below it.

To clarify this description, and make Vigènere encryption precise, we repeat the example given in Chapter 1. Here we suppose that we have a 9 letter plaintext alphabet $A, B, C, D, E, O, S, T, U$. For simplicity in this example we will assume that each row is just a shift of the previous row, but any permutation can be used. We then have the following collection of alphabets.

Key Letters

		A	B	C	D	E	O	S	T	U
a	A	a	b	c	d	e	o	s	t	u
l	B	b	c	d	e	o	s	t	u	a
p	C	c	d	e	o	s	t	u	a	b
h	D	d	e	o	s	t	u	a	b	c
a	E	e	o	s	t	u	a	b	c	d
b	O	o	s	t	u	a	b	c	d	e
e	S	s	t	u	a	b	c	d	e	o
t	T	t	u	a	b	c	d	e	o	s
s	U	u	a	b	c	d	e	o	s	t

Suppose that Bob and Alice are the communicating parties and they have chosen the keyword BET. This is held in secret by the two of them. Suppose that the plaintext message to be encrypted is STAB DOC. The keyword BET is placed repeatedly over the message until the plaintext message if filled as below.

$$B \quad E \quad T \quad B \quad E \quad T \quad B$$
$$S \quad T \quad A \quad B \quad D \quad O \quad C$$

To encrypt we look at B which lies over S. The intersection of the B key letter and the S alphabet is a t so we encrypt the S with T. The next key letter is E which lies over T. The intersection of the E keyletter with the T alphabet is c. Continuing in this manner and ignoring the space we get the encryption

$$\text{STAB DOC} \ \rightarrow \ \text{TCTCTDD}$$

To descrypt we use the keyword BET but now use the method over the cipher text TCTCTDD.

For a long period of time polyalphabetic ciphers were considered unbreakable. In the mid nineteenth century Kasiski developed a technique for the cryptanalysis of Vigènere cryptosystems. Since then there has been substantial work done on the cryptanalysis of Vigènere ciphers. We will describe two of these techniques.

In 1863 the **Kasiski test** was developed. With this method one can try to determine the length of the key word. If the same partial word appears in the plaintext occurs several times then in general it will be encrypted differently. However, if the distance between the first letter of the respective partial word is a multiple of the key word length then the partial word will be encrypted the same. We have to look for partial sequences of length ≥ 3 in the cipher text. There respective distances are probably divisible by the length of the key word.

Then we determine the greatest common divisor as often as possible of such distances. If these are for example $14, 27, 21, 35, \ldots$, then the key word length could be 7.

We now can try to find the key word as follows. The keyword k is represented by a tuple $(n_1, \ldots, n_t)$ where we found the keyword length t. Then form the sets:

$$M_1 : \text{the set of the first}, (t + 1) - th, (2t + 1) - th, \ldots \text{ letters}$$
$$M_2 : \text{the set of the second}, (t + 2) - th, (2t + 2) - th, \ldots \text{ letters}$$
$$M_t : \text{the set of the } t - th, (2t) - th, (3t) - th, \ldots \text{ letters}$$

Now carry out a frequency analysis for the sets $M_1, M_2, \ldots, M_t$ and get possible values for $n_1, \ldots, n_t$ of the keyword.

In 1920, a second test, the **Friedmann test** was further developed. We assume that $N = 26$ and the alphabet is $\{A, B, C, \ldots\}$ but the method is entirely general. Let $(a_n, \ldots, a_n)$ be a sequence of length n of letters from the alphabet. Let n_1 be the number of A's that occur, n_2 the number of B's and so on. Then:

(a) There are $\binom{n_1}{2} = \frac{n_1(n_1-1)}{2}$ pairs of A's in the sequence, $\binom{n_2}{2}$ pairs of B's and so on.

(b) There are altogether $\binom{n_1}{2} + \cdots + \binom{n_{26}}{2}$ pairs of equal letters.

(c) The probability that two randomly chosen letters are equal is therefore given by

$$p_f = \frac{\binom{n_1}{2} + \cdots + \binom{n_{26}}{2}}{\binom{n}{2}}.$$

This number is called the **Friedmann Coincidence Index**.

For German texts the empirically established value for sequences of letters gives $p_F \approx 0.0762$

Theorem 3.4.5. *Given a sequence of letters of length m representing a Vigènere en-crypted cipher text and let p_F be its Friedman Coincidence Index. Let p_L be the Fried-mann Coincidence Index representing the corresponding plaintext (which is unknown but, for example, is usually taken as $p_L = 0.0762$ for German texts). Then the length t of the keyword is*

$$t \approx \frac{(p_L - \frac{1}{26})m}{(n-1)p_F - \frac{m}{26} + p_L}.$$

Proof. We assume approximately that t divides m where m is the length of the se-quence $(\alpha_1, \ldots, \alpha_m)$ representing the cipher reset. We write $(\alpha_1, \ldots, \alpha_m)$ in t columns

s_1	s_2	$\cdots$	s_t
—	—	$\cdots$	—
α_1	α_2	$\cdots$	α_t
α_{t+1}	α_{t+2}	$\cdots$	α_{2t}
$\vdots$	$\vdots$	$\cdots$	$\vdots$

Each column is encrypted with a single alphabet (or a Caesar cipher). We have always the same shift cipher. Hence the coincidence index is about p_L (each column is a shifted plain text). The probability that two letters from different columns are equal is about $\frac{1}{26}$.

Each column has $\frac{m}{t}$ letters. The number of pairs of letters, which are in the same column, is

$$t\binom{\frac{m}{t}}{2} = \frac{m(m-t)}{2t}.$$

The number of pairs of letters that are in different columns is

$$\binom{t}{2}\left(\frac{m}{t}\right)^2 = \frac{m^2(t-1)}{2t}.$$

Hence altogether the number K of pairs of equal letters is

$$K \approx \frac{m(m-t)}{2t}p_L + \frac{m^2(t-1)}{2t}\frac{1}{26}.$$

Hence

$$p_F = \frac{K}{\binom{m}{2}} = \frac{m-t}{t(m-1)}p_L + \frac{m(t-1)}{t(m-1)}\frac{1}{26}.$$

If we solve this for t we get the result

$$t \approx \frac{\left(p_L - \frac{1}{26}\right)n}{(n-1)p_F - \frac{n}{26} + p_L}.\qquad\square$$

3.4.2 Breaking a Protocol

We have defined the general concept of security and defined perfect security, but what exactly is the mathematical formulation of **breaking a code**? This is formally defined in terms of encrypted bits. Let $\epsilon = 0, 1$ be an encrypted bit. In the absence of any information, the determination of ϵ is entirely random, and hence, the probability of correctly determining ϵ is $\frac{1}{2}$. Given a set of security assumptions, then a cryptosystem is said to be **broken**, and an attack **successful**, if by using the attack, the probability of correctly determining an encrypted bit is greater than $\frac{1}{2}$, within a feasible time frame of the encryption. We make this precise.

Definition 3.4.6. Let $\mathcal{S}$ be a set of security assumptions for a cryptographic protocol, $\mathcal{A}$ a given attack, $\mathcal{B}$ the event of correctly determining an encrypted bit from the protocol and finally $\mathcal{T}$ a feasible time frame relative to the encryption. Then the attack $\mathcal{A}$ is successful and the protocol **broken** if

$$P(\mathcal{B}|\mathcal{S}, \mathcal{A}, \mathcal{T}) > \frac{1}{2}.$$

Notice that a cryptosystem may be broken but messages may still be relatively secure since the definition is given solely in terms of encrypted bits. It may still be difficult to then correctly determine a given message. Of course the greater the probability in the previous definition the greater the probability of decrypting the entire message.

A probability, marginally greater than $\frac{1}{2}$ may have a low probability of total message recovery. For example, suppose that $P(\mathcal{B}|\mathcal{S}, \mathcal{A}, \mathcal{T}) = .56$ and the decryption of bits is independent. To then correctly decrypt n bits the probability if $(.56)^n$ which of course can be very small. However, given a set of correct bits the message may be decrypted directly by knowing the gist of the message.

3.5 Complexity of Algorithms: Generic and Average Case

As we have seen, an extremely important aspect of any cryptosystem is its **security**, that is how well the encryption algorithm withstands various type of attacks. Theoretically this involves the **complexity** of the encryption algorithm. Generally an encryption algorithm is based upon some hard to solve problem. In classical cryptography these problems were either number theoretic or combinatorial, however, in group based cryptography the security can be based on group theoretical problems. In practice a problem should be hard to solve both theoretically and practically, like the discrete log problem, to be a valid problem for use in cryptography. Using a hard to solve problem in cryptography involves constructing a trapdoor function whose inverse depends upon the solution to the problem. This trapdoor function would be based on a secret key which makes the hard to solve problem solvable.

Determining how difficult a problem is to solve involves some important aspects of theoretical computer science and in particular complexity theory. In this section we go over some basic concepts.

First we describe a method to compare functions.

Suppose that $f(x), g(x)$ are positive real valued functions. Then

(1) $f(x) = \mathcal{O}(g(x))$ (This is read as $f(x)$ is big $\mathcal{O}$ of $g(x)$.) if there exists a constant A independent of x and an x_0 such that

$$f(x) \leq Ag(x) \quad \text{for all } x \geq x_0.$$

(2) $f(x) = o(g(x))$ (This is read as $f(x)$ is little o of $g(x)$.) if

$$\frac{f(x)}{g(x)} \to 0 \quad \text{as } x \to \infty.$$

In other words $g(x)$ is of a **higher order of magnitude** than $f(x)$.

(3) If $f(x) = \mathcal{O}(g(x))$ and $g(x) = \mathcal{O}(f(x))$, that is there exist constants A_1, A_2 independent of x and an x_0 such that

$$A_1 g(x) \leq f(x) \leq A_2 g(x) \quad \text{for all } x \geq x_0,$$

then we say that $f(x)$ and $g(x)$ are of the **same order of magnitude** and write

$$f(x) \approx g(x).$$

(4) If

$$\frac{f(x)}{g(x)} \to 1 \quad \text{as } x \to \infty$$

then we say that $f(x)$ and $g(x)$ are **asymptotically equal** and we write

$$f(x) \sim g(x).$$

In general we write $\mathcal{O}(g)$ or $o(g)$ to signify an unspecified function f such that $f = \mathcal{O}(g)$ or $f = o(g)$. Hence for example, writing $f = g + o(x)$ means that $\frac{f-g}{x} \to 0$ and saying that f is $o(1)$ means that $f(x) \to 0$ as $x \to \infty$.

It is clear that being $o(g)$ implies being $\mathcal{O}(g)$ but not necessarily the other way around. Further it is easy to see that

$$f \sim g \text{ is equivalent to } f = g + o(g) = g(1 + o(1)).$$

The **length** of an integer is its total number of binary bits. An algorithm is a **polynomial time algorithm** if there exists an integer d such that the number of bit operations required to perform the algorithm on integers of total length k is $\mathcal{O}(k^d)$. The ordinary arithmetic operations, for example, are polynomial time algorithms. An **exponential time algorithm** has a time estimate of $\mathcal{O}(e^{ck})$ on integer inputs of total length k. Factoring an integer or primality testing an integer by trial division is an exponential time algorithm.

Definition 3.5.1. A decision problem $\mathcal{P}$ is a **polynomial time problem** if there exists a constant c and an algorithm such that if an instance of $\mathcal{P}$ has length $\leq n$ then the algorithm answers the problem in time $\mathcal{O}(n^c)$. We denote the class of polynomial time problems by **P**.

Definition 3.5.2. A decision problem is in **non-deterministic polynomial time** if a person with unlimited computing power not only can answer the problem but if the answer is "yes" the solver can supply evidence that another person could use to verify the correctness of the answer in polynomial time. The demonstration that the "yes" answer is correct is called a **polynomial time certificate**. The class of non-deterministic polynomial time problems is denoted by **NP**.

The class **co-NP** consists of problems like those in NP but with "yes" replaced by "no".

If a problem can be solved in polynomial time it can certainly be solved in non-deterministic polynomial time. Therefore $P \subset NP$, that is the class of problems solvable in polynomial time is a subset of the class of problems solvable in non-deterministic polynomial time. However, no example has been found of a problem solvable in non-deterministic polynomial time that cannot be solved in polynomial time. Therefore it has not been proved that $P \neq NP$. The $P = NP$ question is the most important open problem now in computer science.

If $\mathcal{P}_1$ and $\mathcal{P}_2$ are two decision problems then we say that $\mathcal{P}_1$ **reduces** to $\mathcal{P}_2$ in polynomial time if there exists a polynomial time algorithm that constructs an answer for $\mathcal{P}_2$ given a solution to $\mathcal{P}_1$. Hence the problem of breaking the ElGamal cryptosystem reduces to the discrete log problem (see next chapter).

Finally a decision problem is **NP-complete** if any other problem in NP can be reduced to it in polynomial time. A decision problem is **NP-hard** if any NP problem reduces to it.

A good cryptographic decision problem should be at least NP-hard. However, what is also important is **average case complexity** as opposed to **worst case complexity**.

The worst case complexity of a problem is the hardest instance of the problem. Hence an algorithm to solve a problem has exponential worst case complexity if there exists an instance of the problem which is exponential to solve with that algorithm. The average case complexity of an algorithm is the complexity on average of the algorithm where a measure is put on the set of inputs. Finally an algorithm has **generic case complexity** if given a measure on the set of inputs the algorithm solves the problem with measure one.

A more complete discussion of the various complexity classes and the interrelationship between them can be found in the book [MSU1].

3.6 Exercises

3.1 The table below gives the approximate statistical frequency of occurrence of letters in the English language. The following passage is encrypted with a simple substitution cipher without punctuation. Use a frequency analysis to decode it.

ZKIRNVMFNYVIRHZKLHRGREVRMGVTVIDSR
XSSZHZHGHLMOBKLHRGREVWRERHLIHLMVZ
MWRGHVOUKIRNVMFNYVIHKOZBZXIFXRZOI
LOVRMMFNYVIGSVLIBZMWZIVGSVYZHRHUL
IGHSHVMLGVHGSVIVZIVRMURMRGVOBNZMB
KIRNVHZMWGSVBHVIEVZHYFROWRMTYOLXP
HULIZOOGSVKLHRGREVRMGVTVIH

letter	frequency	letter	frequency	letter	frequency
A	.082	B	.015	C	.028
D	.043	E	.127	F	.022
G	.020	H	.061	I	070
J	.002	K	.008	L	.040
M	.024	N	.067	O	.075
P	.019	Q	.001	R	.060
S	.063	T	.091	U	.028
V	.010	W	.023	X	.001
Y	.020	Z	.001		

3.2 A text of Edgar Allen Poe from the "The Gold-Bug" is encrypted with the Vigenère cryptosystem using the keyword *gold*. The ciphertext is

guzrjuwdygtqzvpeogsrvgsryhpoobekkrpyozdvkoe
iufebubpgkuchkglqjhslxhphtatqahpvtccwnslvzoyg
hmyrxhspgwyexoyfngpykbekrwxekodwywohyvzrztcrs
hshrsqwkmprlhshjslwngshgrlekswltsquuaekkhchk
hsuuirkzvpvnceioteblspwuie

Use a frequency analysis to decode the ciphertext.

3.3 Using an encryption with a Vigènere cryptosystem we obtained the ciphertest

TLYIWMSIOVPZKMXKUVQKXCZAOY
BZMGZJCBOYLRKRLSJBBLJZMXCC
ZAOYZVQKWRFVKLSWLOCAYIEDQV
MNLRRBOYGYVOYBZMLPGEMGPFLV
NOYMMBKGTPDPRVVCTCNMSDCUIX
YYLNNPLCMDKRVVDPGCLOCPVQCP

Calculate the key length k with the help of the Kasiski test and the Friedmann test.

3.4 Use the Vigenére code to encode the German sentence

In Hamburg an der Alster traf ich einen ehemaligen Schulkameraden und wir plauderten von alten Zeiten.

with help of the keyword *Schule*.

3.5 The following message is in ciphertext encoding a German language plaintext message. The last ten digits have been lost and don't appear on the list below. What are the missing digits and which cryptosystem was used?

$$10, 4, 13, 13, 12, 4, 8, 20, 1, 16, 14, 15, 12$$
$$20, 4, 12, 15, 17, 4, 19, 1, 8, 20, 1, 19, 8, 20$$
$$18, 2, 4, 19, 2, 14, 2, 15, 4, 14, 12, 15, 6, 16$$
$$13, 2, 15, 1, 5, 8, 13, 10, 4, 8, 4, 13, 12, 4$$
$$14, 4, 18, 2, 12, 1, 4, 15, 12, 1, 8, 19, 1, 12$$
$$4, 14, 18, 2, 12, 12, 17, 20, 16, 19, 2, 14$$
$$13, 12, 15, 10, 2, 4, 6, 8, 13, 1, 4, 8, 15, 15$$
$$20, 1, 19, 4, 10, 4, 13, 13, 12, 4$$

3.6 The following message has been encrypted with the Vernam one-time pad. The ciphertext is:

PITTW

What is the plaintext?

3.7 The plaintext 010101 is encrypted with a Vernam one-time pad to the ciphertext 101010. What is the key?

(a) 000000

(b) 111111

(c) 000111

3.8 Let m be the number of bit operations required to compute 3^n in binary. Estimate using the big-$\mathcal{O}$ notation m as a simple function of n, the input binary digits.

(a) Do the same for n^n.

(b) Estimate in terms of a simple function of n and N the number of bit operations required to compute N^n.

3.9 Consider the ciphertext message

CHGFFBMIFUHFKSYWWOXTSROLO

The following is known about the cryptosystem that constructed this code.

(1) A Vigènere cryptosystem was used with a pseudo random k.

(2) The key was generated with a 4-bit LFSR with elements from $\mathbb{Z}_{26}$ in the cells.

(3) A probable word in the plaintext is BROADCAST.

(a) Decrypt the message.
(b) Encrypt the rest of the text *announced the Allied invasion to be expected within forty-eight hours.*
(c) Calculate the period of this LFSR.

4 Cryptographic Protocols

4.1 Cryptographic Protocols

Up to this point, we have primarily considered encryption systems, that is, methods to encrypt and decrypt secret messages. Recall that cryptosystems can be classified as symmetric key or public (asymmetric) key. In the former, both the encoding and decoding algorithms are supposedly known only to the sender and receiver, while in the latter, the encryption method is public knowledge, but only the receiver knows how to decode.

Secure confidential message transmission is only one type of task that must be done with secrecy and there are many other tasks and procedures that are important in cryptography. What is meant by a cryptographic task will be made a bit more precise via specific examples below.

Suppose that several parties want to perform a cryptographic task. To accomplish this the involved parties must communicate and cooperate. Further each party has to obey certain rules and use certain preassigned algorithms. The set of all methods and algorithms used to perform such a cryptographic task is called a **cryptographic protocol**. A cryptosystem is just one type of cryptographic protocol. Formally:

Definition 4.1.1. Suppose that several parties want to manage a cryptographical task. Then they must communicate with each and cooperate, and hence, each party must follows certain rules and implement certain agreed upon algorithms. The set of all such methods and rules to perform a cryptographical task is called a **cryptographic protocol**.

Several cryptographic tasks are described below. Each will be discussed in more detail later in this chapter. As more techniques are introduced, we will look at further instances of these cryptographic tasks, and cryptographic protocols to handle them.

(1) **Authentication**: Authentication refers to the process of determining that a message, supposedly from a given person, does come from that person and further has not been tampered with. Included in the general topic of authentication are the concepts of **hash functions** and **digital signatures**. Hash functions will be introduced in the next section. Another important usage is **password identification**. We will discuss a secure backup password system using non-commutative methods later in the book.

(2) **Key Exchange and Key Transport**: In a **key exchange protocol** two people, usually called Bob and Alice, exchange a secret shared key to be used in some symmetric encryption. In a **key transport protocol**, one party transports to another, a secret key that is to be used.

(3) **Secret Sharing**: Secret sharing involves methods where some secret is to be shared by k people but not available to any proper subset of them. There are many

ways to accomplish this and it is related to a classical lock and key problem. A beautiful simple solution to the general problem, using polynomial interpolation, is due by Shamir. In Section 4.4 we will describe Shamir's technique as well as a nice geometric alternative to it.

(4) **Zero Knowledge Proof**: A **zero-knowledge proof** is an argument that convinces someone that you have solved a problem, for example a combinatorial problem, without giving away the solution. This is tied to authentication.

4.2 Cryptographic Hash Functions

Before describing individual cryptographic protocols, we introduce the concept of cryptographic hash functions. These functions play a crucial role in many different types of protocols.

Modern cryptography is done via a computer. Therefore all messages, both plaintext and ciphertext, can actually be presented as binary strings, that is strings of binary bits, $\{0, 1\}$'s. We first define a **hash function** as a function on bit strings. Later we will give a more general definition.

Definition 4.2.1. A **cryptographic hash function** is a deterministic function

$$h : S \rightarrow \{0, 1\}^n$$

that returns for each arbitrary block of data, called a **message**, a fixed size bit string. The bit string returned for a given message is called its **hash value**. The function should have the property that a change in the data will change the hash value. The hash value is also called the **digest**.

To satisfy the properties necessary to be used in a cryptographic protocol an ideal hash function has the following properties:
(1) It is easy to compute the hash value for any given message.
(2) It is infeasible to find a message that has a given hash value. This is called **pre-image resistance**.
(3) It is infeasible to modify a message without changing its hash.
(4) It is infeasible to find two different messages with the same hash value. This is called **collision resistance**.

Hash functions can be used with encryption functions to construct authenticated encryption protocols. Suppose that Bob and Alice want to communicate openly. They have exchanged a secret key k that supposedly only they know. Let f_k be an encryption function or encryption algorithm based on the key k. Alice wants to send the message m to Bob and m is given as a binary bit string. Alice sends to Bob

$$f_k(m) \oplus h(k)$$

where $\oplus$ is the XOR operation, that is, addition modulo 2.

Bob knows the key k and hence its hash value $h(k)$. He now computes

$$f_k(m) \oplus h(k) \oplus h(k).$$

Since addition modulo 2 has order 2 we have

$$f_k(m) \oplus h(k) \oplus h(k) = f_k(m).$$

Bob now applies the decryption algorithm g_k to decode the message.

Alice could have just as easily sent $f_K(m) \oplus k$. However, sending the hash has two benefits. Usually the hash is shorter than the key and from the properties of hash functions it gives another level of security. Used in this manner, a cryptographic hash function can serve as a **digital signature**. As we will see, tying the secret key to the actual encryption in this manner is the basis for the ElGamal and Elliptic curve cryptographic methods.

The encryption algorithm f_k is usually a symmetric key encryption so that anyone knowing k can encrypt and decrypt easily. However, it should be resistant to plaintext-ciphertext attacks. That is, if an attacker gains some knowledge of a piece of plaintext together with the corresponding ciphertext, it should not compromise the whole system.

The encryption algorithm can either be a **block cipher** or a **stream cipher**.

For the following, especially the examples, we give a more general definition.

Definition 4.2.2. Let M_1, M_2 be non−empty sets. Then a **(cryptographic) hash function** is a map $f : M_1 \rightarrow M_2$ with the following properties:

(1) f is free of collisions, that is, there is no efficient method to find two elements $x_1, x_2 \in M_1$ with $f(x_1) = f(x_2)$.
(2) f is not efficiently invertible, that is, if $y \in Im(f)$ then there is no efficient way to find $x \in M_1$ with $f(x) = y$.

Although an ideal hash function is not efficiently invertible, to be useful in a cryptographic situation there must be a way to invert such a function or at least partially invert it on the set of messages. This leads us to the concept of a **trapdoor**.

Definition 4.2.3. A hash function with a **trapdoor** is a map $f : M_1 \rightarrow M_2$ satisfying:

(1) f is free of collisions.
(2) There is a secret key by means of which f is efficiently invertible.
(3) If one does not know the secret key then f is not efficiently invertible.

The secret key is the **trapdoor**.

Some Examples of Hash Functions

Here we present several examples of hash functions.

(1) **Quadratic Residues as Hash Functions**: Recall that if n is an integer and $x^2 \equiv b \pmod n$ then b is called a **quadratic residue**. Hence b is a quadratic residue modulo n if $\bar{b}$ has a squareroot in $\mathbb{Z}_n$. If not it is a **quadratic non-residue**. We will discuss quadratic residues in Chapter 4.

Let p, q be two distinct odd primes and let $n = pq$. Determine one quadratic non-residue y modulo n with $y \in \{1, \ldots, n-1\}$ and the greatest common divisor $(y, n) = 1$. Now consider the function

$$f : \{0, 1\} \times \{0, 1, \ldots, N\} \rightarrow \{0, 1, \ldots, n-1\}$$

with $N \geq 1$ and $(\epsilon, m) \mapsto y^\epsilon m^2 \pmod n$.

Then the following hold:

(a) If $\epsilon = 0$ then $y^\epsilon m^2$ is a quadratic residue modulo n.

(b) If $\epsilon = 1$ then $y^\epsilon m^2$ is a quadratic non-residue modulo n.

(c) The function f is collision free in the first argument. That is, we always have $f(0, m) \neq f(1, m_1)$.

(d) If we choose N in such a way that the quadratic residues m^2 modulo n for $m = 0, 1, \ldots, N$ are pairwise distinct then f is free of collisions.

There is no known efficient method to discover whether or not y is a quadratic residue modulo $n = pq$ if p and q are unknown. Therefore the function f is a hash function.

(2) **Modular Squaring as a Hash Function**: Let $p, q \geq 3$ be two distinct odd primes and let $n = pq$. Let $\mathbb{Z}_n^*$ be the unit group modulo n (see Chapter 5 for more details). Now consider the function

$$f : \mathbb{Z}_n^* \rightarrow \mathbb{Z}_n^* \quad \text{given by } x \mapsto x^2.$$

Then the following hold:

(a) If $y \in \text{im}(f)$ then $|f^{-1}(y)| = 4$ (see Chapter 5, Section 5.2).

(b) If we know p and q then for $y \in \text{im}(f)$ we may find efficiently one x such that $y = x^2 = f(x)$.

(c) If we only know n but not p and q there is no efficient algorithm to find an x such that $x^2 = y$ for $y \in \text{im}(f)$. All the known algorithms are of exponential order (see Chapter 3).

(d) If we restrict f to the set

$$\left\{ \bar{0}, \bar{1}, \ldots, \overline{\left[\frac{n}{2} \right]} \right\} \cap \mathbb{Z}_n^*$$

then f is free of collisions and may be used as a hash function with a trapdoor, the secret keys being p and q.

(3) **Discrete Logarithm as Hash Functions**: Recall that if $G = \langle g \rangle$ is a cyclic group and $h \in G$ with $h = g^t$ then t is called a **discrete logarithm** of h. If p is a prime then $\mathbb{Z}_p^*$ is a cyclic group.

Let p be a large odd prime and g a primitive root modulo p, that is $\bar{g}$ is a generator of the unit group $\mathbb{Z}_p^*$. Now consider the function

$$f : \{1, \ldots, p - 1\} \rightarrow \{\overline{1}, \ldots, \overline{p - 1}\} \quad \text{given by } x \mapsto g^x.$$

There is no known efficient algorithm to solve the discrete log problem if $p - 1$ has a large prime factor. Hence in this case f is a hash function with the generator g as the trapdoor.

(4) **Factorization as a Hash Function**: Let $\mathcal{P}$ denote the set of all odd prime numbers. Define the function

$$f : \mathcal{P} \times \mathcal{P} \rightarrow \mathbb{N}$$

given by

$$(p, q) \mapsto pq = n.$$

There is no known efficient algorithm to factorize $n = pq$ if both $p - 1$ and $q - 1$ have large prime factors. Hence in this case the function f is a hash function with a trapdoor.

We note that each of these, especially (3) and (4), can be made the basis of a public key exchange protocol. We will return to these in Chapter 6 when we introduce the Diffie-Hellman and RSA protocols.

4.3 Authentication Protocols

When a confidential message is transmitted there are several important aspects that must be verified. First, there must be a verification to the receiver that the sender is who he or she claims to be. Secondly, there must be a verification to the sender that the receiver is also who he or she claims to be. Next, there should be a verification that the message has not been altered in any way. Finally, there should be in many message transmissions, some form of **undeniability**, that is a procedure that makes it impossible for a sender to claim that he or she did not send the message. All of these verifications are handled by an **authentication protocol**.

There are many specialized types of authentication protocols handling one or more of the verification tasks mentioned in the above paragraph. Here we will discuss just the general ideas in message authentication.

Hash functions provide one technique for message authentication. This was explained briefly in the previous section. Suppose that m is the true message and h is a hash function. What is done in this type of authentication protocol is to verify, using a public key system, $h(m)$ the hash value of the message. From the collision and preimage resistance properties of a hash function, verification of the hash value of the message provides verification of the message itself. Similarly, included with the message could be the hash value of the sender. Again because of the properties of the hash function this provides authentication of the sender.

Any public key encryption system has a type of built-in verification. This will be explained here and explained again in Chapter 7 on public key methods. In that chapter we will also see how to take a specific public key encryption system and build from it an authentication procedure. In Chapter 7 we will introduce how to accomplish this with the RSA encryption system and in Chapter 8 how to do authentication via elliptic curves.

As introduced in Chapter 1, the basic idea in a public key cryptosystem is to have a **one-way function** or **trapdoor function**. That is a function which is easy to implement but very hard to invert. Hence it becomes simple to encrypt a message but very hard, unless you know the inverse, to decrypt. The trapdoor is the simple knowledge that one needs to invert.

The standard model for public key systems is the following. We suppose that Alice's public key is an injective function f_A and her private key is g_A, the left inverse of f_A. It follows that for any message m we have $g_A(f_A(m)) = m$. Here the function f_A indexed for A is Alice's public key. Similarly Bob has a public key f_B.

Alice wants to send a message to Bob. The encrypting map f_A for Alice is public knowledge as well as the encrypting map f_B for Bob. On the other hand, the decryption algorithms g_A and g_B are secret and known only to Alice and Bob respectively. Let m be the message Alice wants to send to Bob. She sends $f_B(m)$. The function f_B is public knowledge and it is easy to encrypt. To decrypt Bob now applies g_B to what he has received. Recall that presumably only he knows his private key g_B. Doing this he obtains

$$g_B\left(f_B(m)\right) = m$$

and hence recovers the message.

If Bob receives a message, how can he be certain that it actually comes from Alice. Within this standard model it is relatively easy to build in **verification** or **authentication**. As above let m be the message Alice wants to send to Bob. Now instead of sending $f_B(m)$ she sends

$$f_B\left(g_A(m)\right).$$

Now to decode Bob applies first g_B, which is his private key so that only he knows. This gives him

$$g_B\left(f_B\left(g_A(m)\right)\right) = g_A(m).$$

He then looks up f_A which is publically available and applies this

$$f_A\left(g_A(m)\right) = m$$

to obtain the message. This supplies **verification** and **authentication** in the following manner. Suppose $\mathcal{P}$ is Alice's verification; signature, social security number etc. If Bob receives $f_B(\mathcal{P})$ it could be sent by anyone since f_B is public. On the other hand, since only Alice supposedly knows g_A getting a reasonable message from $f_A(g_B(f_B(g_A(m))))$ would verify that it is from Alice. Applying g_B alone should result in nonsense.

Important in all authentication protocols is the concept of a **message authentication code** or **MAC**. This is also called a **tag**. A MAC is a short piece of information used to authenticate a message. In using a hash function for authentication the hash value is the tag.

In a MAC protocol, a secret key and an arbitrary length message m to be authenticated, are taken as input. Then a MAC value or tag is outputted. The MAC value protects both the message data integrity as well as its authenticity by allowing verifier's to detect any changes in the message content. A MAC is stronger than just using a hash value.

It is assumed that both sender and receiver have the message m to be verified. The sender generates a secret key k then computes, as a function of both the key and the message, a MAC x. The sender then transmits the secret key and the MAC to the receiver. The receiver then computes in the same manner as a function of the message and the transmitted secret key the MAC value. If the same MAC x is found by the receiver then the message is authenticated and the integrity checked. If they differ something is not right.

4.4 Digital Signatures

A strong form of authentication is given by a **digital signature protocol**. This is to be thought of as the analog of a written signature but for electronic or digital messages. A digital signature is then a mathematical procedure to handle the following authentication needs;
(1) verifying the authenticity of the sender,
(2) verifying that the message has not been tampered with,
(3) providing a means to prevent deniability.

These are the essential requirements of any signature, written or electronic. A digital signature is then tied to a given electronically transmitted message.

In many countries, including the United States and the European Union, digital signatures carry the same legal weight as handwritten signatures.

Mathematically, a **digital signature protocol** consists of three cryptographic algorithms: **a key generation algorithm**, a **digital signing algorithm** and a **digital signature verification algorithm**. In more detail:
(1) **A Key Generation Algorithm** is an algorithm that selects a private key randomly from a set of private keys and then outputs the private key and a corresponding public key.

(2) **A Digital Signing Algorithm** is an algorithm that given a transmitted message m produces a **signature** that will be tied to the message.
(3) **A Digital Signature Verification Algorithm** is an algorithm that accepts three inputs, a message m, a public key k and a signature S and then based on these inputs either accepts or rejects the authenticity of the message.

It is assumed that the authenticity of the given signature for a fixed message and a fixed private key is verified by using the corresponding public key given in the key generation algorithm. It is also required that it is computationally infeasible to generate a valid digital signature for anyone not possessing the private key.

In most applications, the hash value of the message is what is signed rather then the message itself. This is basically for efficiency purposes. Signing algorithms are public key and hence relatively time inefficient. Using the hash value speeds up the process. Further the message often has to be broken up into smaller pieces to be transmitted. If each piece is to be signed there is no verification that all the constituent blocks are included or in the proper order. Signing the hash value of the whole message serves as an authentication of the whole message.

Any public key message transmission system can be used to develop a digital signature protocol. In this section we describe the **Digital Signature Algorithm** or **DSA** which has become the signing standard in many countries where digital signatures are legally binding. This is based on the **ElGamal Signature Method**. Other digital signing protocols, such as the RSA signature method, will be described in later chapters.

In the Digital Signature Algorithm a signed message is to be sent from party A to party B. The first step is to choose the parameters and then the per-user keys, both public and private. These steps are followed by the **signing algorithm** and finally the **verification algorithm**. There is some basic number theory involved in the DSA, and a reader, if desired, may come back and look at this example after reading Chapter 5 on basic number theoretic methods. In the description of DSA below $\mathbb{Z}_p^*$ represents the non-zero elements of the modular ring $\mathbb{Z}_p$. Recall (or see Chapter 5) that if p is a prime then each non-zero element $x \in \mathbb{Z}_p$ has an inverse, that is an element y such that xy is congruent to 1 modulo p.

DSA: Choice of Parameters and Setup
(1) A chooses a large odd prime integer q.
(2) A chooses a second prime integer p with $p \equiv 1 \pmod{q}$. Such primes p exist (see [FR]).
(3) $|\mathbb{Z}_p^*| = p - 1$ an $q \mid (p - 1)$. Then there exists an integer g such that $\bar{g}$ has order q in $\mathbb{Z}_p^*$ (see Chapter 5). To calculate such a g we do as follow:

(i)   Choose a random integer $g_0 \in \{1, \ldots, p-1\}$ and calculate $g_0^{\frac{p-1}{q}}$ modulo p.

(ii) Repeat (i) as long as the result is not congruent to 1 modulo p.

Then define $g = g_0^{\frac{p-1}{q}}$. The order of g is q because q is prime.

DSA: Signature Private and Public Keys

(1) A chooses as a private key a random $x \in \{1, \ldots, p-1\}$.

(2) A calculates $y \equiv g^x \pmod{p}$.

(3) The **public key** for the signing algorithm is (p, q, g, y) while the **private key** is x.

DSA: Signing Algorithm

(1) Let m be the transmitted message and $h(m)$ its hash value for a given hash function h.

(2) A generates randomly per message an integer $k \in \{1, \ldots, q-1\}$ and calculates

$$r \equiv (g^k \pmod{p}) \pmod{q}.$$

(3) If $r = 0$ choose another k.

(4) Calculate $s \equiv k^{-1}(h(m) + xr) \pmod{q}$.

(5) If $s = 0$ choose another k.

(6) The **signature** is (r, s).

DSA: Verification Algorithm

(1) B calculates an integer t with $st \equiv 1 \pmod{(p-1)}$ and then the two integers

$$u_1 \equiv th(m) \pmod{q},$$
$$u_2 \equiv tr \pmod{q}.$$

(2) B then determines the integer

$$g^{u_1} y^{u_2} \pmod{p}$$

and applies the division algorithm with division by q. If the remainder is equal to r then B accepts the signature.

This algorithm is correct, that is the verification algorithm will always accept a genuine signature. From $g^q \equiv 1 \pmod{p}$ we get

$$g^{u_1} y^{u_2} \equiv g^{th}(g^x)^{tr} \equiv g^{t(h+xr)} \equiv g^{tsk} \equiv g^k \pmod{p}$$

and this gives the remainder r modulo q.

As with hand-written signatures there must be security against **forgery**. For digital signatures there are various different levels of forgeries.

(1) A **total break** is where an attacker recovers the signing key.
(2) A **universal forgery** is where an attacker gains the ability to forge signatures for any message.
(3) A **selective forgery** is where an attacker gains the ability to forge a signature on a message of the attacker's choice.
(4) An **existential forgery** is where an attacker recovers a valid message-signature pair not already known to the attacker.

Whereas an ideal digital signature protocol withstands all types of forgeries, a good secure digital signature protocol must be able at least to withstand an existential forgery. In their seminal paper [GMR] Goldwasser, Micali and Rivest describe a hierarchy of possible attacks against digital signature protocols. These mirror the types of attacks on general cryptographic protocols.
(1) A **key-only attack** is where an attacker has access to only the public verification key.
(2) A **known message attack** is where an attacker has access to valid signatures for a variety of messages known but not chosen by the attacker.
(3) An **adaptive chosen message attack** is where an attacker has access to signatures on arbitrary chosen messages.

Closely related to digital signatures is the problem of secure password verification and the preparation of a challenge response system. We describe these in more detail in Section 10.7 in connection with a group based method for challenge response password verification.

4.5 Secret Sharing Schemes

A **secret sharing scheme** is a method to distribute a secret among a group of participants by giving a share of the secret to each. The secret can be recovered only if a sufficient number of participants combine their shares.

Formally we have the following **secret sharing problem**. We have a secret K and a group of n participants. This group is called the **access control group**. A **dealer** allocates shares to each participant under given conditions. If a sufficient number of participants combine their shares then the secret can be recovered. If $t \leq n$ then an (t, n)-**threshold scheme** is one with n total participants and in which any t participants can combine their shares and recover the secret but not fewer than t. The number t is called the **threshold**. It is a **secure secret sharing scheme** if given less than the threshold there is no chance to recover the secret. If a measure is placed on the set of secrets, and on the set of shares, security can be made precise by saying that given less than the threshold all secrets are equally likely but given the threshold there is a unique secret. Secret sharing is an old idea but was formalized mathematically in independent papers in 1979 by Adi Shamir [Sha1] and George Blakely [Bla].

In the 1979 paper Shamir [Sha1] proposed a beautiful (t, n)-threshold scheme, based on polynomial interpolation, that has many desirable properties. It has become the standard method for solving the (t, n)-secret sharing problem, although there are modifications for different situations that we will discuss in this section. Blakely in his original paper [Bla] proposed a geometric solution based on hyperplanes that is less space efficient, for computer storage, than Shamir's. In Blakely's scheme the distributed shares are larger than the secret, whereas in Shamir's scheme they are the same size.

The protection of a private key in an encryption protocol provides strong motivation for the ideas of secret sharing. In an encryption protocol, only the private key used in the encryption scheme is the secret, and not the encryption method itself. When we examined the problem of maintaining sensitive information, we considered two issues: availability and secrecy. If only one person keeps the entire secret, then there is a risk that the person might lose the secret or the person might not be available when the secret is needed. Hence it is often wise to allow several people to have access to the secret. On the other hand, the more people who can access the secret, the higher the chance the secret will be leaked. A secret sharing scheme is designed to solve these issues by splitting a secret into multiple shares and distributing these shares among a group of participants. The secret can only be recovered when the participants of an authorized subset join together to combine their shares.

A secret sharing scheme is a cryptographic primitive with many applications, such as in security protocols, multiparty computation (MPC), Pretty Good Privacy (PGP) key recovering, visual cryptography, threshold cryptography, threshold signature, etc. The paper by Chum, Fine and Zhang [CFZ] provides a strong description of general secret sharing and access security methods.

4.5.1 The Shamir Secret Sharing Scheme

Given a secret K, a (t, n)-**secret sharing threshold scheme** is a cryptographic primitive in which a secret is split into pieces (shares) and distributed among a collection of n participants $\{p_1, p_2, \ldots, p_n\}$ so that any group of t or more participants, with $(t \leq n)$, can recover the secret. Meanwhile, any group of $t - 1$ or fewer participants cannot recover the secret. By sharing a secret in this way the availability and reliability issues can be solved.

Shamir solved the secret sharing problem in a very simple but beautiful manner using polynomial interpolation. The general idea in a Shamir (t, n)-threshold scheme is the following. Let F be any field and $(x_1, y_1), \ldots, (x_n, y_n)$ be n points in F^2 with pairwise distinct x_i. A polynomial $P(x)$ over F **interpolates** these points if $P(x_i) = y_i$ for $i = 1, \ldots, n$. The polynomial $P(x)$ is called an **interpolating polynomial** for the given points. The crucial theoretical result is that for any n points (x_i, y_i) with distinct x_i there always exists a unique interpolating polynomial of degree $\leq n - 1$.

Theorem 4.5.1 (Polynomial Interpolation Theorem). *Let F be any field and $x_1, \ldots, x_n$ be n pairwise distinct elements of F and $y_1, \ldots, y_n$ any elements of F. Then there exists a **unique** polynomial of degree $\leq n - 1$ that interpolates the n points $(x_i, y_i), i = 1, \ldots, n$.*

Using this theorem, the Shamir (t, n)-threshold scheme works in the following manner. A field F with more than n elements is chosen. In general if F is a finite field we assume that the order of F is much much larger than n, the number of participants. The secret K is taken as an element of the field F and a polynomial $P(x)$ of degree $t - 1$ is chosen with the secret K as its constant term. Then pairwise distinct elements of F, $x_1, \ldots, x_n$, are chosen with no $x_i = 0$. The points $(x_i, P(x_i))$ are distributed to each of the n participants. By the polynomial interpolation theorem, Theorem 4.6.1, any t participants can determine the interpolating polynomial $P(x)$ and hence recover the secret K. Given an infinite field and less than t people there are infinitely many polynomials of degree $t - 1$ that can interpolate the given points and hence finding the correct polynomial has probability zero. In a finite field Shamir proved that under random choices for the x_i each secret in F is equally likely so guessing the secret is a random choice from F.

In the next subsection an alternative version to this Shamir scheme will be outlined that uses inner product spaces and the closest vector theorem rather than interpolation.

We now present a more explicit version of the Shamir scheme using the finite field $F = GF(q)$ where $q = p^k$ with $k \geq 1$ and p is a large prime. By using a finite field Shamir was able to place a finite measure on the set of plaintexts and ciphertexts and showed that with this scheme if there are less than t people all secrets are equally likely.

In a secret sharing protocol we assume that there is a **dealer** who fairly and correctly distributes the shares among the participants.

The Shamir (t, n)-Threshold Secret Sharing Scheme
Share distribution: Let K be the secret. The dealer generates a polynomial $P(x)$ of degree at most $t - 1$ over $F = GF(q)$, where q is much larger than n as follows:

$$P(x) = a_0 + a_1 x + \cdots + a_{t-1} x^{t-1} \tag{4.1}$$

where $a_0 = K$ is the secret, $a_1, \ldots, a_{t-1} \in F$ and are generated randomly.

The dealer arbitrarily chooses pairwise distinct $x_i \in F \setminus \{0\}, i = 1, 2, \ldots, n$. Usually, $x_i = i$ will be chosen for simplicity. $x_1, x_2, \ldots, x_n$ are stored in a public area. The dealer calculates $y_i = P(x_i), i = 1, 2, \ldots, n$, and distributes to the n participants via a secure channel so that each participant p_i gets one share y_i. For the rest of the paper, we will not repeat the criteria of the generation of the coefficient a_i of the polynomial $P(x)$ and the calculation of the shares $P(x_i)$.

Secret Recovery (i): When any t participants join together, we have the following system of t equations. For simplicity, we assume the participants $p_1, p_2, \ldots, p_t$ join together.

$$y_1 = P(x_1) = a_0 + a_1 x_1 + \cdots + a_{t-1} x_1^{t-1},$$
$$y_2 = P(x_2) = a_0 + a_1 x_2 + \cdots + a_{t-1} x_2^{t-1},$$
$$\ldots,$$
$$y_t = P(x_t) = a_0 + a_1 x_t + \cdots + a_{t-1} x_t^{t-1}.$$

In matrix representation, it will be:

$$\begin{pmatrix} 1 & x_1 & \cdots & x_1^{t-1} \\ 1 & x_2 & \cdots & x_2^{t-1} \\ \vdots & & \cdots & \vdots \\ 1 & x_t & \cdots & x_t^{t-1} \end{pmatrix} \begin{pmatrix} a_0 \\ a_1 \\ \vdots \\ a_{t-1} \end{pmatrix} = \begin{pmatrix} y_1 \\ y_2 \\ \vdots \\ y_t \end{pmatrix}$$

Let M be the above $t \times t$ Vandermonde matrix. Its determinant is

$$\det(M) = \prod_{1 \le j < k \le t} (x_k - x_j).$$

Since we choose pairwise different points for the participants, that is, pairwise distinct x_i's, $\det(M) \ne 0$, and this guarantees a unique solution. We can solve the system of equations by Gaussian elimination or Cramer's rule. Hence the secret can be recovered.

Secret Recovery (ii): Another method is to use Lagrange interpolation. We can construct the polynomial of degree at most $t - 1$ by any t different points $(x_1, y_1), \ldots, (x_t, y_t)$ as

$$P(x) = \sum_{i=1}^{t} y_i l_i(x),$$

where

$$l_i(x) = \prod_{j=1, j \ne i}^{t} \frac{x - x_j}{x_i - x_j} = \frac{(x - x_1) \cdots (x - x_{i-1})(x - x_{i+1}) \cdots (x - x_t)}{(x_i - x_1) \cdots (x_i - x_{i-1})(x_i - x_{i+1}) \cdots (x_i - x_t)}.$$

So, the secret a_0 will be

$$a_0 = P(0) = \sum_{i=1}^{t} y_i \prod_{j=1, j \ne i}^{t} \frac{-x_j}{x_i - x_j}.$$

Shamir in his original paper was able to prove that this secret sharing scheme is **perfect** in the sense that for $t - 1$ participants any secret $K \in F$ is equally likely. If F is an infinite field, then the probability of correctly guessing the secret is zero.

4.5.2 Alternatives for Secret Sharing Protocols

In this section and the next we describe some alternatives and enhancements to the Shamir Secret Sharing Scheme. The first alternative is a geometric method based on the closest vector theorem and introduced in [CFRZ]. It is similar to the original Blakely scheme but has the advantage that the secret is the same size as the shares. This was not the case in the Blakely protocol. This scheme, like the Shamir scheme, is also perfect in the sense that any t people out of the size n access control group can recover the secret but given less than t each possible secret is equally likely.

The Shamir scheme depends on the uniqueness of interpolating polynomials. The geometric alternative scheme depends on the **closest vector theorem**.

Recall the following facts. Let W be a real inner product space and V a subspace of finite dimension t. Suppose that $w \in W$ and $e_1, e_2, \ldots, e_t$ is an orthonormal basis of V. Then the **closest vector theorem** is the following (see [Atk]).

Theorem 4.5.2 (Closest Vector Theorem). *Let W be a real inner product space and V a subspace of finite dimension t. Suppose that $w \in W$, with w not in V, and $e_1, e_2, \ldots, e_t$ is an orthonormal basis of V. Then the unique vector $w^* \in V$ closest to w is given by*

$$ w^* = < w, e_1 > e_1 + < w, e_2 > e_2 + \cdots + < w, e_t > e_t $$

where $< \ , \ >$ is the inner product on W.

Notice that given any basis for the subspace V, the Gram-Schmidt orthonormalization procedure (see [Atk]) can be used to find an orthonormal basis for V. Hence given $w \in W$ we can algorithmically always find w^*, the unique vector in V closest to w. If a basis for V is not known and we only have knowledge or information on proper subspace spans in V of dimension less than t we cannot do this procedure and any point in W can be the secret. That is, if we do not have complete knowledge of a basis for V, we cannot apply the closest vector theorem. Further, since given a subspace of dimension less than t there are infinitely many subspaces of dimension t properly containing it, there is a probability of zero of obtaining the subspace V with only partial knowledge.

The Secret Sharing Scheme

We start with an inner product space W of dimension m and an access control group of size n. We assume that the dimension m of W is much greater than n, that is $m \gg n$. Within W there is a hidden subspace V of dimension $t < n$. The secret to be shared is given as an element in this hidden subspace, that is the secret $v \in V$ a vector in V.

The dealer distributes to each of the n members of the access control group, $i = 1, \ldots, n$, two vectors, v_i, w, where $v_i \in V$, and w is a vector in the big space W. The common vector w has the property that $w \notin V$ and v is the vector in V closest to w. In general the vector w can be given publically. The set $\{v_1, v_2, \ldots, v_n\}$ has the property that any subset of size t is independent. Hence any subset of size t determines a basis for V.

Suppose t valid users get together. They can determine a basis for V and hence using the Gram-Schmidt procedure (see [Atk]) determine an orthonormal basis. Since w is given, they can determine v by the closest vector theorem and recover the secret.

Given a subset of size less than t the given vectors generate a subspace of V of dimension less than t and hence in W there are infinitely many extensions to subspaces of dimension t. This implies that determining V with less than t elements of a basis has probability zero.

As suggested by Shamir the secret should be altered periodically. In this method it is extremely easy to change the secret v without altering much of the scheme; simply send each user a new w.

This is a general method like the Shamir protocol. In [FMoR], Fine, Moldenhauer and Rosenberger, compared several different secret sharing plans including the classic Shamir plan and the CFRZ plan.

In many instances more involved secret sharing tasks must be handled. As an example consider the following situation. We are in a company that has a directors and vice-directors. The directors and vice-directors are in the access control group but they do not have equal weight. Suppose that a secret can be recovered only if one of the following conditions is satisfied:

(a) two directors of the company cooperate;

(b) three vice-directors of the company cooperate;

(c) one director and two vice-directors of the company cooperate.

Thus here the threshold for the access control group differs depending on the status of the members.

A geometric method to handle this more difficult cryptographic task can done as follows:

(1) Choose a sufficiently large finite field $K = \mathbb{F}_q$, $q = p^n$ with p a prime, $n \geq 1$, or take $K = \mathbb{Q}$.

(2) Consider the 3-space K^3 and choose the point $(0, 0, s)$ on the z-axis where s is the secret.

(3) Choose in K^3 randomly a plane E which contains the point $(0, 0, s)$ but which does not contain the z-axis at all.

(4) Choose randomly in E a straight line G which intersects the z-axis at the point $(0, 0, s)$.

(5) Choose randomly on $G \setminus (0, 0, s)$ pairwise distinct points $D_1, \ldots, D_k$ which are distributed to the directors.

(6) Choose randomly on $E \setminus G$ pairwise distinct points $V_1, \ldots, V_t$ such that points of the V_i are on one straight line and that no point D_k is on a straight line through two distinct points V_i and V_j. The points $V_1, \ldots, V_t$ are distributed to the vice-directors.

This protocol then satisfies the desired conditions for the access control groups.
(1) Any two of the directors can calculate G and hence $(0, 0, s)$.
(2) Any three of the vice-directors can calculate E and hence $(0, 0, s)$.
(3) Any one of the directors combined with any two of the vice-directors can calculate E and hence $(0, 0, s)$.

4.5.3 Verifying Secret Sharing Protocols (VSS)

In a standard secret sharing protocol it is assumed that the dealer and the participants are honest. To ensure the proper behavior of the dealer and the participants we enhance the standard secret sharing scheme. A **verifying secret-sharing protocol**, denoted VSS, is one such enhancement. The aim of a VSS protocol is to be certain that the dealer and the participants behave correctly. A verifiable secret sharing protocol ensures that even if the dealer is dishonest there is a well-defined secret that the participants can recover. Verifiable secret sharing is important in secure multiparty computation.

There are many different verifiable secret sharing protocols (see [CFZ]). We first consider the case of a $(2, 2)$-VSS and then give a generalization. We must construct a protocol where we can be certain that the dealer and the participants behave correctly. This first VSS protocol uses finite group theory. For a formal definition of homomorphism and isomorphism see Chapter 9.
(1) Let G and H be two groups and $\phi : G \rightarrow H$ be a group homomorphism so that

$$\phi(g_1 g_2) = \phi(g_1) \phi(g_2)$$

which is also a hash function.
(2) Let $s \in G$ be the secret. The dealer chooses $g_1, g_2 \in G$ with $g_1 g_2 = s$ and publishes $\phi(g_1), \phi(g_2)$. Nobody can efficiently calculate s from the published data since ϕ is a hash function. However, everyone can verify that $\phi(s) = \phi(g_1) \phi(g_2)$.
(3) The dealer tells g_1 to the participant A_1 and g_2 to the participant A_2. Each A_i can calculate $\phi(g_i)$ from his secret g_i and can check if the correct partial secret was received.
(4) When the participants reconstruct the secret s each A_i can prove if the other participant has exposed his correct partial secret by calculating $\phi(g_i)$ and comparing with the published values.

We now consider a more general (t, n)-VSS protocol. This example requires some elementary number theory. The required material can be found in the next chapter and a reader may want to complete that and then return to this section. Recall that if p is a prime then an integer g is a primitive element modulo p if the order of g is $p - 1$ in the multiplicative group $\mathbb{Z}_p^*$. For an integer g we let $\bar{g}$ denote its residue class modulo p.

In the situation from the $(2, 2)$-VSS protocol above we choose for the group G the additive group modulo $(p - 1)$, that is $(\mathbb{Z}_{p-1}, +)$. For H we choose the multiplicative group $(\mathbb{Z}_p^*, \cdot)$ where p is a sufficiently large prime.

Let g be a primitive element modulo p. As a hash function we choose

$$\phi : \mathbb{Z}_{p-1} \to \mathbb{Z}_p^*$$

by

$$a \mapsto \overline{g}^a \quad \text{for } a \in \{0, 1, \ldots, p - 1\}.$$

The dealer chooses randomly a polynomial

$$f(x) = a_0 + a_1 x + \cdots + a_{t-1} x^{t-1}$$

where $a_0, a_1, \ldots, a_{t-1} \in \mathbb{Z}_{p-1}$ and $a_0 = s$ is the secret.

The dealer then publishes $\overline{g}^{a_i}$ for $i = 0, 1, \ldots, t - 1$. Recall that $\overline{g}^{a_0} = \overline{g}^s$.

The dealer chooses randomly pairwise distinct elements $x_1, \ldots, x_n \in \mathbb{Z}_{p-1}$. He calculates $s_i = f(x_i)$ for $i = 1, \ldots, n$. He then publishes the values x_i and $\overline{g}^{s_i}$ for $i = 1, \ldots, n$.

Then:

(a) Each participant A_i can prove if they received s_i correctly by calculating $\overline{g}^{s_i}$ and comparing with the published values.

(b) Each participant can prove if

$$\prod_{j=0}^{t-1} (\overline{g}^{a_j})^{x_i^j} = \overline{g}^{f(x_i)} = \overline{g}^{s_i}.$$

Practically the dealer cannot cheat. All distributions from the dealer can be proved by the participants.

Finally we present another protocol similar to the one above, due to ElGamal, where the participants cannot cheat but discrepancies from the dealer are hard to discover. This is known as the **ElGamal** (t, n)-threshold signature protocol.

(1) The participants reach an agreement on two large primes p, q with $q|(p - 1)$ and a hash function $h : \mathbb{Z}_p \times \mathbb{Z}_p \to \mathbb{Z}_p$. We assume that $\mathcal{M} = \mathbb{Z}_p$ is the set of plain text units.

(2) There exists a unique cyclic subgroup G_q of $\mathbb{Z}_p^*$ with q elements (see the ElGamal signature method). The participants reach an agreement on a generator $g \in G_q$.

(3) The dealer chooses randomly $a_0, \ldots, a_{t-1} \in \mathbb{Z}_q$ and defines the polynomial

$$f(x) = a_0 + a_1 x + \cdots + a_{t-1} x^{t-1} \in \mathbb{Z}_q[x].$$

(4) Let $s = f(0) = a_0$ be the secret and let $y \equiv g^s \pmod{p}$ be made public.

(5) The dealer chooses randomly pairwise distinct elements $\mu_i, x_i \in \mathbb{Z}_q^*$ for $i = 1, \ldots, n$ and calculates the partial secrets $s_i = \mu_i + f(x_i)$ for the participants A_i. The values x_i are published.

(6) The dealer calculates for each participant A_i the values $y_i \equiv g^{s_i} \pmod p$ and $z_i \equiv g^{\mu_i} \pmod p$ and makes these public.

(7) To sign the protocol, each participant A_i chooses randomly $k_i \in \mathbb{Z}_q^*$ and calculates $r_i \equiv g^{k_i} \pmod p$. The value r_i will be sent to the other participants.

(8) Suppose that t participants have sent their values, say, $A_1, \ldots, A_t$ have sent their values $r_1, \ldots, r_t$. Then participant A_i with $1 \le i \le t$ calculates the value

$$R = r_1 \cdots r_t \equiv g^{k_1 + \cdots + k_t} \pmod p$$

and the value

$$E \equiv h(m, R) \pmod p, \quad m \in \mathcal{M}.$$

Then A_i has his partial signature

$$c_i \equiv s_i \prod_{j \ne i} \frac{-x_j}{x_i - x_j} + k_i E \pmod q.$$

(9) The dealer verifies the partial signatures by checking if

$$y_i^{\prod_{i \ne j} \frac{-x_j}{x_i - x_j}} r_i^E \equiv g^{c_i} \pmod p.$$

(Recall that $r_i \equiv g^{k_i} \pmod p$ and $y_i \equiv g^{s_i} \pmod p$.)

In this case the dealer then calculates $\sigma \equiv (c_1 + \cdots + c_t) \pmod q$. The signature of $m \in \mathcal{M}$ is then

$$(\{A_1, \ldots, A_t\}, R, \sigma).$$

(10) The verification of the signature is to calculate

$$T \equiv \prod_{i=1}^{t} z_i^{\prod_{i \ne j} \frac{-x_j}{x_i - x_j}} \pmod p$$

and check if

$$g^\sigma \equiv y T R^E \pmod p.$$

Notice that the set $\{A_1, \ldots, A_t\}$ has to be given. However, given such a set the participants practically cannot cheat and the dealer can realize this. The signature is correct. Further, in this protocol, which verifies that the participants cannot cheat, it is difficult to discover any violations to the protocol by the dealer.

4.6 Zero-Knowledge Proofs

The final type of cryptographic procedure that we examine in this chapter is a **zero-knowledge proof**. We defer the technical definition to the end of this section. For the present, we say that a **zero-knowledge proof** or **zero-knowledge protocol** is an interactive method for one party to prove to another that a statement is true, without revealing any additional information.

The statement is usually mathematical in nature. In a zero knowledge proof the person proving is called the **prover** and the person who is shown the proof is the **verifier**. In encryption protocols, Alice and Bob, are usually used to stand for the communicating parties. In a prover and verifier protocol, for example a zero-knowledge proof, the prover is normally referred to as Peggy, while the verifier is called Victor. If proving the statement requires knowledge of some secret information on the part of the prover, a zero-knowledge proof implies that the verifier will not be able to prove the statement in turn to anyone else, since the assumption is that the verifier does not possess the secret information and does not learn anything about it. Notice that the statement being proved must include the assertion that the prover has such knowledge. (Otherwise, the statement would not be proved in zero-knowledge, since at the end of the protocol the verifier would gain the additional information that the prover has knowledge of the required secret information). If the statement consists only of the fact that the prover possesses the secret information, it is a special case known as **zero-knowledge proof of knowledge**.

For zero-knowledge proofs of knowledge, the protocol requires interactive input from the verifier, usually in the form of a challenge or challenges such that the responses from the prover will convince the verifier if and only if the prover does have the claimed knowledge. We have looked at this earlier in terms of password verification. In this case Victor is convinced that Peggy has the secret information but does not learn it himself. However, password verification lacks the crucial property of a zero-knowledge proof, namely the *proof* that Victor does not gain any information about Peggy's password.

A zero-knowledge proof is usually not a proof in the mathematical sense but rather a probabilistic proof. For each challenge of the verifier, the prover has a certain probability of being correct purely by guessing. For each correct response by the prover, the probability of consistently correct guessing gets smaller. At the conclusion of the protocol the probability is very high that the prover does know the secret. This will become clearer in the examples below.

It remains to find a way for us to prove that, given a proposed zero-knowledge proof protocol, the verifies does not obtain any information about the secret. This can be achieved as follows: we show that the verifier can produce two recordings of the repeated challenge-and-response experiments. The first recording assumes that Peggy knows the secret and just documents the execution of the protocol. The second recording assumes that Peggy does not know the secret and is guessing her replies. In this case, the verifies records only the successful experiments when Peggy replies correctly and deletes the unsuccessful ones. Then the two recordings are shown to a neutral outside party, and we have to demonstrate that this outside party cannot distinguish the correct recording from the fake one. Then this certifies that the verifier did not receive any extra information during the correct execution of the protocol, as this extra information could be used to distinguish the two recordings.

In the following we present two examples of zero-knowledge proofs.

Example 4.6.1. The following setup is a standard prototype of a zero-knowledge proof of knowledge. We suppose that Peggy, the prover, has knowledge of the secret word used to open a magic door in a cave. She wants to verify to Victor, the verifier, that she does indeed have possession of this magic word without revealing any information about it. The cave is shaped like a rectangle, with the entrance on one side and the magic door blocking the opposite side.

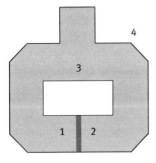

Fig. 4.1: The Zero-Knowledge Cave.

The zero knowledge proof works as follows. Victor waits outside the cave (at point 4) as Peggy goes in. They label the left and right paths from the entrance 1 and 2. Peggy randomly takes either path 1 or 2. Then Victor signals into the cave the name of the path he wants Peggy to use to exit, either 1 or 2, chosen at random. Providing she really does know the magic word, this is easy: if necessary, she opens the door and arrives at point 3 along the desired path. Note that Victor does not know which path she took to get into the cave.

However, suppose that Peggy does not know the word. Then she is only able to return by the desired path if Victor happens to name the same path that she entered by. Since Victor chooses path 1 or 2 at random, she has a 50 % chance of guessing correctly. If they repeat this process many times, her chance of successfully answering all of Victor's requests is extremely small. In fact, after n repetitions it is $\left(\frac{1}{2}\right)^n$. Thus, if Peggy arrives at the correct exit many times, Victor can conclude that she is very likely to know the secret word.

Now let us prove the zero-knowledge property for this protocol. We assume that Victor produces two recordings of the experiments according to the above instructions. If the setup is created correctly, a neutral outside party cannot distinguish the recordings. Therefore we can conclude that Victor does not get any information (such as the time taken by Peggy to arrive at point 3 or the echo of her saying the magic word) from which he could deduce clues about the magic word.

A second example of a zero-knowledge proof of knowledge protocol is based on some number theory and is a numeric version of the cave example.

Example 4.6.2. In the following **Diffie-Hellman Zero-Knowledge Protocol**, Peggy proves to Victor that she knows a certain discrete logarithm (see Section 6.3 for more information on discrete logarithms).
(1) Peggy and Victor agree on a large odd prime number p and a primitive element g modulo p with $g \in \{1, \ldots, p - 1\}$. We then have $\mathbb{Z}_p^* = \langle \overline{g} \rangle$.
(2) Peggy chooses a random integer $a \in \{1, \ldots, p - 1\}$ and calculates $b \equiv g^a \pmod{p}$. She sends b to Victor and will prove to Victor that she knows a.
(3) Peggy chooses a random integer $x \in \{1, \ldots, p - 1\}$ and calculates $y \equiv g^x \pmod{p}$. She sends y to Victor.
(4) Victor chooses (for the question) randomly an $\epsilon \in \{0, 1\}$ and sends it to Peggy.
(5) If $\epsilon = 0$ then Peggy sends $z = x$ to Victor. If $\epsilon = 1$ then Peggy sends $x + a$ modulo $(p - 1)$ to Victor.
(6) If $\epsilon = 0$ then Victor verifies that $g^z \equiv y \pmod{p}$, while if $\epsilon = 1$ then Victor verifies that $g^z \equiv yb \pmod{p}$.

This method proves with probability $\frac{1}{2}$ that Peggy knows a. If $\epsilon = 0$ then

$$g^z \equiv g^x \equiv y \pmod{p}$$

while if $\epsilon = 1$ then

$$g^z \equiv g^{x+a} \equiv g^x g^a \equiv yb \pmod{p}.$$

Hence the probability is $\frac{1}{2}$ that Peggy does not know a. As in the cave example, given n rounds where Peggy is correct, there is only a probability of $(\frac{1}{2})^n$ that she does not know a. The zero-knowledge property of this protocol follows in exactly the same way as above.

To conclude this section, we look more formally at zero-knowledge proofs. Assume that Peggy, the prover, has a secret m and that she has to prove to Victor, the verifier, that she knows m. Further, there is a probability distribution on Peggy's responses to Victor's challenges, so that there is a probability for each of Peggy's responses to be correct. In addition, we assume that Victor's challenges are independent. At the most basic level a zero-knowledge proof must satisfy three properties:
(1) **Completeness:** If the statement is true, the honest verifier (that is, one following the protocol properly) will be convinced of this.
(2) **Soundness:** If the statement is false, no cheating prover can convince the honest verifier that it is true, except with some small probability.
(3) **Zero-Knowledge:** If the statement is true, no cheating verifier learns anything other than this fact. This is formalized by showing that every cheating verifier has some simulator that, given only the statement to be proved (and no access to the prover), can produce a transcript that "looks like" an interaction between an honest prover and the cheating verifier.

The first two properties are present in general interactive proof systems. The third property is what makes the proof zero-knowledge. As can be seen and has been mentioned earlier, zero-knowledge proofs are not proofs in the mathematical sense of being deterministic proofs, but rather probabilistic proofs. In each zero-knowledge protocol, there is some small probability, the **soundness error**, that a cheating prover will be able to convince the verifier of a false statement. The assumption is that repeated applications of the interactive protocol will reduce the soundness error to a negligibly small value.

4.7 Exercises

4.1 Distribute the secret 42 using the Shamir secret sharing scheme evenly among three people such that any two can put together the secret.

4.2 The company *Ruin Invest* has two directors, seven department managers, and 87 further employees. A valuable customer file is protected by a secret key. Develop a procedure for the distribution of the information about the key among the following groups of authorized people:
(1) both directors together,
(2) one director and all seven department managers together,
(3) one director, at least four department managers, and also at least 11 employees.

4.3 A group of at least three employees of a company decides to calculate their average income and then to compare it to their own personal salaries. They are all honest but noone wishes to reveal his salary to the other employees. Describe a protocol to solve this problem.

4.4 Alice and Bob want to decide who among the two of them has the higher salary but neither wants to divulge their salary. Describe a protocol to solve this problem.
Hint: You may assume that there is a finite number of possible salaries.

4.5 Give a fair protocol for an electronic roulette game. You need not deal with the settings in detail.

4.6 Let $p, q \geq 3$ be two large primes and $n = pq$. The integer n is odd so $n = 2k + 1$ for some $k \in N$. Show that

$$f : \{1, \ldots, k + 1\} \to \mathbb{Z}$$

by $x \mapsto x^2$ is a hash function with a trapdoor.

4.7 Two players are playing a game of internet chess. They will decide who plays white. Show how to do this with the help of a hash function,

$$h : \{0, 1\} \times M_1 \to M_2$$

where M_1, M_2 are large sets and so that no partner can alter the decision at a later time.

4.8 This is the **Shamir-No-Key Protocol**.

(a) Consider the following protocol.

    (1) A and B agree publically on a large odd prime p.

    (2) A generates a pair $(a, a') \in \mathbb{Z} \times \mathbb{Z}$ and $aa' \equiv 1 \pmod{(p-1)}$.

    (3) B generates a pair $(b, b') \in \mathbb{Z} \times \mathbb{Z}$ and $bb' \equiv 1 \pmod{(p-1)}$.

    (4) A wants to send the plaintext $m \in \{1, 2, \ldots, p-1\}$ to B. To do this he calculates $x \equiv m^a \pmod{p}$ and sends this to B.

    (5) B calculates $y \equiv x^b \pmod{p}$ and sends this to A.

    (6) A calculates $z \equiv y^{a'} \pmod{p}$ and sends this to B.

    (7) B calculates $m \equiv z^{b'} \pmod{p}$.

    Show that this protocol is correct.

(b) Considering the protocol from part (a) show that if an attacker can find from x and y a $d \in \mathbb{Z}$ with $y^d \equiv x \pmod{p}$, then the attacker can determine m.

5 Elementary Number Theoretic Techniques

5.1 Cryptography and Number Theory

Number theory plays a prominent role in many areas of cryptography. In the simplest case of applying a cryptosystem to an N letter alphabet, as explained in Chapter 1, we consider the letters as integers modulo N. The integers modulo N form a ring called the **modular ring**, $\mathbb{Z}_N$, and hence encryption is done within this ring. Operations within the various modular rings are called **modular arithmetic**. The encryption algorithms then apply number theoretic functions and use modular arithmetic on these integers. It follows that encryption maps on k-length-message units are functions

$$f : \mathbb{Z}_N^k \to \mathbb{Z}_N^k.$$

In addition to providing a framework for the encrypted alphabets, number theoretic problems provide the security in the basic public key protocols. Recall from the last chapter that the security of a cryptographic protocol is usually based on the difficulty of solving some "hard" problem. For example in the Diffie-Hellman method, one of the main techniques in public key cryptography, the hard problem is the **discrete log problem**. We will discuss the discrete log problem and some attempted solutions in Chapter 6. The discrete log problem was used in one of our zero-knowledge proof examples (see Section 4.6). The other most commonly used public key method, the RSA method, is based on the factorization problem. The factorization problem is the relative difficulty of factoring large integers. This will be discussed also in Chapter 6.

In the next two chapters we introduce and discuss all the relevant material from number theory and related abstract algebra, especially finite group theory, that will be necessary for cryptographic applications. For more material and more complete discussions on these mathematical area see the book [FR]. For further references on abstract algebra we refer to [CFR1] or [FGR].

We first discuss modular arithmetic.

5.2 Modular Arithmetic

For each natural number n, we will construct a ring, called the **ring of integers modulo n**, which we will denote by $\mathbb{Z}_n$, and which will be finite with n elements. Each of these rings will be a commutative ring with an identity. Collectively the rings $\mathbb{Z}_n$ are called the **modular rings** and operations within these rings are called **modular arithmetic**. We assume that the reader has a basic familiarity with primes, divisibility, the fundamental theorem of arithmetic and the concepts of a greatest common divisor or gcd and a least common multiple or lcm. Given integers a, b we denote their gcd by

(a, b). If $(a, b) = 1$ then the integers a, b are **relatively prime** or **coprime**. For more information on basic number theory we refer to [FR].

Given $n \geq 1$, in order to construct the ring $\mathbb{Z}_n$ we must first introduce the relation of **congruence modulo n** on the integers $\mathbb{Z}$.

Definition 5.2.1. Suppose that $n \geq 1$ is a positive integer. If x, y are integers such that $x - y$ is a multiple of n or equivalently that n divides $x - y$ we say that x is **congruent to** y **modulo** n and denote this by $x \equiv y \pmod{n}$. If n does not divide $x - y$ then x and y are **incongruent** modulo n.

If $x \equiv y \pmod{n}$, then y is called a **residue** of x modulo n. Given $x \in \mathbb{Z}$ the set of integers

$$\{y \in Z \mid x \equiv y \pmod{n}\}$$

is called the **residue class** for x modulo n. We denote this by $[x]$. Notice that $x \equiv 0 \pmod{n}$ is equivalent to x being a multiple of n. We first show that the residue classes partition $\mathbb{Z}$, that is that each integer falls in one and only one residue class.

Theorem 5.2.2. *Given $n \geq 1$, an integer, then congruence modulo n is an equivalence relation on the integers. Therefore the residue classes partition the integers.*

Proof. Recall that a relation $\sim$ on a set S is an **equivalence relation** if it is **reflexive**, that is $s \sim s$ for all $s \in S$; **symmetric**, that is if $s_1 \sim s_2$ then $s_2 \sim s_1$; and **transitive**, that is if $s_1 \sim s_2$ and $s_2 \sim s_3$, then $s_1 \sim s_3$. If $\sim$ is an equivalence relation then the equivalence classes $[s] = \{s_1 \in S \mid s_1 \sim s\}$ partition S.

We recall the notation $m \mid n$ indicates that m divides n, so that $m = nk$ for some integer k. Consider $\equiv \pmod{n}$ on $\mathbb{Z}$. Given $x \in \mathbb{Z}$, $x - x = 0 = 0 \cdot n$ so $n \mid (x - x)$ and so $x \equiv x \pmod{n}$. Therefore $\equiv \pmod{n}$ is reflexive.

Suppose that $x \equiv y \pmod{n}$. Then

$$n \mid (x - y) \Longrightarrow x - y = an$$

for some $a \in \mathbb{Z}$. Then $y - x = -an$ so $n \mid (y - x)$ and so $y \equiv x \pmod{n}$. Therefore the relation $\equiv \pmod{n}$ is symmetric.

Finally, suppose that $x \equiv y \pmod{n}$ and $y \equiv z \pmod{n}$. Then $x - y = a_1 n$ and $y - z = a_2 n$. But then $x - z = (x - y) + (y - z) = a_1 n + a_2 n = (a_1 + a_2)n$. Therefore $n \mid (x - z)$ and $x \equiv z \pmod{n}$. Therefore $\equiv \pmod{n}$ is transitive and the theorem is proved. $\square$

Hence given $n > 0$ every integer falls into one and only one residue class. We now show that there are exactly n residue classes modulo n.

Theorem 5.2.3. *Given an integer $n > 0$ there exist exactly n residue classes. In particular, $[0], [1], \ldots, [n - 1]$ gives a complete set of residue classes.*

Proof. Given $x \in \mathbb{Z}$, x must be congruent modulo n to one of $0, 1, 2, \ldots, n - 1$ since the remainder when we divide x by n is less than n. Further these are all incongruent modulo n. As a consequence

$$[0], [1], \ldots, [n - 1]$$

gives a complete set of residue classes modulo n and hence there are n of them. □

Recall that a **ring** and a **field** are defined as follows (see [CFR1]).

Definition 5.2.4. A **ring** is a set R with two binary operations defined on it. These are usually called addition, denoted by $+$, and multiplication denoted by $\cdot$ or just by juxtaposition, satisfying the following six axioms:
(1) Addition is commutative: $a + b = b + a$ for all $a, b \in R$.
(2) Addition is associative: $a + (b + c) = (a + b) + c$ for $a, b, c \in R$.
(3) There exists an additive identity, denoted by 0 such that $a + 0 = a$ for each $a \in R$.
(4) For each $a \in R$ there exists an additive inverse denoted by $-a$ such that

$$a + (-a) = 0.$$

(5) Multiplication is associative: $a(bc) = (ab)c$ for $a, b, c \in R$.
(6) Multiplication is left and right distributive over addition:

$$a(b + c) = ab + ac$$
$$(b + c)a = ba + ca$$

for $a, b, c \in R$.
(7) If in addition multiplication is commutative: $ab = ba$ for all a, b in R, then R is a **commutative ring**.
(8) Further if there exists an multiplicative identity denoted by 1 such that $a \cdot 1 = a$ and $1 \cdot a = a$ for each a in R then R is a **ring with identity** or a **ring with unity** if R satisfies (1) through (6) and (8).
(9) If R satisfies (1) through (8) then R is a **commutative ring with an identity** or **with a unity**.
Finally, a **field** F is a commutative ring with an identity such that each non-zero element of F has a multiplicative inverse. That is, if $x \in F$ with $x \neq 0$ then there exists an $x_1 \in F$ such that $xx_1 = 1$.

We now construct the ring $\mathbb{Z}_n$ on the set $\{[0], [1], \ldots, [n-1]\}$ of residue classes modulo n. We first need the following whose proof is straightforward.

Lemma 5.2.5. *If $a \equiv b \pmod{n}$ and $c \equiv d \pmod{n}$, then*
(1) $a + c \equiv b + d \pmod{n}$,
(2) $ac \equiv bd \pmod{n}$.

We now define operations on the set of residue classes.

Definition 5.2.6. Consider the complete residue system $\{[0], \ldots, [n-1]\}$ modulo n. On this set of residue classes define

(1) $[x_i] + [x_j] = [x_i + x_j]$,

(2) $[x_i][x_j] = [x_i x_j]$.

From this we get the following result. The proof is a direct verification of the ring axioms.

Theorem 5.2.7. *Given a positive integer $n \geq 1$, the set of residue classes*

$$\{[0], \ldots, [n-1]\}$$

*forms a commutative ring with an identity under the operations defined above. This is called the **ring of integers modulo n** and is denoted by $\mathbb{Z}_n$. The zero element is $[0]$ and the identity element is $[1]$.*

We usually consider $\mathbb{Z}_n$ as consisting of $0, 1, \ldots, n-1$ with addition and multiplication **modulo n**. For an integer a we will denote the residue class for a and hence the element of $\mathbb{Z}_n$ by $\bar{a}$. When there is no confusion we will denote the element $\bar{a}$ in $\mathbb{Z}_n$ as just a. We emphasize, however, that even though we are denoting the elements of $\mathbb{Z}_n$ by $0, 1, 2, \ldots, n-1$, they are still residue classes or equivalence classes. This is just to simplify our notation. Below we give the addition and multiplication tables modulo 5, that is in $\mathbb{Z}_5$. We note that, in line with the comment that these elements are different from usual integers. Consider $3 + 4$ and $3 \cdot 4$. They are both equal to 2 in $\mathbb{Z}_5$. This is certainly not what happens if we add and multiply 3 and 4 in $\mathbb{Z}$.

Example 5.2.8. Addition and Multiplication Tables for $\mathbb{Z}_5$

+	0	1	2	3	4
0	0	1	1	3	4
1	1	2	3	4	0
2	2	3	4	0	1
3	3	4	0	1	2
4	4	0	1	2	3

.	0	1	2	3	4
0	0	0	0	0	0
1	0	1	2	3	4
2	0	2	4	1	3
3	0	3	1	4	2
4	0	4	3	2	1

Notice for example that modulo 5, $3 \cdot 4 = 12 \equiv 2 \pmod 5$ so that in $\mathbb{Z}_5$, $3 \cdot 4 = 2$. Similarly $4 + 2 = 6 \equiv 1 \pmod 5$ so in $\mathbb{Z}_5$, $4 + 2 = 1$.

5.3 Units and the Multiplicative Group $\mathbb{Z}_n^*$

In a commutative ring with an identity, (like each of the modular rings $\mathbb{Z}_n$) a **unit** is an element that has a multiplicative inverse. For example in the integers $\mathbb{Z}$ only 1 and -1 are units. On the other hand, in a field, such as the rationals, by definition, every non-zero element is a unit. The set of units in such a ring always is non-empty since the multiplicative identity is a unit. We denote the set of units in a ring R by $U(R)$.

Recall next that a **group** G is a set with one binary operation that we denote by $\cdot$ or just by juxtaposition, defined on it, satisfying the following three axioms:

(1) The operation is associative. That is, we have $g_1(g_2 g_3) = (g_1 g_2)g_3$ for all $g_1, g_2, g_3 \in G$.

(2) There is an identity, denoted 1, for the operation. Hence for each $g \in G$ we have $1g = g1 = g$.

(3) Each $g \in G$ has a inverse. That is there exists a g^{-1} for each g such that $gg^{-1} = g^{-1}g = 1$.

If the operation is also commutative then G is called an **abelian group**. The additive part of any ring is an abelian group and hence the additive parts of all the modular rings are finite abelian groups.

We need that the set of units in a commutative ring forms an abelian group.

Lemma 5.3.1. *For a commutative ring R with identity the set of units $U(R)$ forms an abelian group under the ring multiplication. We call this the **unit group** of R.*

We usually denote the unit group in a ring R by R^*. Hence $\mathbb{Z}_n^*$ will denote the unit group in the modular ring $\mathbb{Z}_n$. Given an integer a it is easy to determine if $\bar{a}$, the residue class of a modulo n, is a unit and hence is in $\mathbb{Z}_n^*$. This will be true if and only if a is relatively prime to n. Recall that the gcd of two integers a, b can be expressed as a linear combination of them. Further, a and b are relatively prime if and only if 1 is expressible as a linear combination of a and b.

Lemma 5.3.2. *If $n > 1$ and $a \in \mathbb{Z}$ then $\bar{a}$ is a unit in $\mathbb{Z}_n^*$ if and only if $(a, n) = 1$.*

Proof. Suppose $(a, n) = 1$. Then there exists $x, y \in \mathbb{Z}$ such that $ax + ny = 1$. This implies that $ax \equiv 1 \pmod{n}$ which in turn implies that $\overline{ax} = 1$ in $\mathbb{Z}_n^*$ and therefore a is a unit.

Conversely, suppose that a is a unit in $\mathbb{Z}_n^*$. Then there is an $x \in \mathbb{Z}_n^*$ with $ax = 1$. In terms of congruence then

$$ax \equiv 1 \pmod{n} \implies n | ax - 1 \implies ax - 1 = ny \implies ax - ny = 1.$$

Therefore 1 is a linear combination of a and n and so $(a, n) = 1$. $\square$

The next result, called the **Euclidean algorithm**, provides a technique for both finding the gcd of two integers and expressing the gcd as a linear combination of the two integers. If an element $a \in \mathbb{Z}_n$ is a unit this will also provide a method to find its inverse in $\mathbb{Z}_n$.

Theorem 5.3.3 (The Euclidean Algorithm). *Given integers b and $a > 0$ form the repeated divisions*

$$b = q_1 a + r_1, \qquad \text{with } 0 < r_1 < a,$$
$$a = q_2 r_1 + r_2, \qquad \text{with } 0 < r_2 < r_1,$$

$$\vdots$$

$$r_{n-2} = q_n r_{n-1} + r_n, \qquad \text{with } 0 < r_n < r_{n-1},$$
$$r_{n-1} = q_{n+1} r_n.$$

The last non-zero remainder r_n is the gcd of a, b. Further r_n can be expressed as a linear combination of a and b by successively eliminating the r_i's in the intermediate equations.

Example 5.3.4. Find the gcd of 270 and 2412 and express it as a linear combination of 270 and 2412.

We apply the Euclidean algorithm

$$2412 = (8)(270) + 252,$$
$$270 = (1)(252) + 18,$$
$$252 = (14)(18).$$

Therefore the last non-zero remainder is 18 which is the gcd. We now must express 18 as a linear combination of 270 and 2412.

From the first equation

$$252 = 2412 - (8)(270)$$

which gives in the second equation

$$270 = (2412) - (8)(270) + 18 \implies 18 = (-1)(2412) + (9)(270)$$

which is the desired linear combination.

The following lemma whose proof can be found in [FR] provides a technique to find the lcm of two integers. If a, b are two integers we denote their least common multiple by $\mathrm{lcm}(a, b)$ or $[a, b]$.

Lemma 5.3.5. *If $a, b \in \mathbb{Z}$ then $(a, b)[a, b] = ab$.*

Example 5.3.6. Find the lcm of 270 and 2412.

From the previous example we found that $\gcd(270, 2412) = 18$. Therefore $\mathrm{lcm}(270, 2412) = \frac{(270)(2412)}{18} = 36\,180$.

The same method can be applied to finding the inverse of a unit and to solve linear equations in a modular ring. Recall from Chapter 1 that a decryption of an affine cipher requires solving a linear equation in a modular ring.

Example 5.3.7. Find the multiplicative inverse of 5 in $\mathbb{Z}_{17}$.

Notice that $\mathbb{Z}_{17}$ is a field so every non-zero element is a unit. Now $(5, 17) = 1$ so we use the Euclidean algorithm to express 1 as a linear combination of 5 and 17. We get

$$1 = (7)(5) - (2)(17).$$

Considering these in $\mathbb{Z}_{17}$ and using that $\overline{17} = \overline{0}$ we get

$$\overline{1} = (\overline{7})(\overline{5}).$$

Hence $\overline{7}$ is the inverse of $\overline{5}$ in $\mathbb{Z}_{17}$.

If a is a unit in $\mathbb{Z}_n$ then a linear equation

$$ax + b = c$$

can always be solved with a unique solution given by $x = a^{-1}(c - b)$.

Example 5.3.8. Solve $5x + 4 = 2$ in $\mathbb{Z}_6$.

Since $(5, 6) = 1$ it follows that 5 is a unit in $\mathbb{Z}_6$. Therefore $x = 5^{-1}(2 - 4)$. Now $2 - 4 = -2 = 4$ in $\mathbb{Z}_6$. Further $5 = -1$ so $5^{-1} = -1^{-1} = -1$. Then we have

$$x = 5^{-1}(2 - 4) = -1(4) = -4 = 2.$$

Thus the unique solution in $\mathbb{Z}_6$ is $x = 2$.

Since an element $\overline{a}$ is a unit in $\mathbb{Z}_n$ if and only if $(a, n) = 1$ it follows that the number of units in $\mathbb{Z}_n$ is equal to the number of positive integers less than or equal to n and relatively prime to n. This number is given by what is called the **Euler Phi Function**.

Definition 5.3.9. For any $n > 0$,

$$\phi(n) = \text{number of positive integers less than or equal to } n$$
$$\text{and relatively prime to } n.$$

Example 5.3.10. $\phi(6) = 2$ since among $1, 2, 3, 4, 5, 6$ only $1, 5$ are relatively prime to 6.

The following is immediate from our characterization of units.

Lemma 5.3.11. *The number of units in $\mathbb{Z}_n$, which is the order of the unit group $U(\mathbb{Z}_n)$, is $\phi(n)$.*

Definition 5.3.12. Given $n > 0$ a **reduced residue system** modulo n is a set of integers $x_1, \ldots, x_k$ such that each x_i is relatively prime to n, $x_i \neq x_j$ modulo n unless $i = j$ and if $(x, n) = 1$ for some integer x then $x \equiv x_i \pmod{n}$ for some i.

Hence a reduced residue system is a complete collection of representatives of those residue classes of integers relatively prime to n. Hence it is a complete collection of units (up to congruence modulo n) in $\mathbb{Z}_n$. It follows that any reduced residue system modulo n has $\phi(n)$ elements.

Example 5.3.13. A reduced residue system modulo 6 would be $\{1, 5\}$.

We now develop a formula for $\phi(n)$. We first determine a formula for prime powers and then paste back together via the fundamental theorem of arithmetic to develop our general formula.

Lemma 5.3.14. *For any prime p and $m > 0$,*

$$\phi(p^m) = p^m - p^{m-1} = p^m \left(1 - \frac{1}{p}\right).$$

Proof. The numbers a, $1 \le a \le p^e$ with $(a, p^e) > 1$ are precisely the multiples of p, that is

$$p, 2p, 3p, \ldots, p^{e-1}p.$$

All the others are relatively prime to p^e. Therefore $\phi(p^e) = p^e - p^{e-1}$. $\qquad\square$

The following lemma says that the Euler Phi Function is a **multiplicative function**. For a proof of this lemma we refer to [FR].

Lemma 5.3.15. *If $(a, b) = 1$ then $\phi(ab) = \phi(a)\phi(b)$.*

Based on this lemma and the fundamental theorem of arithmetic we get the general formula for $\phi(n)$.

Theorem 5.3.16. *Suppose $n = p_1^{e_1} \cdots p_k^{e_k}$ where the p_i are distinct primes, then*

$$\phi(n) = (p_1^{e_1} - p_1^{e_1-1})(p_2^{e_2} - p_2^{e_2-1}) \cdots (p_k^{e_k} - p_k^{e_k-1}) = n \prod_{i=1}^{k} \left(1 - \frac{1}{p_i}\right).$$

Proof. From the previous Lemmas we have since $\gcd(p_i^{e_i}, p_j^{e_j}) = 1$ for $i \ne j$

$$\begin{aligned}
\phi(n) &= \phi(p_1^{e_1})\phi(p_2^{e_2}) \cdots \phi(p_k^{e_k}) \\
&= (p_1^{e_1} - p_1^{e_1-1})(p_2^{e_2} - p_2^{e_2-1}) \cdots (p_k^{e_k} - p_k^{e_k-1}) \\
&= p_1^{e_1}\left(1 - \frac{1}{p_1}\right) \cdots p_k^{e_k}\left(1 - \frac{1}{p_k}\right) \\
&= p_1^{e_1} \cdots p_k^{e_k}\left(1 - \frac{1}{p_1}\right) \cdots \left(1 - \frac{1}{p_k}\right).
\end{aligned}$$
$\qquad\square$

Example 5.3.17. Determine $\phi(126)$. Now

$$126 = 2 \cdot 3^2 \cdot 7 \implies \phi(126) = \phi(2)\phi(3^2)\phi(7) = (1)(3^2 - 3)(6) = 36.$$

Hence there are 36 units in $\mathbb{Z}_{126}$.

5.4 The Field $\mathbb{Z}_p$ and Finite Fields

In the previous section we saw that an element $\bar{a}$ in the modular ring $\mathbb{Z}_n$ is a unit if and only if $(a, n) = 1$. If $n = p$ with p a prime then every integer except multiples of p are relatively prime to p and therefore become units in $\mathbb{Z}_p$. The multiples of the prime p correspond to the zero element of $\mathbb{Z}_p$ and hence it follows that every non-zero element of $\mathbb{Z}_p$ is a unit. Recall that a **field** is a commutative ring with an identity where every non-zero element is a unit. Since for any natural number n the modular ring $\mathbb{Z}_n$ is a commutative ring with an identity it follows that if p is a prime then $\mathbb{Z}_p$ is a field. This proves the first part of the following theorem.

Theorem 5.4.1. *The modular ring $\mathbb{Z}_n$ is a field if and only if $n = p$ is a prime. For the field $\mathbb{Z}_p$ the non-zero elements $\mathbb{Z}_p^*$ form a group, namely the unit group.*

Proof. If p is a prime then as explained above $\mathbb{Z}_p$ forms a field.

Conversely, suppose that $\mathbb{Z}_n$ is a field. First recall that a field must be an integral domain, that is there can be no zero divisors, that is no non-zero elements a, b with $ab = 0$. Now suppose that n is not prime so that $n = mk$ with $1 < m, k < n$. Then neither $\overline{m}$ nor $\overline{k}$ are 0 in $\mathbb{Z}_n$ but $\overline{mk} = 0$ in $\mathbb{Z}_n$. Hence $\mathbb{Z}_n$ has zero divisors which cannot occur in a field. Therefore n must be prime. □

Hence for each prime p we have a finite field $\mathbb{Z}_p$. We next show that if F is any finite field, that is a field with finitely many elements, then F must have size p^n for some prime p and natural number n. Further given a prime power p^n there exists a unique field (up to isomorphism) of order p^n. We call this unique field $GF(p^n)$ denoting the Galois field of order p^n. We also denote this by $F = \mathbb{F}_q, q = p^n$. Finite fields are used quite often in cryptography. In the AES and DES symmetric key systems computations are done in the Galois fields $GF(2^n)$.

Theorem 5.4.2. *Any finite field F has size p^n for some prime p and natural number n.*

Proof. Let F be a finite field. If its characteristic were 0 then F would be infinite and its characteristic (see [CFR1]) must be a prime p. It follows that F is a finite dimensional vector space over the finite field $\mathbb{Z}_p$. Suppose that $v_1, \ldots, v_n$ is a finite basis. Then every element of F can be written uniquely as a linear combination

$$c_1 v_1 + \cdots + c_n v_n \quad \text{with } c_i \in \mathbb{Z}_p.$$

Since there are p choices for each c_i there are p^n choices for these linear combinations. □

A proof of the following theorem can be found in [CFR1].

Theorem 5.4.3. *Given a prime p a and natural number n there exists a unique (up to isomorphism) field of size p^n.*

As noted above this unique field is denoted by $GF(p^n)$ or by $\mathbb{F}_q$ with $q = p^n$.

We close this subsection with several other results in elementary number theory involving primes that appear in cryptography. The first is called **Fermat's Theorem**.

Theorem 5.4.4 (Fermat). *Let p be a prime. Then*

$$a^p \equiv a \pmod{p}.$$

for each $a \in \mathbb{Z}$, that is $p|(a^p - a)$. Hence in the ring $\mathbb{Z}_p$ we have $\bar{a}^p = \bar{a}$ and if $\bar{a} \neq \bar{0}$ then $\bar{a}^{p-1} = \bar{1}$.

Proof. Here we give a purely number theoretic proof. In the next section we give a proof based on elementary group theory. If $a \equiv 0 \pmod{p}$ then $a^p \equiv 0 \pmod{p}$ so $a^p \equiv a \pmod{p}$.

If $a \not\equiv 0 \pmod{p}$ so that $(a, p) = 1$, let $\{r_1, \ldots, r_p\}$ be a complete residue system modulo p with $r_p = \bar{0}$. Since $(a, p) = 1$ then $\{ar_1, ar_2, \ldots, ar_p\}$ is also a complete residue system modulo p and $ar_p \equiv 0 \pmod{p}$. Then we have

$$(ar_1)(ar_2)\cdots(ar_{p-1}) \equiv r_1 r_2 \cdots r_{p-1} \pmod{p}$$
$$\implies a^{p-1}(r_1 \cdots r_{p-1}) \equiv (r_1 \cdots r_{p-1}) \pmod{p}.$$

However, $(r_1 \cdots r_{p-1})$ is not divisible by p so it is a unit in $\mathbb{Z}_p$ and hence we can divide through by it to get the result;

$$a^{p-1} \equiv 1 \pmod{p} \implies a^p \equiv a \pmod{p}. \qquad \Box$$

Theorem 5.4.5 (Wilson). *Let p be a prime. Then*

$$(p - 1)! \equiv -1 \pmod{p}.$$

Conversely if $(p - 1)! \equiv -1 \pmod{p}$ then p is a prime.

Proof. Now $(p - 1)! = (p - 1)(p - 2) \cdots 1$. Since $\mathbb{Z}_p$ is a field each

$$x \in \{1, 2, \ldots, p - 1\}$$

has a multiplicative inverse modulo p. Further suppose $x = x^{-1}$ in $\mathbb{Z}_p$. Then $x^2 = 1$ which implies $(x - 1)(x + 1) = 0$ in $\mathbb{Z}_p$ and hence either $x = 1$ or $x = -1$ since $\mathbb{Z}_p$ is an integral domain. Therefore in $\mathbb{Z}_p$ only $1, -1$ are their own multiplicative inverses. Further $-1 = p - 1$ since $p - 1 \equiv -1 \pmod{p}$.

Hence in the product $(p - 1)(p - 2) \cdots 1$ considered in the field $\mathbb{Z}_p$ each element is paired up with its distinct multiplicative inverse except 1 and $p-1$. Further the product of each with its inverse is 1. Therefore in $\mathbb{Z}_p$ we have $(p - 1)(p - 2) \cdots 1 = p - 1$. Written as a congruence then

$$(p - 1)! \equiv p - 1 \equiv -1 \pmod{p}.$$

Conversely suppose $(n - 1)! \equiv -1 \pmod{n}$. If n were composite then $n = mk$ with $1 < m < n - 1$ and $1 < k < n - 1$. Hence, if $m \neq k$ then both m and k are included in $(n - 1)!$. It follows that $(n - 1)!$ is divisible by n so that $(n - 1)! \equiv 0 \pmod{n}$ contradicting the assertion that $(n - 1)! \equiv -1 \pmod{n}$. If $m = k$ then $n = m^2$. If $m = 2$ then $n = 4$ and $(n - 1)! \equiv 2 \pmod 4$ contradicting the assertion $(n - 1)! \equiv -1 \pmod 4$. If $m \geq 3$ then $(n - 1)! \equiv 0 \pmod{m}$ but $-1 \not\equiv 0 \pmod{m}$.

Therefore n must be prime. $\qquad \Box$

5.5 Finite Abelian Groups

The modular rings $\mathbb{Z}_n$ form finite abelian groups under addition, while their unit groups $U(\mathbb{Z}_n)$ form finite abelian groups under multiplication. Here we discuss some important results from finite group theory that are applicable to both number theory and cryptography.

Recall that the number of elements in a finite group G is called the **order** of G denoted $|G|$. Each element in a group G generates a subgroup called the **cyclic subgroup** generated by g. This consists of the powers of the element g. The cyclic subgroup generated by g is denoted $\langle g \rangle$ and the order of this subgroup is called the **order** of g. We will denote the order of an element $g \in G$ by $o(g)$ or $|g|$. The first result called **Lagrange's Theorem** restricts the order of subgroups. For proofs of the results in this section we refer the reader to the book [CFR1].

Theorem 5.5.1 (Lagrange's Theorem). *Let G be a finite group of order n. Then the order of any subgroup divides n. In particular, the order of any element divides n and moreover $g^n = 1$ for each $g \in G$.*

Lagrange's theorem provides an alternative proof of Fermat's Theorem and an extension of it called **Euler's Theorem**.

Consider a unit $a \in \mathbb{Z}_n$. Then $a \in U(\mathbb{Z}_n)$ and hence a has a **multiplicative order**, that is there is an integer m with $a^m = 1$ in $\mathbb{Z}_n$. In terms of congruences this means that $a^m \equiv 1 \pmod n$. If $a \in \mathbb{Z}_n$ is not a unit then there cannot exist a power $m \geq 1$ such that $a^m \equiv 1 \pmod n$, if such an m existed then a^{m-1} would be an inverse for a. The unit group $U(\mathbb{Z}_n)$ is a finite group of order $\phi(n)$ where ϕ is the Euler phi-function. Applying Lagrange's Theorem to the group $U(\mathbb{Z}_n)$ we get the following result, known as **Euler's Theorem**, which is an extension of Fermat's Theorem.

Theorem 5.5.2 (Euler's Theorem). *If $(a, n) = 1$ then*

$$a^{\phi(n)} \equiv 1 \pmod n.$$

If p is a prime $\phi(p) = p - 1$ so applying Euler's Theorem to the modular fields $\mathbb{Z}_p$ we recover as a corollary Fermat's Theorem.

Corollary 5.5.3 (Fermat). *If $a \in \mathbb{Z}_p$ and $a \neq 0$ then $a^{p-1} \equiv 1 \pmod p$.*

In general for finite abelian groups we have the following. Recall that a finite group G is the direct product

$$G = H_1 \times H_2 \times \cdots \times H_k$$

of subgroups $H_1, \ldots, H_k$ if
(1) each $g \in G$ can be written uniquely in the form $g = h_1 h_2 \cdots h_k$ with $h_i \in H_i$ for $i = 1, \ldots, k$ (unique up to the ordering of the factors);
(2) $h_i h_j = h_j h_i$ for $h_i \in H_i, h_j \in H_j$ if $i \neq j$.

Theorem 5.5.4. *Let G be a finite abelian group with $|G| = n$ then*
(1) *if $g_1, g_2 \in G$ with $|g_1| = a$, $|g_2| = b$ then $(g_1 g_2)^{\text{lcm}(a,b)} = 1$;*
(2) *if $g_1, g_2 \in G$ with $|g_1| = a$, $|g_2| = b$ and $(a, b) = 1$ then $|g_1 g_2| = ab$;*
(3) *if $n = p_1^{e_1} p_2^{e_2} \cdots p_k^{e_k}$ is the prime factorization of n then*

$$G = H_1 \times H_2 \times \cdots \times H_k$$

where $|H_i| = p_i^{e_i}$.

Corollary 5.5.5. *Let G be a finite abelian group and let $H_1, \ldots, H_k$ be subgroups of G with pairwise relatively prime orders $|H_i|$. Suppose that $|G| = |H_1||H_2| \cdots |H_k|$. Then $G = H_1 \times \cdots \times H_k$.*

5.6 Cyclic Groups and Primitive Elements

It follows from Euler's Theorem that any unit a in $\mathbb{Z}_n$ has finite multiplicative order. For $a \in \mathbb{Z}_n$ we denote this order by $o(a)$ and further this natural number divides $\phi(n)$. The order of a is the order of the cyclic multiplicative subgroup of $U(\mathbb{Z}_n)$ generated by $\bar{a}$. If $(a, n) = 1$ and the order of a is exactly $\phi(n)$ then a is called a **primitive element** modulo n. In this case the unit group is cyclic with a as a generator. We will prove that for $n = p$ a prime there is always a primitive element. First we review some material about cyclic groups.

Recall that a cyclic group G is a group with a single generator g. G then consists of all the powers of g, that is $G = \{1, g^{\pm 1}, g^{\pm 2}, \ldots\}$. If G is finite of order n then $g^n = 1$ and n is the least positive integer x such that $g^x = 1$. It is then clear that if $g^m = 1$ for some power m it must follow that $m \equiv 0 \pmod{n}$, and if $g^k = g^l$ then $k \equiv l \pmod{n}$.

Let $H = (\mathbb{Z}_n, +)$ denote the additive subgroup of $\mathbb{Z}_n$. Then H is cyclic of order n with generator 1. If $G = \langle g \rangle$ is also cyclic of order n then since multiplication of group elements is done via addition of exponents it is fairly straightforward that the homomorphism $f : G \to (\mathbb{Z}_n, +)$ given by $g \to 1$ is actually an isomorphism (see Chapter 9 for a formal definition of homomorphism and isomorphism and see the exercises). Further if $G = \langle g \rangle$ is cyclic of infinite order then $g \to 1$ gives an isomorphism from G to the additive group of $\mathbb{Z}$.

Lemma 5.6.1. *Let G be a cyclic group,*
(1) *If G is finite of order n then G is isomorphic to $(\mathbb{Z}_n, +)$. In particular, all finite cyclic groups of a given order are isomorphic.*
(2) *If G is an infinite cyclic group then G is isomorphic to $(\mathbb{Z}, +)$.*

Cyclic groups are abelian and hence their subgroups are also abelian. However, as an almost direct consequence of the division algorithm we get that any subgroup of a cyclic group must also be cyclic.

Lemma 5.6.2. *Let G be a cyclic group. Then any subgroup of G is also cyclic.*

Proof. Suppose $G = \langle g \rangle$ and $H \subset G$ is a subgroup. Since G consists of powers of g, H also consists of certain powers of g. Let k be the least positive integer such that $g^k \in H$. We show that $H = \langle g^k \rangle$ that is H is the cyclic subgroup generated by g^k. This is clearly equivalent to showing that every $h \in H$ must be a power of g^k.

Suppose $g^t \in H$. We may assume that $t > 0$ and that $t > k$ since k is the least positive integer such that $g^k \in H$. If $t < 0$ work with $-t$. By the division algorithm we then have

$$t = qk + r \quad \text{with } r = 0 \quad \text{or} \quad 0 < r < k.$$

If $r \neq 0$ then $0 < r < k$ and $r = t - k$. Hence $g^r = g^{t-k} = g^t g^{-k}$. Now $g^t \in H$ and $g^k \in H$ and since H is a subgroup it follows that $g^{t-k} \in H$. But then $g^r \in H$ which is a contradiction since $0 < r < k$ and k is the least power of g in H. Therefore $r = 0$ and $t = qk$. We then have

$$g^t = g^{qk} = (g^k)^q$$

completing the proof. $\qquad\square$

Each element of a cyclic group G generates its own cyclic subgroup. The question is when does this cyclic subgroup coincide with all of G. In particular, which powers g^k are generators of G. The answer is purely number theoretic.

Lemma 5.6.3. *Let G be a cyclic group.*
(1) *Let $G = \langle g \rangle$ be a finite cyclic group of order n. Then g^k with $k > 0$ is a generator of G if and only if $(k, n) = 1$, that is k and n are relatively prime.*
(2) *If $G = \langle g \rangle$ is an infinite cyclic group then g, g^{-1} are the only generators.*

Proof. Suppose first that $G = \langle g \rangle$ is finite cyclic of order n and suppose that $(k, n) = 1$. Then there exists integers x, y such that $kx + ny = 1$. It follows then that

$$g = g^1 = g^{kx+ny} = g^{kx} g^{ny} = (g^k)^x (g^n)^y.$$

But $g^n = 1$ so $(g^n)^y = 1$ and therefore

$$g = (g^k)^x.$$

Therefore g is a power of g^k and hence every power of g is also a power of g^k. The whole group g then consists of powers of g^k and hence g^k is a generator for G.

Conversely, suppose that g^k is also a generator for G. Then there exists a power x such that $g = (g^k)^x = g^{kx}$. Hence $kx \equiv 1 \pmod{n}$ and so k is a unit modulo n which implies from the last section that $(k, n) = 1$.

Suppose next that $G = \langle g \rangle$ is infinite cyclic. Then there is no power of g which is the identity. Suppose g^k is also a generator with $k > 1$. Then there exists a power x such that $g = (g^k)^x = g^{kx}$. But this implies that $g^{kx-1} = 1$ contradicting that no power of g is the identity. Hence $k = 1$. $\qquad\square$

Recall that the Euler phi function $\phi(n)$ is equal to the number of positive integers less than n which are relatively prime to n. This is then the number of generators of a cyclic group of order n.

Corollary 5.6.4. *Let G be a finite cyclic group of order n. Then there are $\phi(n)$ generators for G.*

By Lagrange's Theorem, for any finite group, the order of a subgroup divides the order of a group, that is if $|G| = n$ and $|H| = d$ with H a subgroup of G then $d|n$. However, the converse in general is not true, that is if $|G| = n$ and $d|n$ there need not be a subgroup of order d. Further if there is a subgroup of order d, there may or may not be other subgroups of order d. For a finite cyclic group G of order n however, there is, for each $d|n$, a **unique** subgroup of order d.

Theorem 5.6.5. *Let G be a finite cyclic group of order n. Then for each d|n with d $\geq$ 1 there exists a unique subgroup H of order d.*

Proof. Let $G = \langle g \rangle$ and $|G| = n$. Suppose $d|n$, then $n = kd$. Consider the element g^k. Then $(g^k)^d = g^{kd} = g^n = 1$. Further if $0 < t < d$ then $0 < kt < kd$ so $kt \not\equiv 0 \pmod{n}$ and hence $g^{kt} = (g^k)^t \neq 1$. Therefore d is the least power of g^k which is the identity and hence g^k has order d and generates a cyclic subgroup of order d. We must show that this is unique.

Suppose $H = \langle g^t \rangle$ is another cyclic subgroup of order d (recall that all subgroups of G are also cyclic). We may assume that $t > 0$ and we show that g^t is a power of g^k and hence the subgroups coincide.

Since H has order d we have $g^{td} = 1$ which implies that $td \equiv 0 \pmod{n}$. Since $n = kd$ it follows that $t > k$. Apply the division algorithm

$$t = qk + r \quad \text{with } 0 \leq r < k.$$

If $r \neq 0$ then $0 < r < k$ and $r = t - qk$. Then

$$r = t - qk \implies rd = td - qkd \equiv 0 \pmod{n}.$$

Hence $n \mid rd$ which is impossible since $rd < kd = n$. Therefore $r = 0$ and $t = qk$. From this

$$g^t = g^{qk} = (g^k)^q.$$

Therefore g^t is a power of g^k and $H = \langle g^k \rangle$. $\qquad\square$

From this result we get the following concerning the Euler Phi Function.

Theorem 5.6.6. *For n > 1 and for d $\geq$ 1*

$$\sum_{d|n} \phi(d) = n.$$

Proof. Consider a cyclic group G of order n. For each $d|n$, $d \geq 1$ there is a unique cyclic subgroup H of order d. H then has $\phi(d)$ generators. Each element in G generates its own cyclic subgroup H_1, say of order d and hence must be included in the $\phi(d)$ generators of H_1. Therefore

$$\sum_{d|n} \phi(d) = \text{sum of the numbers of generators of the cyclic subgroups of } G.$$

But this must be the whole group and hence this sum is n. □

We now return to primitive elements in the modular rings.

Theorem 5.6.7. *For a prime p there is always an element a of order $\phi(p) = p - 1$, that is a primitive element. Equivalently the unit group of $\mathbb{Z}_p$ is always cyclic.*

Since every non-zero element in $\mathbb{Z}_p$ is a unit, the unit group $U(\mathbb{Z}_p)$ is precisely the multiplicative group of the field $\mathbb{Z}_p$. The fact that $U(\mathbb{Z}_p)$ is cyclic follows from the following more general result, which also shows that the multiplicative group of any finite field is cyclic.

Theorem 5.6.8. *Let F be a field. Then any finite subgroup of the multiplicative group of F must be cyclic.*

Proof. Suppose $G \subset F$ is a finite multiplicative subgroup of the multiplicative group of F. Suppose $|G| = n$. As has been our general mode of approaching results we will prove it for n a power of a prime and then paste the result together via the fundamental theorem of arithmetic.

Suppose $n = p^k$ for some k. Then the order of any element in G is p^α with $\alpha \leq k$. Suppose the maximal order is p^t with $t < k$. Then the lcm of the orders is p^t. It follows that for every $g \in G$ we have $g^{p^t} = 1$. Therefore every $g \in G$ is a root of the polynomial equation

$$x^{p^t} - 1 = 0.$$

However, over a field, a polynomial cannot have more roots than its degree. Since G has $n = p^k$ elements and $p^t < p^k$ this is a contradiction. Therefore the maximal order must be $p^k = n$. Therefore G has an element of order $n = p^k$ and hence this element generates G and G must be cyclic.

We now do an induction on the number of distinct prime factors in $n = |G|$. The above argument handles the case where there is only one distinct prime factor. Assume the result is true if the order of G has less than k distinct prime factors. Suppose $n = p_1^{e_1} \cdots p_k^{e_k}$. Then $n = p^e c$ where c has less than k distinct prime factors. Since G is a finite abelian group with

$$|G| = n = p^e c \implies G = H \times K \qquad \text{with } |H| = p^e, \quad |K| = c.$$

By the inductive hypothesis H and K are both cyclic so H has an element h of order p^e and K has an element k of order c. Since $(p^e, c) = 1$ the element hk has order $p^e c = n$ completing the proof. □

Corollary 5.6.9. *The multiplicative group of any finite field is cyclic.*

This fact will play a role in the implementation of the Diffie-Hellman protocol (see Chapter 6).

Example 5.6.10. Determine a primitive element modulo 7.

This is equivalent to finding a generator for the multiplicative group of $\mathbb{Z}_7$. The non-zero elements are 0, 1, 2, 3, 4, 5, 6 and we are looking for an element of order 6.

The table below list these elements and their orders

x	1	2	3	4	5	6
$|x|$	1	3	6	3	6	2

Therefore there are 2 primitive roots 3 and 5 modulo 7. To see how these were determined powers were taken modulo 7 until a value of 1 was obtained. For example

$$3^2 = 9 = 2, \quad 3^3 = 2 \cdot 3 = 6, \quad 3^4 = 3 \cdot 6 = 18 = 4, \quad 3^5 = 3 \cdot 4 = 12 = 5,$$
$$\text{and} \quad 3^6 = 3 \cdot 5 = 15 = 1.$$

Example 5.6.11. Show that there is no primitive element modulo 15.

The units in $\mathbb{Z}_{15}$ are $\{1, 2, 4, 7, 8, 11, 13, 14\}$. Since $\phi(15) = 8$ we must show that there is no element of order 8. The table below gives the units and their respective orders

x	1	2	4	7	8	11	13	14
$|x|$	1	4	2	4	4	2	4	2

Therefore there is no element of order 8.

We have the following result which determines when there exists a primitive element in $\mathbb{Z}_n^*$. A proof can be found in [FR].

Theorem 5.6.12. *There is a primitive element for $\mathbb{Z}_n$ if and only if $n = 2, 4, p, 2p, p^n$ for some odd prime p.*

5.7 The Chinese Remainder Theorem

Systems of linear congruences are handled by the next result which is called the **Chinese Remainder Theorem**. This will also provide an important result describing the structure of the modular rings.

Theorem 5.7.1 (Chinese Remainder Theorem). *Suppose that $m_1, m_2, \ldots, m_k$ are k positive integers that are relatively prime in pairs. If $a_1, \ldots, a_k$ are any integers then the simultaneous congruences*

$$x \equiv a_i \ (\text{mod } m_i), \quad i = 1, \ldots, k$$

have a common solution which is unique modulo $m_1 m_2 \cdots m_k$.

Proof. The proof we give not only provides a verification but also provides a technique for finding the common solution.

Let $m = m_1 m_2 \cdots m_k$. Since the m_i are relatively prime in pairs we have $\left(\frac{m}{m_i}, m_i\right) = 1$. Therefore there is a solution x_i to the reduced congruence

$$\frac{m}{m_i} x_i \equiv 1 \pmod{m_i}.$$

Further for x_i we clearly have

$$\frac{m}{m_j} x_i \equiv 0 \pmod{m_i} \quad \text{if } i \neq j.$$

Now let

$$x_0 = \sum_{i=1}^{k} \frac{m}{m_i} x_i a_i.$$

We claim that x_0 is a solution to the simultaneous congruences and that it is unique modulo m.

Now

$$x_0 = \sum_{i=1}^{k} \frac{m}{m_i} x_i a_i \equiv \frac{m}{m_j} x_j a_j \pmod{m_j}$$

since $\frac{m}{m_i} x_i \equiv 0 \pmod{m_j}$ if $i \neq j$. It follows then that

$$x_0 \equiv \frac{m}{m_j} x_j a_j \pmod{m_j} \equiv a_j \pmod{m_j}$$

since $\frac{m}{m_j} x_j \equiv 1 \pmod{m_j}$. Therefore x_0 is a common solution. We must show the uniqueness part.

If $\tilde{x}_0$ is another common solution then $\tilde{x}_0 \equiv x_0 \pmod{m_i}$ for $i = 1, \ldots, k$. Therefore $\tilde{x}_0 \equiv x_0 \pmod{m}$.

We note that if the integers m_i are not relatively prime in pairs there may be no solution to the simultaneous congruences. □

Example 5.7.2. Solve the simultaneous congruences

$$x \equiv 6 \pmod{13}$$
$$x \equiv 9 \pmod{45}$$
$$x \equiv 12 \pmod{17}.$$

Here $m_1 = 13, m_2 = 45, m_3 = 17$ so $m = 13 \cdot 45 \cdot 17$. We first solve

$$(17)(45)x = 1 \pmod{13} \implies x \equiv \ 6 \pmod{13}$$
$$(13)(17)x = 1 \pmod{45} \implies x \equiv 11 \pmod{45}$$
$$(13)(45)x = 1 \pmod{17} \implies x \equiv \ 5 \pmod{17}.$$

To see how these solutions are found lets look at the second one:

$$(13)(17)x \equiv 1 \ (\text{mod } 45) \implies 221x \equiv 1 \ (\text{mod } 45) \implies 41x \equiv 1 \ (\text{mod } 45)$$

since $221 \equiv 41 \ (\text{mod } 45)$. We now use the Euclidean algorithm,

$$45 = 1 \cdot 41 + 4, \ 41 = 10 \cdot 4 + 1 \implies 1 = (11)(41) - (10)(45)$$
$$\implies 41^{-1} \equiv 11 \ (\text{mod } 45).$$

Therefore, using these solutions, the common solution is

$$x_0 = \frac{13 \cdot 45 \cdot 17}{13}(6)(6) + \frac{13 \cdot 45 \cdot 17}{45}(11)(9) + \frac{13 \cdot 45 \cdot 17}{17}(5)(12)$$
$$\implies x_0 = 27\,540 + 21\,879 + 35\,100 = 84\,519 \equiv 4959 \ (\text{mod } 9945)$$
$$\implies x_0 = 4959.$$

The following theorem describes the structure of modular rings in terms of the multiplicative decomposition of the modulus and is based directly on the Chinese Remainder Theorem.

Theorem 5.7.3. *Suppose that $m_1, m_2, \ldots, m_k$ are k positive integers that are relatively prime in pairs and $n = m_1 m_2 \cdots m_k$. Then the map*

$$\phi : \mathbb{Z}_n \to \mathbb{Z}_{m_1} \times \mathbb{Z}_{m_2} \times \cdots \times \mathbb{Z}_{m_k}$$

given by

$$\bar{a} \mapsto (\bar{a}, \ldots, \bar{a})$$

is a ring isomorphism.
The $\bar{a}$'s in the tuple are the corresponding residue classes for each m_i.

5.8 Exercises

5.1 Verify that the following are rings. Indicate which are commutative and which have identities. Which are integral domains?
 (a) The set of rational numbers.
 (b) The set of continuous functions on a closed interval $[a, b]$ under ordinary addition and multiplication of functions.
 (c) The set of 2×2 matrices with integral entries.
 (d) The set $n\,\mathbb{Z}$ consisting of all integers which are multiples of the fixed integer n.

5.2 (a) Show that in an ordered ring, squares must be positive. Conclude that in an ordered ring with identity the multiplicative identity must be positive.

(b) Show that the complex numbers under the ordinary operations cannot be ordered.

5.3 Show that any ordered ring must be infinite. (Hint: Suppose $a > 0$ then $a + a > 0$, $a + a + a > 0$ and continue.)

5.4 Find the gcd and lcm of the following pairs of integers and then express the gcd as a linear combination
(a) 78 and 30;
(b) 175 and 35;
(c) 380 and 127.

5.5 Prove that if $a = qb + r$ then $(a, b) = (b, r)$.

5.6 Prove that if $d = (a, b)$ then $\frac{a}{d}$ and $\frac{b}{d}$ are relatively prime.

5.7 Show that if $(a, b) = c$ then $(a^2, b^2) = c^2$. (Hint: The easiest method is to use the fundamental theorem of arithmetic.)

5.8 Show that an integer is divisible by 3 if and only if the sum of its digits (in decimal expansion) is divisible by 3. (Hint: Write out the decimal expansion and take everything modulo 3.)

5.9 Find the multiplicative inverse if it exists
(a) of 13 in $\mathbb{Z}_{47}$;
(b) of 17 in $\mathbb{Z}_{22}$;
(c) of 6 in $\mathbb{Z}_{30}$.

5.10 Solve the linear equations in the given rings
(a) $4x + 6 = 2$ in $\mathbb{Z}_7$;
(b) $5x + 9 = 12$ in $\mathbb{Z}_{47}$;
(c) $3x + 18 = 27$ in $\mathbb{Z}_{40}$.

5.11 Find $\phi(n)$ for
(a) $n = 17$;
(b) $n = 526$;
(c) $n = 138$.

5.12 Determine the units and write down the group table for the unit group $U(\mathbb{Z}_n)$ for
(a) $\mathbb{Z}_{12}$;
(b) $\mathbb{Z}_{26}$.

5.13 Suppose that $G = \langle g \rangle$ is also cyclic of order n. Show that the map $f : G \to (\mathbb{Z}_n, +)$ given by $g \mapsto 1$ is actually a group isomorphism.

5.14 For any natural number m, let $(\mathbb{Z}_m, +)$ denote the additive group of $\mathbb{Z}_m$, and let $U(\mathbb{Z}_m)$ be the group of units of $\mathbb{Z}_m$. Let $n = n_1 n_2 \cdots n_k$ be a factorization of n with pairwise relatively prime factors. Then prove that

$$(\mathbb{Z}_n, +) \cong (\mathbb{Z}_{n_1}, +) \times (\mathbb{Z}_{n_2}, +) \times \cdots \times (\mathbb{Z}_{n_k}, +)$$
$$U(\mathbb{Z}_n) = U(\mathbb{Z}_{n_1}) \times \cdots \times U(\mathbb{Z}_{n_k}).$$

5.15 Prove that if an integer is congruent to 2 modulo 3 then it must have a prime factor congruent to 2 modulo 3.

5.16 Prove that if bc is a perfect square for integers b, c and $(b, c) = 1$ then both b and c are perfect squares.

5.17 Determine a primitive root modulo 11.

5.18 We outline a proof of the theorem: An integer n will have a primitive root modulo n if and only if

$$n = 2, 4, p^k, 2p^k$$

where p is an odd prime.

(a) Show that if $(m, n) = 1$ with $m > 2$, $n > 2$ then there is no primitive root modulo mn

(b) Show that there is no primitive root modulo 2^k for $k > 2$.

(c) Prove: If p is an odd prime then there exists a primitive root a modulo p such that a^{p-1} is not congruent to 1 modulo p^2. (Hint: Let a be a primitive root modulo p. Then $a + p$ is also a primitive root modulo p. Show that either a or $(a + p)$ satisfy the result.)

(d) Prove: If p is an odd prime then there exists a primitive root modulo p^k for any $k \geq 2$. (Hint: Let a be the primitive root modulo p from part (c). Then this is a primitive root modulo p^k for any $k \geq 2$.)

(e) Prove: If p is an odd prime and if a is a primitive root modulo p^k then, if a is odd, a is also a primitive root modulo $2p^k$. If a is even then $a + p^k$ is a primitive root modulo $2p^k$.

5.19 If $m > 2$ show that $\phi(m)$ is even.

5.20 Prove that $\phi(n^2) = n\phi(n)$ for any positive integer n.

5.21 Prove that if $n \geq 2$ then

$$\sum_{(m,n)=1, 0<m<n} m = \frac{n\phi(n)}{2}.$$

5.22 Prove that if n has k distinct odd factors then $2^k \mid \phi(n)$.

6 Some Number Theoretic Algorithms

6.1 Number Theoretic Algorithms for Public Key Cryptography

Number theory plays a prominent role in many areas of cryptography. In the simplest case, to develop a cryptosystem for an N letter alphabet, we consider the letters as integers modulo N. As described in Chapter 5, the integers modulo N form a ring called the **modular ring** $\mathbb{Z}_N$ and encryption is done using algebraic operations on this ring. The totality of operations within the various modular rings is called **modular arithmetic**. The encryption algorithms then apply number theoretic functions and use modular arithmetic on these integers. Hence encryption maps on k-length message units are functions

$$f : \mathbb{Z}_N^k \rightarrow \mathbb{Z}_N^k.$$

In addition to providing a framework for the encrypted alphabets, number theoretic problems provide the security in the basic public key protocols. Recall that the security of a cryptographic protocol is usually based on the difficulty of solving some "hard" problems. The two most commonly used public key methods in cryptography are the Diffie-Hellman method and its variations and the RSA method and its variations. For the Diffie-Hellman method, the hard problem used for cryptography is the discrete log problem that we will discuss in depth in Section 6.5. The hard problem for the RSA method is the factorization problem, that is the relative difficulty of factoring large integers.

In this chapter we discuss these hard number theoretic problems and several important number theoretic algorithms related to them. We then discuss primality testing, that is determining whether or not an integer is prime. The relative difficulty of finding large primes versus factoring large integers provides the security in RSA and related cryptographic methods.

6.2 Quadratic Residues and Square Roots

If $a \in \mathbb{Z}$ then we write $\bar{a}$ for the residue class for a modulo n. Sometimes, especially in calculations, it is convenient to write just a as an element of $\mathbb{Z}_n$ without any misunderstanding.

Definition 6.2.1. If $(a, m) = 1$ and and $x^2 \equiv a \pmod{m}$ has a solution then a is called a **quadratic residue** modulo m. If $x^2 \equiv a \pmod{m}$ has no solution then a is a **quadratic non-residue**.

Further if $x^2 \equiv a \pmod{m}$ then x is a **square root** of a modulo m.

The basic idea in using quadratic residues in cryptography stems from the following. If n is a prime, then we can decide effectively if a is a quadratic residue modulo n or not. However if $n = pq$, the product of two distinct odd primes p and q, then in general there is no known probabilistic algorithm to decide if a is a quadratic residue modulo n unless p and q are known.

The following result sets up a probabilistic method for determining modular square roots by counting the number of quadratic residues and quadratic non-residues modulo a prime.

Theorem 6.2.2. *Let $p \geq 3$ be an odd prime.*
(1) *The set $\{1, 2, \ldots, p - 1\}$ contains exactly $\frac{p-1}{2}$ quadratic residues and hence exactly $\frac{p-1}{2}$ quadratic non-residues.*
(2) *If $a \in \mathbb{Z}$ and $(a, p) = 1$ then a has 0 or 2 square roots modulo p, that is the equation $x^2 \equiv a \pmod{p}$ has 0 or 2 solutions modulo p.*
(3) *If p is an odd prime and $(a, p) = 1$ then a is a quadratic residue modulo p if and only if $a^{\frac{p-1}{2}} \equiv 1 \pmod{p}$. If a is a quadratic non-residue then $a^{\frac{p-1}{2}} \equiv -1 \pmod{p}$.*
(4) *If $a, b \in \mathbb{Z}$ with $(a, p) = (b, p) = 1$ then ab is a quadratic residue modulo p if both a and b are quadratic residues modulo p or if both a and b are quadratic non-residues modulo p.*

Proof. (2) Consider the polynomial $p(x) = x^2 - \overline{a} \in \mathbb{Z}_p[x]$. If x_0 is one solution of $p(x) = 0$ then $-x_0$ is another. Since a quadratic polynomial over a field can have at most 2 zeros there are the only square roots of a. On the other hand if there is no solution the polynomial is irreducible and there are no square roots. (For more details see [CFR1].)

(1) Now $x^2 \equiv (-x)^2 \pmod{p}$. Hence

$$\overline{1} = \overline{p-1}^2, \overline{2}^2 = \overline{p-2}^2, \ldots, \left(\overline{\frac{p+1}{2}}\right)^2 = \left(\overline{\frac{p-1}{2}}\right)^2$$

in $\mathbb{Z}_p^*$ and $1^2, 2^2, \ldots, (\frac{p-1}{2})^2$ are quadratic residues modulo p.

(3) Suppose $(a, p) = 1$. We do the computations in the field $\mathbb{Z}_p$. Since $a \neq 0$ then from Fermat's Theorem $a^{p-1} = 1$ in $\mathbb{Z}_p$. This implies that

$$(a^{\frac{p-1}{2}} - 1)(a^{\frac{p-1}{2}} + 1) = 0$$

in $\mathbb{Z}_p$. Since $\mathbb{Z}_p$ is a field it has no zero divisors and this implies that either $a^{\frac{p-1}{2}} = 1$ or $a^{\frac{p-1}{2}} = -1$. Hence either $a^{\frac{p-1}{2}} \equiv 1 \pmod{p}$ or $a^{\frac{p-1}{2}} \equiv -1 \pmod{p}$. We show that in the former case and only in the former case is a a quadratic residue.

Suppose that $x^2 = a$ has a solution say x_0 in $\mathbb{Z}_p$. Then

$$a^{\frac{p-1}{2}} = (x_0^2)^{\frac{p-1}{2}} = x_0^{p-1} = 1.$$

It follows further that if $a^{\frac{p-1}{2}} = -1$ there can be no solution.

Conversely suppose $a^{\frac{p-1}{2}} = 1$. Since the multiplicative group of $\mathbb{Z}_p$ is cyclic (see the last section) it follows that there is a $g \in \mathbb{Z}_p$ which generates this cyclic group and $a = g^t$ for some t. Hence $g^{\frac{t(p-1)}{2}} = 1$. However the order of the multiplicative group of

$\mathbb{Z}_p$ is $p - 1$ and therefore this implies that

$$\frac{t(p-1)}{2} \equiv 0 \ (\text{mod}(p-1)).$$

Therefore t must be even $t = 2k$. Hence $a = g^{2k} = (g^k)^2$ and there is a solution to $x^2 = a$.

(4) This is a direct consequence of (3). $\qquad\square$

We now introduce the Legendre symbol and use this to show how we can effectively decide, for $(a, p) = 1$ and p an odd prime, whether or not a is a quadratic residue. The restriction to odd primes is obvious because $a \equiv 1 \ (\text{mod } 2)$ for each odd integer a.

Definition 6.2.3. If p is an odd prime and $(a, p) = 1$ then the **Legendre symbol** $\left(\frac{a}{p}\right)$ is defined by

(1) $\left(\frac{a}{p}\right) = 1$ if a is a quadratic residue modulo p;

(2) $\left(\frac{a}{p}\right) = -1$ if a is a quadratic non-residue modulo p.

Therefore the value of the Legendre symbol distinguishes quadratic residues from quadratic non-residues. The next lemma establishes the basic properties of $\left(\frac{a}{b}\right)$.

Theorem 6.2.4. *If p is an odd prime and $(a, p) = (b, p) = 1$ then*

(1) $\left(\frac{a^2}{p}\right) = 1$;

(2) *If $a \equiv b \ (\text{mod } p)$ then $\left(\frac{a}{p}\right) = \left(\frac{b}{p}\right)$;*

(3) $\left(\frac{a}{p}\right) \equiv a^{\frac{p-1}{2}} \ (\text{mod } p)$;

(4) $\left(\frac{ab}{p}\right) = \left(\frac{a}{p}\right)\left(\frac{b}{p}\right)$;

(5) *If $a = p_1 p_2 \cdots p_k$ is a prime factorization of a then*

$$\left(\frac{a}{p}\right) = \left(\frac{p_1}{p}\right) \cdots \left(\frac{p_k}{p}\right);$$

(6) $\left(\frac{2}{p}\right) = (-1)^{\frac{p^2-1}{8}}$;

(7) $\left(\frac{-1}{p}\right) = (-1)^{\frac{p-1}{2}}$ *and* $\left(\frac{1}{p}\right) = 1$.

Proof. Parts (1) and (2) are immediate from the definition of the Legendre symbol. Part (3) is a direct consequence of Lemma 2.6.1.

To see part (4) notice that $(ab)^{\frac{p-1}{2}} = a^{\frac{p-1}{2}} b^{\frac{p-1}{2}}$ and use part (3). $\qquad\square$

From part (4) of this last lemma we see that to compute $\left(\frac{a}{p}\right)$ we can use the prime factorization of a and then restrict to $\left(\frac{q}{p}\right)$ where q is a prime distinct from p. The quadratic reciprocity law which is presented next will allow us to compute this for odd primes and then use the result for $\left(\frac{2}{p}\right)$. We present examples after the theorem on how to do this. There are many proofs of the quadratic reciprocity law. For more information we refer to [FR]. We now give the theorem.

Theorem 6.2.5 (Law of Quadratic Reciprocity). *If p, q are distinct odd primes then*

$$\left(\frac{p}{q}\right)\left(\frac{q}{p}\right) = (-1)^{(\frac{p-1}{2})(\frac{q-1}{2})}.$$

Alternatively if p, q are distinct odd primes then
(1) *If at least one of p, q is congruent to 1 modulo 4 then*

$$x^2 \equiv q \,(\mathrm{mod}\, p) \quad and \quad x^2 \equiv p \,(\mathrm{mod}\, q)$$

are either both solvable or both unsolvable.
(2) *If both p and q are congruent to 3 modulo 4 then one of*

$$x^2 \equiv q \,(\mathrm{mod}\, p) \quad and \quad x^2 \equiv p \,(\mathrm{mod}\, q)$$

is solvable and the other is unsolvable.

For a proof of the quadratic reciprocity law and more details we refer to [FR].

Example 6.2.6. Let $p = 1993$ and $q = 65\,537 = 2^{16} + 1$. We show that p is a quadratic non-residue modulo q.

$$\left(\frac{1993}{65\,537}\right) = \left(\frac{65\,537}{1993}\right) = \left(\frac{1761}{1993}\right) = \left(\frac{232}{1761}\right) = \left(\frac{2}{1761}\right)^3 \left(\frac{29}{1761}\right)$$
$$= \left(\frac{1761}{29}\right) = \left(\frac{21}{29}\right) = \left(\frac{3}{29}\right)\left(\frac{7}{29}\right) = \left(\frac{29}{3}\right)\left(\frac{29}{7}\right)$$
$$= \left(\frac{2}{3}\right)\left(\frac{1}{7}\right) = \left(\frac{2}{3}\right) = -1.$$

Since the value of the Legendre symbol is -1 it follows that p is a quadratic non-residue modulo q.

Example 6.2.7. Let $p = 7$ and $a = 870$. We show that a is a quadratic residue modulo p.

$$\left(\frac{870}{7}\right) = \left(\frac{2}{7}\right)\left(\frac{3}{7}\right)\left(\frac{5}{7}\right)\left(\frac{29}{7}\right) = -\left(\frac{7}{3}\right)\left(\frac{7}{5}\right) = -\left(\frac{1}{3}\right)\left(\frac{2}{5}\right) = 1.$$

Since the value of the Legendre symbol is 1 if follows that a is a quadratic residue modulo p.

6.3 Modular Square Roots

We now describe a probabilistic algorithm to calculate square roots modulo an odd prime p. By a **probabilistic algorithm**, we mean an algorithm that will return either that an inputted integer a with $(a, p) = 1$ does not have a square root modulo p, or that a has a given degree of probability of having a square root.

Suppose first that $p \equiv 3 \pmod 4$. Then $\frac{p+1}{4} \in \mathbb{N}$. If $\overline{a} = \overline{b}^2$ in $\mathbb{Z}_p^*$ then

$$\overline{a}^{\frac{p+1}{4}} = \overline{b}^{\frac{p+1}{2}} = \overline{b}^{\frac{p-1}{2}+1} = \overline{b}^{\frac{p-1}{2}}\overline{b}.$$

Since $\overline{b}^{\frac{p-1}{2}} = \pm\overline{1}$ it follows that the whole expression equals $\pm\overline{b}$. Therefore

$$\overline{a}_p = \overline{a}^{\frac{p+1}{4}} \in \mathbb{Z}_p^*$$

is a square root of $\overline{a}$ and $\pm\overline{a}_p$ are the only square roots of $\overline{a}$. Therefore in the case $p \equiv 3 \pmod 4$ we get the square roots of a modulo p by simple exponentiation.

Theorem 6.3.1. *Let p be an odd prime and $a \in \{1,\ldots,p-1\}$ a quadratic residue modulo p. Then the following algorithm produces with arbitrary probability $\geq 1 - (\frac{1}{2})^k$ a square root of a modulo p.*
(1) *If $p \equiv 3 \pmod 4$ then give $a^{\frac{p+1}{4}}$ modulo p and stop.*
(2) *If $p \equiv 1 \pmod 4$ write $p - 1 = 2^s t$ with $s \geq 2$ and t odd.*
    (a) *Choose a random $\overline{u} \in \mathbb{Z}_p^*$ and compute $\overline{u}^{\frac{p_1}{2}}$. Repeat this as long as $\overline{u}^{\frac{p-1}{2}} = -1$ up to a maximum of k times. Now calculate $\overline{v} = \overline{u}^t$.*
    (b) *Determine $\overline{y} = \overline{a}^{\frac{t+1}{2}}$*
    (c) *Calculate $\overline{a}^{2^{s-i}t}$ for $i = 2, 3, \ldots, s$. Then define $b_0, b_1, \ldots, b_{s-2} \in \{0,1\}$ recursively by*

$$\overline{a}^{2^{s-i}t} = \overline{v}^{b_0 2^{s-i+1} + b_1 2^{s-i+2} + \cdots + b_{s-2} 2^{s-1}}. \qquad (\ast)$$

    Define

$$b = b_0 + 2b_1 + \cdots + 2^{s-2} b_{s-2}$$

    and give $\overline{x} = \overline{y}\,\overline{v}^{-b}$.

Proof. First we show that in part (c) we get a square root modulo p for a.
    For $i = s$ we get

$$\overline{a}^{2^s t} = \overline{a}^t = \overline{v}^{2b_0 + 2^2 b_1 + \cdots + 2^{s-1} b_{s-2}} = \overline{v}^{2b}$$

and hence

$$\overline{x}^2 = \overline{y}^2 \overline{v}^{-2b} = \overline{a}^{t+1} \overline{a}^{-t} = \overline{a}.$$

We now show that we can realize the equation $(\ast)$.
a is a quadratic residue modulo p so that $a^{\frac{p-1}{2}} = a^{2^{s-1}t} \equiv 1 \pmod p$. Then

$$a^{\frac{p-1}{4}} = a^{2^{s-2}t} \equiv \pm 1 \pmod p.$$

On the other side

$$\overline{v}^{2^{s-1}} = -\overline{1} \quad \text{because } \overline{v} = \overline{u}^t \quad \text{and} \quad \overline{u}^{2^{s-1}t} = -\overline{1}.$$

If $a^{2^{s-2}t} \equiv 1 \pmod p$ then choose $b_0 = 0$ and if $a^{2^{s-2}t} \equiv -1 \pmod p$ then choose $b_0 = 1$. From this we get

$$\overline{a}^{2^{s-2}t} = \overline{v}^{2^{s-1}b_0} \implies \overline{a}^{2^{s-2}t}\overline{v}^{-2^{s-1}b_0} = \overline{1}.$$

Now we may construct $b_1, b_2, \ldots, b_{s-2}$ step by step. Recall that if we construct b_j for $j \in \{1, \ldots, s-2\}$ then $i = j + 2$ and

$$\bar{v}^{2^{s-i+j+1}} = \bar{v}^{2^{s-1}}.$$ $\square$

We now consider the situation $n = pq$ with p, q two distinct primes.

Theorem 6.3.2. *Let $p, q \geq 2$ be two distinct primes. Let $a \in \mathbb{Z}$ with $(a, p) = (a, q) = 1$. Then a is a quadratic residue modulo pq if and only if a is a quadratic residue modulo both p and q.*

Proof. Suppose that $x^2 \equiv a \pmod{pq}$. Then directly $x^2 \equiv a \pmod{p}$ and $x^2 \equiv a \pmod{q}$.

Conversely, suppose then $x^2 \equiv a \pmod{p}$ and $y^2 \equiv a \pmod{q}$. Then by the Chinese Remainder Theorem (see Chapter 5) there exists an integer z with $z \equiv x \pmod{p}$ and $z \equiv y \pmod{q}$. Then $z^2 \equiv a \pmod{p}$ and $z^2 \equiv a \pmod{q}$ and hence also $z^2 \equiv a \pmod{pq}$. $\square$

Theorem 6.3.3. *Let $n = pq$ with p, q two distinct odd primes. Let a be an integer with $(a, p) = (b, p) = 1$. Then the congruence $x^2 \equiv a \pmod{n}$ has either 0 or 4 solutions modulo n.*

If the prime factorization $n = pq$ is known and if a is a quadratic residue modulo n then there exists a probabilistic algorithm to efficiently calculate the four solutions of $x^2 \equiv a \pmod{n}$.

Proof. If the number a is a quadratic non-residue modulo n then the congruence $x^2 \equiv a \pmod{n}$ has no solutions modulo n.

Now let a be a quadratic residue modulo n. We apply the Chinese Remainder Theorem to obtain a bijection

$$\phi : \mathbb{Z}_n^* \to \mathbb{Z}_p^* \times \mathbb{Z}_q^* \quad \text{with } \bar{a} \mapsto (\bar{b}, \bar{c}).$$

Then $\bar{b} \in \mathbb{Z}_p^*$ and $\bar{c} \in \mathbb{Z}_q^*$ are quadratic residues modulo p or modulo q respectively.

Hence there exist $\bar{y} \in \mathbb{Z}_p^*$ and $\bar{z} \in \mathbb{Z}_q^*$ with $\bar{b} = \bar{y}^2$ and $\bar{c} = \bar{z}^2$.

The residue classes

$$\bar{x}_1 = \phi^{-1}((\bar{y}, \bar{z})), \quad \bar{x}_2 = \phi^{-1}((-\bar{y}, \bar{z}))$$
$$\bar{x}_3 = \phi^{-1}((\bar{y}, -\bar{z})) \quad \text{and} \quad \bar{x}_4 = \phi^{-1}((-\bar{y}, -\bar{z}))$$

are pairwise distinct and satisfy

$$\phi(\bar{x}_i^2) = (\bar{y}, \bar{z})^2 = (\bar{b}, \bar{c}).$$

Hence

$$(\bar{x}_i)^2 = \phi^{-1}(\bar{b}, \bar{c}) \implies (\bar{x}_i)^2 = \bar{a} \in \mathbb{Z}_n^*$$

for $i = 1, 2, 3, 4$.

For the second part of the theorem we apply Theorem 6.3.1 to determine $\bar{y}$ and $\bar{z}$. We can then apply the Chinese Remainder Theorem to calculate $\bar{x}_1, \bar{x}_2, \bar{x}_3, \bar{x}_4$. $\square$

If $n = pq$, with p, q distinct large odd primes, then there is no known general probabilistic algorithm which can be applied to check efficiently if an integer a is a quadratic
residue modulo n or not, unless the prime factors p and q are known.

We should mention the case where $p \mid a$ or $q \mid a$. Here consider $n = pq$ with p, q
two distinct odd primes and suppose that $a \in \mathbb{Z}$.

(1) If we have $p \mid a$ and $(a, q) = 1$, or if we have $q|a$ and $(a, p) = 1$, then the congruence
$x^2 \equiv a \pmod{n}$ has either 0 or 2 solutions modulo n.

(2) If $p \mid a$ and $q \mid a$ then the congruence $x^2 \equiv a \pmod{n}$ has only the solution 0
modulo n.

(3) Let $n = 2p$ with $p \geq 3$ and let a be an integer.
   (a) If $2 \mid a$ and $(a, p) = 1$ then the congruence $x^2 \equiv a \pmod{n}$ has either 0 or 2
   solutions modulo n.
   (b) If $(2, a) = 1$ and $p|a$ then the congruence $x^2 \equiv a \pmod{n}$ has either 0 or 1
   solution modulo n.
   (c) If $2 \mid a$ and $p \mid a$ then the congruence $x^2 \equiv a \pmod{n}$ has only the solution 0
   modulo n.

If p and q are known in (2) and (3) then the solutions can be calculated as in Theorem 6.3.2 by applying the Chinese Remainder Theorem.

6.4 Products of Two Primes

The two most commonly used methods in constructing public key cryptosystems are
the RSA method and the Diffie-Hellman method. The Diffie-Hellman method depends
for its security on the supposed difficulty of the discrete log problem, which we will
discuss in Section 6.5. The RSA method depends on the difficulty of factoring large
integers of the form $n = pq$, with p, q distinct odd large primes. Here we show that
there are important connections between the square root problem and the factorization problem.

Theorem 6.4.1. *Let p and q be two distinct primes and $n = pq$. If there exists a probabilistic algorithm to efficiently determine a square root b modulo n for every quadratic
residue a modulo n then there exists a probabilistic algorithm to factorize n.*

Proof. We randomly choose an integer $x \in \{1, \ldots, n-1\}$. We may assume that $(x, n) = 1$,
otherwise we may factorize n. Hence we may also assume that p, q are odd primes with
$(x, p) = (x, q) = 1$. Let $c \equiv x^2 \pmod{n}$. Determine a square root a for x modulo n. The
integer a is modulo n one of the four square roots of c (see last section). We have

$$\bar{a}^2 = \bar{x}^2 \in \mathbb{Z}_n^* \quad \text{that is} \quad n = pq \mid (a^2 - x^2) = (a - x)(a + x).$$

Here $|a - x| \in \{0, 1, \ldots, n-1\}$ because $a, x \in \{1, \ldots, n-1\}$. There are four possibilities
(1) $p \mid (a - x), q \mid (a - x) \implies (a - x, n) = n,$

(2) $p \mid (a + x), q \mid (a + x) \implies (a - x, n) = 1$,
(3) $p \mid (a - x), q \mid (a + x) \implies (a - x, n) = p$,
(4) $p \mid (a + x), q \mid (a - x) \implies (a - x, n) = q$.

Since each of these four cases occurs with equal probability we get that $d = (a - x, n)$ is a proper divisor of n with probability $\frac{1}{2}$. Hence after k repetitions we may factorize n with probability $1 - (\frac{1}{2})^k$ in its prime factors. □

The preceding theorem, together with Theorem 6.5, show that the problem of factorizing $n = pq$ and to determine square roots modulo n are probabilistically algorithmically equivalent.

We may extend the previous result as follows.

Theorem 6.4.2. *Let n be a composite positive integer. Let $x, y \in \mathbb{Z}$ with $x^2 \equiv y^2 \pmod{n}$ and x not congruent to $\pm y$ modulo n. Then $(x + y, n)$ and $(x - y, n)$ are proper divisors of n.*

Proof. From $x^2 \equiv y^2 \pmod{n}$ we get that $x^2 - y^2 \equiv 0 \pmod{n}$ and hence $(x-y)(x + y) = kn$ for some integer k. Note that both x and y are not congruent to 0 modulo n and $k \neq 0$ because x is not congruent to y or $-y$ modulo n.

Hence n is a divisor of $(x + y)(x - y)$ but neither a divisor of $x + y$ nor $x - y$, again because x is not congruent to $\pm y$ modulo n. Therefore $(x - y, n)$ and $(x + y, n)$ are proper divisors of n. □

Corollary 6.4.3. *Let $n = pq$ with p, q two distinct primes. Let $x, y \in \mathbb{Z}$ with $x^2 \equiv y^2 \pmod{n}$ and x not congruent to $\pm y$ modulo n. Then $|(x + y, n)|$ and $|(x - y, n)|$ are the prime factors of n.*

Example 6.4.4.
(a) Let $n = 7429, x = 227, y = 210$. Then $x^2 - y^2 = n$ and $x - y = 17, x + y = 437$.
 Therefore $(x - y, n) = 17$ is a proper divisor of n.
(b) Let $n = 253, x = 17, y = 6$. Then $17^2 \equiv 6^2 \pmod{253}$ and $x - y = 11, x + y = 23$.
Therefore $(11, 253) = 11, (23, 253) = 23$ and $253 = (23)(11)$.

Note that if $n = pq$ is the product of two distinct primes and we can find integers x, y with $x^2 \equiv y^2 \pmod{n}$ then we can factorize n. The problem of course is to find such x and y. There is a quite efficient probabilistic algorithm called the **quadratic sieve method** to find such x and y values.

The method works as follows. Let $m = [\sqrt{n}]$ where $[a]$ denotes the greatest integer function, that is the greatest integer less than or equal to a.

Then consider the polynomial

$$f(t) = (t + m)^2 - n$$

and evaluate $f(a)$ for $a \in \mathbb{Z}$. We then try to find the prime factorization

$$f(a) = \pm p_1^{e_1} \cdots p_k^{e_k}.$$

If we find $a_1, \ldots, a_r$ such that $f(a_1) \cdots f(a_r) \equiv y^2 \pmod{n}$ for some integer y, then we can define

$$x \equiv (a_1 + m) \cdots (a_r + m) \pmod{n}$$

and test if x is not congruent to y modulo n and x is not congruent to $-y$ modulo n.
For more details see [LL1], [LL2] and [KKi].

Example 6.4.5 (This is an example from [Buc]). Let $n = 7429$ so that $m = 86$ and $f(t) = (t + 86)^2 - 7429$.
We have $f(1) = 87^2 - 7429 = 140 = 2^2 \cdot 5 \cdot 7$ and $f(2) = 88^2 - 7429 = 315 = 3^2 \cdot 5 \cdot 7$.
It follows that

$$(87 \cdot 88)^2 \equiv (2 \cdot 3 \cdot 5 \cdot 7)^2 \pmod{n}$$

and we define $x = (87)(88) \equiv 227 \pmod{n}$ and $y \equiv (2)(3)(5)(7) = 210 \pmod{n}$. Now take $x = 227$ and $y = 210$ and we have the two integers from the quadratic sieve.

A variation of the quadratic sieve method is the **Fermat factorization method**. Here let $n = pq$ with p, q distinct odd primes. Then $t = \frac{p+q}{2}$ and $s = \frac{p-q}{2}$ are integers satisfying

$$n = pq = \left(\frac{p+q}{2}\right)^2 - \left(\frac{p-q}{2}\right)^2.$$

Suppose that we want to factor $n = pq$. Start with $t = [\sqrt{n}] + 1$. Then we determine $t^2 - n$. If $t^2 - n = s^2$ for some s, then from the above equation we determine $p = t + s$ and $q = t - s$.

If $t^2 - n$ is not a square we replace t by $t + 1$ and repeat the process. We continue until we get a square.

This methods works quite efficiently if $|p - q|$ is small. In this case $t = \frac{p+q}{2}$ is only slightly bigger than $\sqrt{n}$. If $|p - q|$ is very large, then many iterations are needed to get the desired result.

A variation of the Fermat factorization method works with a small positive integer k, the starting value $t = [\sqrt{kn}] + 1$ and the values $t^2 - kn$. Here if $t^2 - kn = s^2$ is a square then check $(t + s, n)$ and $(t - s, n)$.

Another probabilistic algorithm to factorize a composite number $n \in \mathbb{N}$ that uses congruences is based on the following idea.

Let $n \in \mathbb{N}$ be a composite integer. Assume that the prime number p divides n and that we have two integers x, y with x not congruent to y modulo n and $x \equiv y \pmod{p}$. Let $d = (x - y, n)$.

From $p \mid (x - y)$ and $p | n$ we obtain that $p | d$ and so $d \neq 1$. From n not being a divisor of $x - y$ we get that $d \neq n$. Hence $1 < d < n$ and d is a proper divisor of n.

Now if $n = pq$, for two distinct primes, then we have the factorization. Therefore we want to find such x, y for $n = pq$ with two distinct odd primes. This can be done probabilistically applying Pollard's ρ-algorithm.

We take an integer x_0 to start ρ and a function $f(x)$. A suitable function is $f(x) = x^2 + c$ with an integer c and $c \neq 0, -2$. Then we recursively define $x_{i+1} = f(x_i)$. Define $y_i \in \{0, 1, \ldots, n - 1\}$ with $x_i \equiv y_i \pmod{n}$ for $i = 0, 1, \ldots$.

If p is a prime divisor (which so far we do not know) then there are $i \neq j$ with $y_i \equiv y_j \pmod{p}$, that is $d = (y_i - y_j, n) > 1$. If in addition $y_i \neq y_j$ then $d \neq n$ because $y_i, y_j \in \{0, 1, \ldots, n-1\}$. Hence we must determine the values $(y_i - y_j, n)$ for $i \neq j$.

Example 6.4.6. Let $n = 1517$ and $x_0 = y_0 = 70$ with $f(x) = x^2 + 1$.
  Then $y_7 = 1358$ and $y_{14} = 825$ and we get $(1358 - 825, 1517) = 41$.

There are also factorization methods utilizing the Theorems of Fermat and Euler. The next theorem shows the connection between the Euler phi function $\phi(n)$ and the factorization of $n = pq$ with p, q distinct odd primes.

Theorem 6.4.7. Let $n = pq$ with p, q distinct odd primes. Then the pair (p, q) can be efficiently calculated if and only if $\phi(n)$ can be efficiently calculated.

Proof. We have $\phi(n) = (p-1)(q-1) = n - p - q + 1$. Hence $p + q = n + 1 - \phi(n)$ and $pq = n$. Therefore p and q are exactly the two solutions of the quadratic equation

$$x^2 - (n + 1 - \phi(n))x + n = 0. \qquad \square$$

The following is necessary for our next factorization method.

Lemma 6.4.8. Let $n = pq$ with p, q distinct odd primes. Suppose that there exists an even integer $m \geq 2$ with $\bar{a}^m = \bar{1}$ in $\mathbb{Z}_n^*$ for all $\bar{a} \in \mathbb{Z}_n^*$. Assume that there exists a $\bar{b} \in \mathbb{Z}_n^*$ with $\bar{b}^{\frac{m}{2}} \neq \bar{1}$. Then
(1) For at least 50 % of the residue classes $\bar{a}$ in $\mathbb{Z}_n^*$ we have $\bar{a}^{\frac{m}{2}} \neq \bar{1}$.
(2) Suppose that there exists an $\bar{a} \in \mathbb{Z}_p^*$ with $\bar{a}^{\frac{m}{2}} \neq \bar{1}$ in $\mathbb{Z}_p^*$. Then $\bar{b}^{\frac{m}{2}} = \bar{1}$ for exactly
     50 % of the residue classes $\bar{b} \in \mathbb{Z}_p^*$ and $\bar{b}^{\frac{m}{2}} = -\bar{1}$ for the other 50 %.
The analogous result holds for q.

Proof.
(a) Let $\{\bar{a}_1, \ldots, \bar{a}_s\}$ be the different residue classes in $\mathbb{Z}_n^*$ with $\bar{a}^{\frac{m}{2}} = \bar{1}$.
     Then the residue classes $\bar{a}_i \bar{b}$ are pairwise distinct and satisfy

$$(\bar{a}_i \bar{b})^{\frac{m}{2}} = (\bar{a}_i)^{\frac{m}{2}}(\bar{b})^{\frac{m}{2}} = \bar{b}^{\frac{m}{2}} \neq \bar{1}.$$

(b) If $\bar{a}^k = \bar{1}$ in $\mathbb{Z}_n^*$ then $\bar{a}^k = \bar{1}$ in $\mathbb{Z}_p^*$ for an integer k. The proof now follows from
     Theorem 6.2. $\qquad \square$

Now we describe a probabilistic algorithm for the factorization of $n = pq$ based on Fermat's and Euler's theorems.

Theorem 6.4.9 (Pollard's $p - 1$ method). Let $n = pq$ with p, q distinct odd primes. Suppose that we know an integer $m \geq 1$ with $\bar{a}^m = \bar{1}$ for all $\bar{a} \in \mathbb{Z}_n^*$. In particular we have such an m if we know $\phi(n)$. Then we can efficiently determine the prime factors p and q with the following probabilistic algorithm:
(1) Check for many, say 100, randomly chosen integers $a \in \{1, \ldots, n-1\}$ with $(a, n) = 1$
     if $\bar{a}^{\frac{m}{2}} = \bar{1}$. If this is the case then replace m by $\frac{m}{2}$.

(2) *Repeat* (1) *as often as one has an a with* $\overline{a}^{\frac{m}{2}} \neq \overline{1}$.
(3) *For a randomly chosen* $a \in \{1, \ldots, n-1\}$ *with* $(a, n) = 1$ *and* $\overline{a}^{\frac{m}{2}} \neq \overline{1}$, *calculate*
  $p_1 = (n, a^{\frac{m}{2}} - 1)$ *and check if* $p_1 > 1$. *If this is the case then n is factored. Define*
  $q_1 = \frac{n}{p_1}$ *and return the pair* (p_1, q_1).
(4) *Repeat* (3) *as long as there is a* p_1 *with* $p_1 > 1$.
This is a probabilistic algorithm that finds the factorization $n = pq$ *in finitely many steps*
with a probability close to 1.

Proof. We first remark that m must be even. If m were odd we would have

$$(-1)^m \equiv (n-1)^m \equiv -1 \pmod{n}$$

contradicting the fact that $\overline{a}^m = \overline{1}$ for all $\overline{a} \in \mathbb{Z}_n^*$. By Lemma 6.4.1 part (a) there are at
least 50 % of the residue classes $\overline{a} \in \mathbb{Z}_n^*$ with $\overline{a}^{\frac{m}{2}} \neq \overline{1}$ if there exists one such $\overline{a}$. If we
check N times in step (1) whether $\overline{a}^{\frac{m}{2}} = \overline{1}$ or $\overline{a}^{\frac{m}{2}} \neq \overline{1}$ in $\mathbb{Z}_n^*$ then we have a probability
$\leq \left(\frac{1}{2}\right)^N$ that always $\overline{a}^{\frac{m}{2}} = \overline{1}$.

Hence the loop (1)–(2) gives with high probability an even integer m such that
there is an $\overline{a} \in \mathbb{Z}_n^*$ with $\overline{a}^{\frac{m}{2}} \neq \overline{1}$.

Now we have two cases.

Case 1: $(p-1) \mid \frac{m}{2}$ or $(q-1) \mid \frac{m}{2}$.

Suppose without loss of generality that $(p-1) \mid \frac{m}{2}$. Then $(q-1)$ cannot divide $\frac{m}{2}$
because otherwise $\phi(n) = (p-1)(q-1)$ is a divisor of $\frac{m}{2}$ contradicting $\overline{a}^{\frac{m}{2}} \neq \overline{1}$ in
$\mathbb{Z}_n^*$. Recall form Euler's Theorem that $\overline{a}^{\phi(n)} = \overline{1}$ for all $\overline{a} \in \mathbb{Z}_n^*$.
Hence $c^{\frac{m}{2}} \equiv 1 \pmod{p}$ for all $c \in \{1, \ldots, n-1\}$ with $(c, n) = 1$ and there exists $b \in$
$\{1, 2, \ldots, n-1\}$ with $(b, n) = 1$ and $b^{\frac{m}{2}}$ not congruent to 1 modulo q. By Lemma 6.4.2
part (b) there is therefore a $b \in \{1, \ldots, n-1\}$ with $(b, n) = 1$ and $b^{\frac{m}{2}} \equiv -1 \pmod{q}$.
We cannot have $(n, b^{\frac{m}{2}} - 1) = 1$ or q or n since $p \mid (b^{\frac{m}{2}} - 1)$ and q does not divide
$b^{\frac{m}{2}} - 1$. Hence $(n, b^{\frac{m}{2}} - 1) = p$ and n is factorized.

Case 2: $(p-1)$ does not divide $\frac{m}{2}$ and $(q-1)$ does not divide $\frac{m}{2}$.

Then for $b \in \{1, \ldots, n-1\}$ with $(b, n) = 1$ we have four possibilities:
(a) $b^{\frac{m}{2}} \equiv 1 \pmod{p}$ and $b^{\frac{m}{2}} \equiv 1 \pmod{q}$;
(b) $b^{\frac{m}{2}} \equiv -1 \pmod{p}$ and $b^{\frac{m}{2}} \equiv 1 \pmod{q}$;
(c) $b^{\frac{m}{2}} \equiv 1 \pmod{p}$ and $b^{\frac{m}{2}} \equiv -1 \pmod{q}$;
(d) $b^{\frac{m}{2}} \equiv -1 \pmod{p}$ and $b^{\frac{m}{2}} \equiv -1 \pmod{q}$.
Hence with probability $\frac{1}{2}$ we have $b \in \{1, \ldots, n-1\}$ with $(b, n) = 1$ and $b^{\frac{m}{2}} \equiv$
$-1 \pmod{p}$ and $b^{\frac{m}{2}} \equiv 1 \pmod{q}$ or $b^{\frac{m}{2}} \equiv 1 \pmod{p}$ and $b^{\frac{m}{2}} \equiv -1 \pmod{q}$ and
we may factorize n as in Case (1). Recall that there is an $a \in \mathbb{Z}$ with $\overline{a}^{\frac{m}{2}} \neq \overline{1}$ in
$\mathbb{Z}_n^*$ and hence $\overline{a}^{\frac{m}{2}} = -\overline{1}$ in $\mathbb{Z}_p^*$ or in $\mathbb{Z}_q^*$ by the Chinese Remainder Theorem. Now
apply Lemma 6.4.1(b). In cases (a) and (d) we move to another b and repeat the
calculations. $\qquad\square$

Corollary 6.4.10. *Let $n = pq$ with p, q distinct odd primes. Then the following conditions are equivalent.*

(1) *If we know an integer e with $(e, \phi(n)) = 1$ then we can efficiently find an integer d with $ed \equiv 1 \pmod{\phi(n)}$.*

(2) *We can efficiently find the prime factors p and q of n.*

Proof. Suppose that we assume (1), so we suppose that we know such a d. Then $ed - 1$ is divisible by $\phi(n) = (p-1)(q-1)$ which is even. We then get $a^{ed-1} \equiv 1 \pmod{n}$ for all integers $a \in \{1, \ldots, n-1\}$ with $(a, n) = 1$. This can be seen as follows.

There exists an integer r with $ed = 1 + r(p-1)(q-1)$. Then

$$a^{ed} = a^{1+r(p-1)(q-1)} = a\left(a^{(p-1)(q-1)}\right)^r$$

and $a^{ed} \equiv a \pmod{n}$ by Fermat's theorem and the Chinese remainder theorem. With $n = ed - 1$, which is even, we can calculate p and q with the probabilistic algorithm from the theorem. Hence we get (2).

Now suppose that we have (2) so that we know p and q.
Then $\phi(n) = (p-1)(q-1)$ and we can find an integer d with $de \equiv 1 \pmod{\phi(n)}$ using the Euclidean algorithm. $\qquad\square$

6.5 Discrete Logarithms and the Discrete Log Problem

Diffie-Hellman in 1976 developed the original public key cryptographic protocol using the **discrete log problem**. In this section we introduce this and also discuss two algorithms for attempted solutions of the problem. The basic idea is that in modular arithmetic it is easy to raise an element to a power but difficult to determine, given an element, if it is a power of another element. We define the problem in general.

Definition 6.5.1. If G is a finite group, such as the cyclic multiplicative group of $\mathbb{Z}_p$ where p is a prime, and $h = g^k$ for some k, then a **discrete logarithm** or **discrete log** of h to the base g is any integer t with $h = g^t$. We denote this by

$$k = \log_g^G(h) \quad \text{or simply} \quad k = \log_g(h) \quad \text{if } G \text{ is fixed.}$$

In the Diffie-Hellman method, that we will discuss in Chapter 7, this is usually applied to the cyclic group $\mathbb{Z}_p^*$ where p is a prime.

Example 6.5.2. If $G = \mathbb{Z}_{19}^*$ then $6 = \log_2(7)$ since $2^6 = 64 \equiv 7 \pmod{19}$.

The discrete log does not have to be unique as the next example shows.

Example 6.5.3. Let $G = \mathbb{Z}_8^*$. Then $2 = \log_3(1)$ since $3^2 \equiv 1 \pmod 8$. But also $4 = \log_3(1)$ since $3^4 \equiv 1 \pmod 8$ and 3 and 4 are incongruent modulo 8.

Notice that if $G = \langle g \rangle$ is of order n and $h = g^i$ with $1 \leq i \leq n$ then the discrete log of h is uniquely defined for the generator g, that is i is unique modulo n for g.

The **discrete logarithm problem** denoted also by DLP is then the following. Let $G = \langle g \rangle$ and $h \in G$. Find $x \in \mathbb{Z}$ such that $h = g^x$.

The smallest non-negative x with this property is often called the **discrete logarithm** of h relative to the generator g.

Other than an exhaustive search there is no known algorithm to solve the discrete log problem. This makes it ideal then as a method to construct a one-way function. In the next two subsections we look at algorithms to attempt to find a solution.

6.5.1 Shank's Baby Step Giant Step Algorithm (BSGS)

In this section we describe an algorithm, due to Shanks, to find a discrete log. This is called the **baby step giant step** algorithm which is abbreviated BSGS. Let $G = \langle b \rangle$ be a finite cyclic group and suppose that $a \in G$. We will try to find algorithmically an integer x such that $a = b^x$ and hence $x = \log_b a$. We need an upper bound $S \in \mathbb{N}$ for the order of G, that is $|G| \leq S$. Define

$$n = \min\{k \in \mathbb{N} \mid k^2 \geq S\}.$$

Then $n^2 \geq S$ and in particular $x = \log_b a \leq n^2$. Then $x = qn + r$ with $r \in \{0, 1, \ldots, n-1\}$ and $q \in \{0, 1, \ldots, n\}$ by the division algorithm.

We get $a = b^x = b^{qn+r}$ so that $ab^{-r} = b^{qn}$. From this last equation we must find r and q to obtain the discrete logarithm $x = qn + r$. There are n possibilities for r and $n + 1$ possibilities for q. Hence we have to make two list of group elements. The first is the **baby step list** given by

$$(1) \quad \begin{array}{c|ccccc} r & 0 & 1 & 2 & \cdots & n-1 \\ \hline ab^{-r} & a & ab^{-1} & ab^{-2} & \cdots & ab^{-(n-1)} \end{array}$$

and the **giant step list** given by

$$(2) \quad \begin{array}{c|ccccc} q & 0 & 1 & 2 & \cdots & n \\ \hline b^{qn} & 1 & b & b^{2n} & \cdots & b^{n^2} \end{array}$$

Now we compare both lists. A common element $ab^{-r} = b^{qn}$ in both lists gives the integers q, r and hence the discrete logarithm

$$x = qn + r = \log_b(a).$$

There must be a common element because $|G| \leq n^2$.

Example 6.5.4 (see [Buc]). Let $G = \mathbb{Z}_p^*$ with $p = 2017$. The integer 5 is a primitive element in G that is $\mathbb{Z}_p^* = \langle \overline{5} \rangle$. Suppose that we want to find $x = \log_5(\overline{3})$. We can take $n = 45$ because $45^2 > 2016 = |G|$.

We write down the baby step list $r = 0, 1, \ldots, 44$ and the giant step list for $q = 0, 1, \ldots, 45$ and find a common element for $r = 40$ and $q = 22$. Hence

$$x = (22)(45) + 40 = 1030 = \log_5(\overline{3}).$$

For practical implementation of the BSGS one can first calculate the giant step list because it is independent of the integer a and can be used several times. Then we calculate the baby step list and compare each value step by step to the giant step list. We need $\leq 2n$ group operations and we must store n group elements. Therefore we get the following result describing the complexity of the BSGS algorithm (see Chapter 3).

Theorem 6.5.5. *The complexity of the baby-step giant-step algorithm in a cyclic group G is $\mathcal{O}(\sqrt{|G|})$.*

It follows that if $|G| > 2^{160}$ the BSGS algorithm is not very helpful due to the size of the necessary storage space. The problem of storage space is handled in the next algorithm we discuss, the Pollard ρ-method.

6.5.2 Pollard's ρ-Algorithm

The second method we describe for finding discrete logarithms is called the **Pollard ρ-algorithm**. This method will handle some of the storage space problems of the BSGS algorithms. We first need the following.

Lemma 6.5.6. *Let $a, c \in \mathbb{Z}$ and $n \in \mathbb{N}$. Then the congruence $ax \equiv c \pmod{n}$ with $(a, n) = d$ is solvable if and only if $d|c$. In this case there are exactly d solutions modulo n that are given by*

$$x = x_0 + \frac{tn}{d}, \quad t = 0, 1, \ldots, d - 1$$

where x_0 is any solution of the reduced equation

$$\frac{a}{d}x \equiv \frac{c}{d} \equiv 0 \left(\bmod \frac{n}{d}\right).$$

For a proof of this we refer to the book [FR].

We can now give the Pollard ρ-algorithm for the discrete log problem. If $G = \langle b \rangle$ is a finite cyclic group and $a \in G$ then this algorithm consists of three steps.

Step (1): Divide G into three pairwise disjoint subsets G_1, G_2, G_3 with $1 \notin G_2$ and each of about the same size.

Step (2): Determine two integers s, t with $a^s = b^t$ in the following manner. Produce a sequence inductively with $x_0 = 1$ and

$$x_{i+1} = ax_i \quad \text{if } x_i \in G_1;$$
$$x_{i+1} = x_i^2 \quad \text{if } x_i \in G_2;$$
$$x_{i+1} = bx_i \quad \text{if } x_i \in G_3.$$

Then $x_1 \in \{a, b\}$, $x_2 \in \{a^2, ab, b^2\}$ and so on. Altogether we get that $x_i = a^{\alpha_i} b^{\beta_i}$ for some $\alpha_i, \beta_i \in \mathbb{N} \cup \{0\}$.

Since G is finite there are indices i, j with $i < j$ and $x_i = x_j$. We then have

$$a^{\alpha_i} b^{\beta_i} = a^{\alpha_j} b^{\beta_j} \quad \text{so that} \quad a^{\alpha_i - \alpha_j} = b^{\beta_j - \beta_i}.$$

Setting

$$s = \alpha_i - \alpha_j \quad \text{and} \quad t = \beta_j - \beta_i$$

produces the desired integers.

Step (3): Now determine the discrete log $x = \log_b(a)$ as follows. Let s, t be as found in step (2). Then

$$b^t = a^s = b^{sx} \quad \text{since} \quad a = b^x \quad \text{and so} \quad sx \equiv t \pmod{n}.$$

Hence x is a solution of the congruence

$$sx \equiv t \pmod{n}. \tag{**}$$

Suppose that $d = (s, n)$. By Lemma 6.5.2.1 the set of integer solutions to this congruence is

$$y = x + \frac{tn}{d}, \quad t = 0, 1, \ldots, d - 1$$

where x is the constructed solution.

Without loss of generality suppose that v is the smallest positive solution to (**) which we can find using the Euclidean algorithm. Since we may assume that the discrete logarithm is smaller than n we may assume that

$$x = v + \frac{kn}{d}, \quad k = 0, 1, \ldots, d - 1.$$

Determine all values $b^{v + \frac{kn}{d}}$, $k = 0, 1, \ldots, d - 1$ and compare the results with a. This produces $x = \log_b(a)$.

Example 6.5.7 (see [Buc]). As before let $G = \mathbb{Z}_p$ with $p = 2017$. The integer 5 is a primitive element so that $\mathbb{Z}_p^* = \langle 5 \rangle$. We want to find $x = \log_5(3)$. Take

$$G_1 = \{1, \ldots, 672\}, \quad G_2 = \{673, \ldots, 1344\}, \quad G_3 = \{1345, \ldots, 2016\}.$$

If we start with $x_0 = 1023$ we get $5^{-800} = 3^{-128}$. To get x we have to solve the congruence $128x \equiv 800 \pmod{2016}$. Doing this we get $v = 22$ and $x = 22 + (16)(63)$. Recall that $(128, 2016) = 32$ and $63 = \frac{2016}{32}$.

The complexity of the Pollard ρ-algorithm in the finite cyclic group G is the same as that of the BSGS algorithm, namely $\mathcal{O}(\sqrt{|G|})$. However this method is more storage efficient because we must only save one of the sets G_1, G_2, G_3.

The two algorithms to solve the DLP can also be used to determine the order of an element in a finite abelian group.

Theorem 6.5.8. *Let G be a finite abelian group and* $a \in G$. *Suppose that* $B \in \mathbb{N}$ *is a bound for the order of a, that is* $|a| < B$ *(we could take* $B = |G| + 1$*). Then the following algorithm determines* $|a|$.

Step (1): *Let* $\beta = \min\{k \in \mathbb{N} \mid k^2 \geq B\}$. *Then* $\beta^2 \geq B$. *Determine* $a, a^2, a^3, \ldots, a^\beta$. *Define*
$$a_1 = a^\beta.$$

Step (2): *Compute* a_1^α *for* $\alpha = 1, \ldots, \beta$ *and check if* $a_1^\alpha \in \{a, a^2, \ldots, a^\beta\}$. *As soon as this is the case go to Step (3).*

Step (3): *If* $a_1^\alpha = a^\gamma$ *with* $\alpha, \gamma \in \{1, \ldots, \beta\}$ *then factorize* $\alpha\beta - \gamma$ *and find the smallest positive divisor* δ *with* $a^\delta = 1$. *The value* δ *gives the order* $|a|$ *of a.*

Proof. Let $n = |a| = o_G(a)$ be the order of the element a in the group G. Write $n = \alpha_1\beta + \gamma_1$ with $\alpha_1, \gamma_1 \in \{0, \ldots, \beta - 1\}$ by the division algorithm. This is possible since $n < B \leq \beta^2$. Then

$$1 = a^n = a^{\alpha_1\beta + \gamma_1} = a_1^{\alpha_1} a^{\gamma_1} = a_1^\alpha a^{-\gamma} = a^{\alpha\beta - \gamma}$$

with $\alpha = \alpha_1 + 1 \in \{1, \ldots, \beta\}$ and $\gamma = \beta - \gamma_1$ because $a_1 = a^\beta$. In particular $\gamma \in \{1, \ldots, \beta\}$. Hence in (2) we must have a match $a_1^\alpha = a^\gamma$. $\qquad\square$

Finally we show how to reduce the DLP in a finite cyclic group to the DLP for cyclic groups of prime order.

Theorem 6.5.9. *Let* $G_1 = \langle g_1 \rangle$ *and* $G_2 = \langle g_2 \rangle$ *be finite cyclic groups with* $(|G_1|, |G_2|) = 1$. *Let* $G = G_1 \times G_2 = \langle g \rangle$ *with* $g = (g_1, g_2)$. *Let* $a = (a_1, a_2) \in G$. *If* $x = \log_g(a)$ *then*

$$x \equiv \log_{g_i}(a_i) \pmod{|G_i|} \quad \text{for } i = 1, 2.$$

Proof. Let $x = \log_g(a)$ so that $g^x = a$ and hence $(g_1^x, g_2^x) = (a_1, a_2)$. Then $g_i^x = a_i = g_i^y$ with $x \equiv y \pmod{|G_i|}$, $0 \leq y < |G_i|$ for $i = 1, 2$. Hence $x \equiv \log_{g_i}(a_i) \pmod{|G_i|}$ for $i = 1, 2$. $\qquad\square$

Corollary 6.5.10. *Let* G, G_1, G_2, g *be as in the above theorem. If we know the discrete logarithms* $x_i = \log_{g_i}(a_i), i = 1, 2$ *then we can determine* $x = \log_g(a)$.

Proof. Since $(|G_1|, |G_2|) = 1$ we can apply the Chinese Remainder Theorem and obtain the result. $\qquad\square$

This provides the reduction of the DLP to cyclic groups of prime power orders.

Example 6.5.11 (see [KKi]). Let $G = \mathbb{Z}_p^*$ with $p = 73$. A primitive element modulo 73 is 5 so that $G = \langle \bar{5} \rangle$.

We have the decomposition $p - 1 = 72 = 2^3 3^2$.

We want to determine $x = \log_{\bar{5}}(\overline{58})$.

We get $G = G_1 \times G_2$ with

$$G_1 = \langle \bar{5}^9 \rangle = \langle \overline{10} \rangle \quad \text{with } |G_1| = 8$$

and

$$G_2 = \langle \overline{5}^8 \rangle = \langle \overline{2} \rangle \quad \text{with } |G_2| = 9.$$

For $a = \overline{58}$ we get the decomposition $a = \overline{58} = (\overline{58}^9, \overline{58}^8) = (\overline{51}, \overline{55})$. Now

$$x_1 = \log_{\overline{10}}(\overline{51}) = 3 \text{ in } G_1 \quad \text{and}$$
$$x_2 = \log_{\overline{5}}(\overline{55}) = 7 \text{ in } G_2$$

using the BSGS algorithm.

Now we apply the Chinese remainder theorem for

$$x \equiv 3 \pmod 8$$
$$x \equiv 7 \pmod 9.$$

This gives the solution $x \equiv 43 \pmod{72}$. Hence $\log_{\overline{5}}(\overline{58}) = 43$ and $\overline{5}^{43} = \overline{58}$.

Finally we can reduce the DLP to cyclic groups of prime order.

Theorem 6.5.12. *Let $G = \langle g \rangle$ be a cyclic group with $|G| = p^n$ with p a prime and $n \geq 1$. By taking the logarithm n times we can reduce the DLP for G to the DLP for groups of order p.*

Proof. If $n = 1$ there is nothing to prove. Suppose $n \geq 2$. For an induction suppose that the result holds for cyclic groups of order p^{n-1}.

Let $U = < g^p >$. Then U is a subgroup of G of order p^{n-1}. The factor group G/U is a cyclic group of order p (see [CFR1]) with $G/U = \langle gU \rangle$. Let $a \in G$. Determine $y \in \mathbb{N}$ with $a = (gU)^y = g^y U$. Then $ag^{-y} \in U = \langle g^p \rangle$. From the inductive hypothesis we may determine an integer x by taking the logarithm $(n-1)$ times such that $ag^{-y} = g^{px}$. Hence $a = g^{y+px}$.

We now describe the procedure. Let $G = \langle g \rangle$ be a finite cyclic group of order p^n with p prime and $n \geq 1$. Let $a \in G$ and suppose we want to find the discrete $\log x = \log_g(a)$. We write

$$x = x_0 + x_1 p + \cdots + x_{n-1} p^{n-1}$$

with $x_0, \ldots, x_{n-1} \in \{0, 1, \ldots, p-1\}$. We determine $x_0, \ldots, x_{n-1}$ step by step from the equation

$$g^x = g^{x_0 + x_1 p + \cdots + x_{n-1} p^{n-1}} = a \qquad (*)$$

and get $x = \log_g(a)$ this way.

Recall that $g^{p^n} = 1$. We first determine x_0.
From $g^x = a$ we get $g^{p^{n-1}x} = (g^{p^{n-1}})^{x_0} = a^{p^{n-1}}$ because $g^{p^n} = 1$.
Now $g^{p^{n-1}}$ has order p and we determine

$$x_0 = \log_{g^{p^{n-1}}}(a^{p^{n-1}}).$$

We then determine x_1 from the equation

$$ag^{-x_0} = g^{x_1 p + \cdots + x_{n-1} p^{n-1}}.$$

from this we get

$$(ag^{-x_0})^{p^{n-2}} = g^{p^{n-1}x_1} = (g^{p^{n-1}})^{x_1}$$

and we determine

$$x_1 = \log_{g^{p^{n-1}}}(a^{p^{n-2}} g^{-p^{n-2}x_0}).$$

Now continue in the analogous manner to determine $x_2, \ldots, x_{n-1}$ if necessary. $\square$

Example 6.5.13 (see [KKi]). Let $G = \langle g \rangle$ with $|G| = 5^3$. We realize G as a subgroup of index 2 in $\mathbb{Z}_{251}^*$.

We have $\mathbb{Z}_{251}^* = \langle \overline{6} \rangle$ and hence $G = \langle \overline{6}^2 \rangle = \langle \overline{36} \rangle$. We determine $x = \log_{\overline{36}}(\overline{4})$.
Write $x = x_0 + 5x_1 + 5^2 x_2$ with $x_0, x_1, x_2 \in \{0, 1, 2, 3, 4\}$ and determine x_0, x_1, x_2 in steps. From $\overline{36} = \overline{4}$ we get that $(\overline{36}^{25})^{x_0} = \overline{4}^{25} = \overline{1}$, that is $x_0 = 0$.
From $\overline{36}^{-5x_1+5^2x_2} = \overline{4} \cdot \overline{36}^{-0} = \overline{4}$ we get $(\overline{36}^{25})^{x_1} = \overline{4}^{-5} = \overline{20}$, that is $x_1 = \log_{\overline{36}^{25}}(\overline{20}) = 2$.

Therefore we obtain

$$\overline{36}^{-5^2 x_2} = \overline{4} \cdot \overline{36}^{-0-(2)(5)} = \overline{149}.$$

Then $(\overline{36}^{25})^{x_2} = \overline{149}$ and $x_2 = \log_{\overline{36}^{25}}(\overline{149}) = 4$. Altogether

$$x = \log_{\overline{36}}(\overline{4}) = 0 + (2)(5) + (4)(5^2) = 110.$$

In general, this algorithm determines n discrete logarithms $x_0, x_1, \ldots, x_{n-1}$ in the cyclic group $\langle g^{p^{n-1}} \rangle$ of order p, and finally the desired discrete logarithm x in $\langle g \rangle$ of order p^n.
Summarizing all these reductions we have the following.

Theorem 6.5.14. *Let G be a finite cyclic group of order $p_1^{n_1} p_2^{n_2} \cdots p_k^{n_k}$ with this being the prime decomposition. Then the DLP for G can be reduced to the DLP in cyclic groups of orders $p_1, \ldots, p_k$.*

Note that if $|G|$ has only small prime factors then the DLP for G is relatively easy to solve. For example suppose that $|G| = 2^{104} 3^{44} 5^{49} 7^{47}$ and is cyclic. Then the DLP is solved in cyclic groups of orders $2, 3, 5, 7$, respectively.

6.5.3 The Index Calculus Method (ICM) for $\mathbb{Z}_p^*$

Given a prime number $p \geq 3$, the **index calculus method** or **ICM** is a method to determine probabilistically the discrete logarithm k of b where $b = \overline{g}^k$ and g is a primitive root modulo p. The method is done with the following steps.

Step (1): Choose a good bound B with $B \geq 2$. For a positive integer B the **factor base** is given by

$$F(B) = \{q \mid q \text{ is a prime number with } q \leq B\}.$$

Let $v = |F(B)|$ and $p_1 < p_2 < \cdots < p_v$ be primes with $p_1 = 2$ so that $F(B) = \{p_1, \ldots, p_v\}$.

Step (2): Choose randomly and equally distributed an integer

$$a \in \{1,\ldots,p-1\}$$

and factor g^a modulo p, that is, factor the integer $c \in \{1,\ldots,p-1\}$ with $g^a \equiv c \pmod{p}$.

We say that g^a modulo p is B-smooth if each prime factor of g^a modulo p is in $F(B)$. If this is the case, then save a, g^a modulo p and the prime factorization of g^a modulo p.

Repeat this step as long as there are at least v piecewise different B-smooth values and each $p_i \in F(B)$ occurs at least once.

Step (3): Let $a_1,\ldots,a_v$ be the exponents of the pairwise different B-smooth values g^{a_i} modulo $p,\ldots,g^{a_v}$ modulo p found in Step 2. Then

$$g^{a_i} \equiv \prod_{j=1}^{v} p_j^{e_{ij}} \pmod{p}$$

with

$$e_{ij} \in \mathbb{Z}, e_{ij} \geq 0, \quad \text{for each } i \in \{1,\ldots,v\}.$$

From this we get the system of congruences

$$a_i \equiv \sum_{j=1}^{v} e_{ij} \log_{\bar{g}}(p_j) \pmod{(p-1)}, \quad i = 1,\ldots,v.$$

If all $\log_{\bar{g}}(p_i)$, $i = 1,\ldots,v$, occur, then solve the system for these $\log_{\bar{g}}(p_j)$. If this system is not solvable, then determine more B-smooth values as in Step 2 and continue as long as the system is solvable.

Step (4): Choose randomly and equally distributed $s \in \{1,\ldots,p-1\}$ as long as bg^s modulo p is B-smooth. Then

$$bg^s \equiv \prod_{j=1}^{v} p_j^{c_j} \pmod{p} \quad \text{for non-negative integers } c_1,\ldots,c_v.$$

Take the discrete logarithms, $\log_{\bar{g}}(p_j), j = 1,\ldots,v$, as found in Step 2, and put them into the congruence

$$\log_{\bar{g}}(b) + s \equiv \sum_{j=1}^{v} c_j \log_{\bar{g}}(p_j) \pmod{(p-1)}$$

and solve this for $k = \log_{\bar{g}}(b)$.

This is a probabilistic algorithm which gives uniquely the discrete log, $k = \log_{\bar{g}}(b)$ modulo $(p-1)$ of b, for the generator $\bar{g}$ of $\mathbb{Z}_p^*$.

The first step in the ICM algorithm is to find an optimal bound B. Then, for the complexity of the ICM algorithm most significant is the complexity of a good factorization algorithm to factor the values g^a modulo p which is in general subexponential in $\log p$. Having this in mind and combining this with the described steps we get that the ICM algorithm is subexponential in $\log p$. More details can be found in the book by S. S. Wagstaff Jr. [Wag].

6.6 Primality Testing

We now consider the question of determining whether a particular given positive integer n is prime or not prime. The methods used in solving this problem are called **primality testing** and consist of algorithms to determine whether or not an inputted positive integer is prime. Primality testing is extremely important due to its close ties to **public key cryptography**. We will be looking at public key methods carefully in the next chapter. Both the Diffie-Hellman method and the RSA method, the two main public key techniques, require the user to determine large primes. The security of these methods depends on the fact that it is easier to determine large primes than to factor large integers.

At first glance, the problem of determining if a positive integer n is prime, seems like an easy one. If n is not prime it must have a divisor m with $1 < m < n$. Therefore test all integers $2, \ldots, \frac{n}{2}$ to see if they divide n or not. If there is such a divisor then n is composite. If not, then n is prime.

Of course this can be improved on in several ways. First of all, if $n = mk$ is a non-trivial factorization, that is $1 < m < n, 1 < k < n$, then one of m or k must be $\leq \sqrt{n}$. Hence to check for primality we need only check divisibility of n by integers from 2 to $\sqrt{n}$ rather than from 2 to $\frac{n}{2}$. Further if n has a divisor m with $1 < m \leq \sqrt{n}$, then n must have a prime divisor p with $1 < p \leq \sqrt{n}$. Therefore it is only necessary to check the primes $\leq \sqrt{n}$. Therefore knowing all the primes $\leq \sqrt{n}$ provides the necessary knowledge to test all the integers $\leq n$ for primality. We summarize all these comments to give an elementary general algorithm for primality testing.

General Algorithm for Primality Testing
Given $n > 0$, test all primes p with $p \leq \sqrt{n}$. The integer n is prime if and only if none of these primes divides n.

Example 6.6.1. Test whether the integer 83 is prime.

Now $9 < \sqrt{83} < 10$ so we must test all the primes less than 9. Hence we must test $2, 3, 5, 7$. None of these divides 83 and therefore 83 is prime.

This general algorithm is simple and always works. However it becomes computationally infeasible for large integers. Therefore other methods become necessary to determine primality. Most of these methods rely on a number theoretic property, such as Fermat's Theorem, which is true for all primes but may not be true for all composites. Recall that Fermat's Theorem says that $a^{p-1} \equiv 1 \pmod{p}$ for any prime p and for any a with $1 < a < p$. We will return to these shortly.

6.6.1 Sieving Methods

Another set of techniques for determining primes are called **sieving methods**. In ordinary language a sieve is a device to separate, or sift, finer particles from coarser particles. This idea has been applied to number theory via numerical sieving methods. A **sieve** in number theory is a method or procedure to find numbers with desired properties (for example primes) by sifting through all the positive integers up to a certain bound, successively eliminating invalid candidates until only numbers with the particular attributes desired are left. Sieving methods are quite effective for obtaining lists of primes (and numbers with other characteristics) up to a reasonably small limit.

Relative to generating lists of primes, sieving methods originated with the **Sieve of Eratosthenes**. This is a straightforward method to obtain all the primes less than or equal to a fixed bound x. It is ascribed (as the name suggests) to Eratosthenes (276–194 B. C.) who was the chief librarian of the great ancient library in Alexandria. Besides the sieve method he was an influential scientist and scholar in the ancient world, developing a chronology of ancient history (up to that point) and helping to obtain an accurate measure (within the measurement errors of his time) of the dimensions of the Earth.

The method of the Sieve of Eratosthenes is direct and works as follows. Given $x > 0$ list all the positive integers less than or equal to x. Starting with 2, which is prime, cross out all multiples of 2 on the list. The next number on the list, not crossed out, which is 3, is prime. Now cross out all the multiples of 3 not already eliminated. The next number left uneliminated, 5, is prime. Continue in this manner. As explained for the primality test described in the previous section the elimination must only be done for numbers $\leq \sqrt{x}$. Upon completion of this process, any number not crossed out must be a prime.

Below we exhibit the Sieve of Eratosthenes for numbers ≤ 100. In beginning each round of elimination we must only consider numbers $\leq \sqrt{100} = 10$.

1	2	3	4	5	6	7	8	9	10
11	12	13	14	15	16	17	18	19	20
21	22	23	24	25	26	27	28	29	30
31	32	33	34	35	36	37	38	39	40
41	42	43	44	45	46	47	48	49	50
51	52	53	54	55	56	57	58	59	60
61	62	63	64	65	66	67	68	69	70
71	72	73	74	75	76	77	78	79	80
81	82	83	84	85	86	87	88	89	90
91	92	93	94	95	96	97	98	99	100

After completing the sieving operation we obtain the list

$$\{2, 3, 5, 7, 11, 13, 17, 19, 23, 29, 31, 37, 41, 43, 47, 53, 59, 61, 67, 71, 73,$$
$$79, 83, 89, 97\}$$

which comprises all the primes less than or equal to 100.

6.6.2 Fermat's Primality Testing

A **primality test** is an algorithm which inputs a positive integer n and outputs whether it is prime or not. These tests can be subclassified as either **deterministic primality tests** or **probabilistic primality tests**. In a deterministic test an integer n is inputted and the output is, yes the integer is prime, or, no the integer is not prime. Hence both the direct method of trial division and the Sieve of Eratosthenes are deterministic tests.

A non-deterministic primality test takes an inputted integer n and returns either no it is not prime or it may be a prime. A **probabilistic primality test** is a non-deterministic test that returns either the inputted integer is not a prime or is probably a prime to some given degree of accuracy. There are various tests which can give this accuracy to as high a probability as desired. Numbers that pass a probabilistic primality test are called **probable primes**. For use in cryptography, knowing if an integer is prime to a high probability, is often just as good as knowing if it is definitely prime. For this reason probable primes with a high degree of probability are called **industrial grade primes**, a term originally coined by M. Cohen.

The majority of probabilistic tests are based on either Fermat's Theorem or some variation of it. If n is an integer and suppose that a is relatively prime to n with a^{n-1} not congruent to 1 modulo n. Then n cannot be prime. This is usually called the **Fermat Probable Prime Test**. Basically given n we find an a with $(a, n) = 1$ and compute a^{n-1} modulo n. If this value is not 1 modulo n, then n is not prime. If it is congruent to 1 modulo n, then n may be prime. In the latter case, by trying different values for a we can assign a probability value. We will make this precise. The basic **Fermat Probable Prime Test** is the following.

The Fermat Probable Prime Test

Suppose n is an inputted integer. Find an a with $(a, n) = 1$. Compute a^{n-1} modulo n. If this value is not 1 modulo n then n is not prime. If this value is 1 modulo n then n may be prime.

Example 6.6.2. Test whether 11387 is prime.

This integer is relatively small so even by trial division determining whether it is prime is easy. We use the Fermat method just to illustrate the technique.

Start with $a = 2$ and we test $2^{11\,386}$ modulo $11\,387$. The basic idea is to use repeated squarings to reduce the congruence. All the equivalences are modulo $11\,387$.

$$2^{13} = 8192 \equiv -3195 \implies 2^{26} \equiv 10\,208\,025 \equiv 5273$$
$$\implies 2^{52} \equiv 8862 \equiv 2525 \implies 2^{104} \equiv 10\,292 \equiv -1095$$
$$\implies 2^{208} \equiv 3390 \implies 2^{416} \equiv 2617 \implies 2^{832} \equiv 5102$$

Continuing in this manner we eventually get

$$2^{11\,388} \equiv 8642 \implies 2^{11387} \equiv 4321.$$

From Fermat's theorem if n is prime we would have $a^{n-1} \equiv 1 \pmod n$ and therefore $a^n \equiv a \pmod n$. Here 4321 is not congruent to 2 modulo $11\,387$. Therefore $11\,387$ is not prime.

For this integer using trial division it is easy to obtain the factorization

$$11\,387 = (59)(193).$$

However even with an integer this size a calculator at least is necessary.

There are several methods to make Fermat testing deterministic. In 1891 Lucas gave the following extension of Fermat's Theorem.

Theorem 6.6.3 (Lucas). *Let $n > 1$. If for every prime factor p of $n - 1$ there exists an integer a such that*
(1) $a^{n-1} \equiv 1 \pmod n$ *and*
(2) $a^{\frac{n-1}{p}}$ *is not congruent to 1 modulo n.*
Then n is prime.

Proof. Suppose n satisfies the conditions of the theorem. To show that n is prime we will show that $\phi(n) = n - 1$ where ϕ is the Euler phi function. Since in general $\phi(n) < n - 1$, to show equality we will show that under the above conditions $n - 1$ divides $\phi(n)$. Suppose not. Then there exists a prime p such that p^r divides $n - 1$, but p^r does not divide $\phi(n)$ for some exponent $r \geq 1$. For this prime p, there exists an integer a satisfying the conditions of the theorem. Let m be the order of a modulo n. Then m divides $n - 1$ since the order of an element divides any power which equals 1 (see Chapter 5). However by the second condition in the theorem and for the same reason, m does not divide $\frac{n-1}{p}$. Therefore p^r divides m which divides $\phi(n)$ contradicting our assumption. Hence $n - 1 = \phi(n)$ and therefore n is prime. □

The following result is related.

Theorem 6.6.4. *Suppose that $a, k, n, q \in \mathbb{N}$ with $n - 1 = qk$ and $q > k$. Suppose further that*
(1) *q is prime;*
(2) *$a^{n-1} \equiv 1 \pmod n$;*

(3) $(a^k - 1, n) = 1$.

Then n is prime.

Proof. Suppose that n is not prime. Then there exists a prime divisor p of n with $p \leq \sqrt{n}$. Let d be the order of a in $\mathbb{Z}_p^*$. From (2) we get that $a^{n-1} \equiv 1 \pmod{p}$, that is $d \mid (n-1)$.

From (3) we get that d does not divide k for otherwise $p \mid (a^k - 1)$ and $p \mid n$.

From (1) we not get that $q \mid d$. Suppose that q does not divide d. Then $q \mid m$ f $n - 1 = dm - qk$. Hence $k = d\frac{m}{q}$ and $d \mid k$ which gives a contradiction. Therefore $q \mid d$.

Altogether we have

$$\sqrt{n} > p - 1 \geq d \geq q$$

since $d \mid (p - 1)$. But $q \geq \sqrt{n}$ from $q > k$ which gives a contradiction. It follows that n is a prime number. □

Example 6.6.5. Show that 2922 259 is a prime number.

We get the factorization $2922\,259 - 1 = 1721 \cdot 1698$. We find that 1721 is prime and apply the theorem with $k = 1698$. Computing we find that

$$2^{2922259-1} \equiv 1 \pmod{2923\,359} \quad \text{and} \quad (2^{1698-1}, 2922\,259) = 1$$

and hence applying the theorem the number is prime.

Although these tests are deterministic, they are, in most cases, no more computationally feasible than trial division or sieving, since the tests depend on the factorization of $n-1$. In general, factorization is even more difficult than solely testing for primality. Therefore, even here, further methods are necessary. We note that the idea in the Lucas Test has been quite effective in developing methods for testing Fermat and Mersenne numbers for primality.

Primality testing is essentially a computational problem. Therefore a primality test raises questions about the accompanying algorithm's computational speed and computational complexity. For these types of number theoretic algorithms the computational complexity is measured in terms of functions of the input length, which here is roughly the number of digits of the inputted integer. The Sieve of Eratosthenes requires, for an inputted integer n, roughly the same order n of operations. If n has $\log_{10} n$ digits Then the Sieve requires $\mathcal{O}(10^{\log_{10} n})$ operations to prove primality. We say that this algorithm is of **exponential time** in terms of the input length. The big open question was whether there existed a deterministic algorithm which was of **polynomial time** in the input length. This means, that for this algorithm there is a positive integer d such that the number of operations in the algorithm to prove primality is $\mathcal{O}((\log n)^d)$. Earlier, Miller and Rabin had shown that the Miller-Rabin Test, which we will describe later in this section, can be made deterministic. Further it is of polynomial time **if** one accepts as true the extended Riemann hypothesis (see [FR]). However, prior to 2003 it was an open question whether there was a deterministic algorithm for primality which could be shown to be of polynomial time without using any unproved conjectures.

In 2003, M. Agrawal and two of his students, N. Kayal and N. Saxena, developed an algorithm, now called the **AKS Algorithm**, which was deterministic and could be proved to be of polynomial time. The result was even more spectacular since it was accomplished with relatively elementary methods. The basic algorithm depends on two rather straightforward extensions of Fermat's Theorem. This result has of course generated a great deal of attention and much has already been written about it. We refer the reader to the articles [Bor] and [Ber] for a more complete discussion of the algorithm and its development and to [FR] for a more complete discussion of the algorithm as well as its proof.

To handle the computational problems probabilistic tests were introduced. To repeat what we defined earlier, a non-deterministic primality test takes an inputted integer n and returns either no it is not prime or it may be a prime. A **probabilistic primality test** is a non-deterministic test that returns either the inputted integer is not a prime or is probably a prime to some given degree of accuracy.

6.6.3 Pseudoprimes and Probabilistic Primality Testing

In a probabilistic primality test a number that can possibly be a prime is called a **probable prime**. Here we look at some probable primes for the Fermat probabilistic test.

Definition 6.6.6. Let n be a composite integer. If $b > 1$ with $(n, b) = 1$ Then n is a **pseudoprime** to the base b if $b^{n-1} \equiv 1 \pmod{n}$.

If n is a pseudoprime to the base b then it passes the Fermat Test and hence is a probable prime.

Example 6.6.7. 25 is a pseudoprime to the base 7. To see this notice that

$$7^2 = 49 \equiv -1 \pmod{25}.$$

This implies that $7^4 \equiv 1 \pmod{25}$ and hence $7^{24} \equiv 1^6 \equiv 1 \pmod{25}$.

Notice that 25 is not a pseuodprime modulo 2 or 3.

Theorem 6.6.8. *For each base $b > 1$ there exists infinitely many pseudoprimes to the base b.*

Proof. Suppose $b > 1$. We show that if p is any odd prime not dividing $b^2 - 1$ then the integer $n = \frac{b^{2p}-1}{b^2-1}$ is a pseudoprime to the base b. Note that for this n we have

$$n = \frac{b^{2p} - 1}{b^2 - 1} = \frac{b^p - 1}{b - 1} \cdot \frac{b^p + 1}{b + 1}$$

so that n is composite.

Given b, from Fermat's Theorem we have $b^p \equiv b \pmod{p}$ and hence $b^{2p} \equiv b^2 \pmod{p}$. Now $n - 1 = \frac{b^{2p}-b^2}{b^2-1}$ and since p does not divide $b^2 - 1$ by assumption it follows that p divides $n - 1$.

Further

$$n - 1 = b^{2p-2} + b^{2p-4} + \cdots + b^2.$$

Therefore $n - 1$ is a sum of an even number of terms of the same parity so $n - 1$ must be even. It follows that $2p$ divides $n - 1$. Hence $b^{2p} - 1$ is a divisor of $b^{n-1} - 1$. However

$$b^{2p} - 1 \equiv 0 \pmod{n} \implies b^{n-1} - 1 \equiv 0 \pmod{n}.$$

Therefore n is a pseudoprime to the base b proving the theorem. □

Although there are infinitely many pseudoprimes they are not that common. It has been shown for example that there are only $21\,853$ pseudoprimes to the base 2 among the first $25\,000\,000\,000$ integers. Hence there is a good chance that if a number, especially a large number, passes a test as a pseudoprime, then it is really a prime. The question becomes how to make this chance, or probability, precise. Lists of many pseudoprimes can be found on various internet websites (see [PP]).

From simple congruences it is clear that if n is a pseudoprime to the base b_1 and also a pseudoprime to the base b_2 then it is a pseudoprime to the base $b_1 b_2$.

Probabilistic methods proceed by testing n to a base b_1. If it is not a pseudoprime, then it is composite and we are done. If it is a pseudoprime, test a second base b_2 and so on, in the hope of finding a base where it is not a pseudoprime. However there do exist numbers which are pseudoprimes to every possible base.

Definition 6.6.9. A composite integer n is a **Carmichael number** if n is a pseudoprime to each base $b > 1$ with $(n, b) = 1$.

The Carmichael numbers can be completely classified. Interestingly this was done even before the existence of Carmichael numbers was shown. The following is called the **Korselt criterion** after A. Korselt.

Theorem 6.6.10. *An odd composite number n is a Carmichael number if and only if n is squarefree and $(p - 1) \mid (n - 1)$ for every prime p dividing n.*

Proof. We first show that if a number n is not squarefree then it cannot be a Carmichael number.

Suppose that n is not squarefree. Then there exists a prime p with $p^2 \mid n$. The multiplicative group in $\mathbb{Z}_{p^2}$ is cyclic, that is there exists a primitive element, (see [FR]), and hence, there is a multiplicative generator g modulo p^2. Since $\phi(p^2) = p(p - 1)$ we have $g^{p(p-1)} \equiv 1 \pmod{p^2}$ and this is the least power of g that is congruent to 1 modulo p^2. Now let $m = p_1 \cdots p_k$ where $p_1, \ldots, p_k$ are the other primes besides p dividing n. Notice that p^k is not a Carmichael number so these primes exist. Choose a solution b to the pair of congruences

$$b \equiv g \pmod{p^2}$$
$$b \equiv 1 \pmod{m}$$

which exists from the Chinese Remainder Theorem. Since $b \equiv g \pmod{p^2}$ it follows that b also has multiplicative order $p(p - 1)$ modulo p^2. Suppose n was a Carmichael number. Then n would be a pseudoprime to the base b and hence

$$b^{n-1} \equiv 1 \pmod{n}.$$

This implies that $p(p - 1) \mid n$ from the multiplicative order of b. However since $p \mid n$ we have $n - 1 \equiv -1 \pmod{p}$. On the other hand if $p(p - 1) \mid n - 1$ we have $n - 1 \equiv 0 \pmod{p}$ a contradiction. Therefore n cannot be a pseudoprime to the base b and hence is not a Carmichael number.

Now suppose that n is squarefree so that $n = p_1 p_2 \cdots p_k$ with $k \geq 2$ and the p_i distinct primes. Suppose first that $(p_1 - 1) \mid (n - 1)$ for $i = 1, \ldots, k$ and suppose that $(b, n) = 1$. Then

$$b^{n-1} \equiv b^{(p_1 - 1)k} \equiv 1^k \equiv 1 \pmod{p_i}, \quad i = 1, \ldots, k.$$

Hence

$$b^{n-1} \equiv 1 \pmod{p_1 \cdots p_k}.$$

Therefore n is a pseudoprime to the base b and since b was arbitrary with $(b, n) = 1$ it follows that n is a Carmichael number.

Conversely, suppose that $n = p_1 \cdots p_k$ is a Carmichael number. Let p_i be one of these primes and suppose that g is a generator of the multiplicative group of $\mathbb{Z}_{p_i}$. Recall, as in the proof of the squarefree property, that this group is cyclic. Hence g has multiplicative order $p_i - 1$ modulo p_i. Now let b be a solution to the pair of congruences

$$b \equiv g \pmod{p_i}$$

$$b \equiv 1 \left(\text{mod } \frac{n}{p_i}\right).$$

Then b also has multiplicative order $p_1 - 1$ modulo p_i. Further since $(b, p_1) = 1$ and $(b, \frac{n}{p_i}) = 1$ it follows that $(b, n) = 1$. Since n is a Carmichael number it is a pseudoprime to the base b and hence

$$b^{n-1} \equiv 1 \pmod{n} \implies b^{n-1} \equiv 1 \pmod{p_i}.$$

It follows that $(p_1 - 1) \mid (n - 1)$ proving the theorem. $\square$

Corollary 6.6.11. *A Carmichael number must be divisible by at least 3 primes.*

Proof. Suppose that n is a Carmichael number. Then from the proof of the previous theorem $n = p_1 \cdots p_k$ with $k \geq 2$ and the p_i distinct primes. We must show that $k > 2$. Suppose that $n = pq$ with $p < q$ primes. Since n is a Carmichael number we get from the previous theorem $(q - 1) \mid (n - 1)$. However

$$n - 1 = pq - 1 = p(q - 1 + 1) - 1 \equiv p - 1 \;(\text{mod}(q - 1)).$$

Since $(q - 1) \mid (n - 1)$ this would imply that $(q - 1) \mid (p - 1)$ which is impossible since $p < q$. Therefore if $n = pq$ it cannot be a Carmichael number and hence $k > 2$ so that n must be divisible by at least 3 distinct primes. □

Using the Korselt criterion we can present an example of a Carmichael number.

Example 6.6.12. The integer $n = 561 = 3 \cdot 11 \cdot 17$ is a Carmichael number. Here $n - 1 = 560$ which is divisible by 2, 10 and 16 and hence by the Korselt criterion it is a Carmichael number. This is well known as the smallest Carmichael number (see exercises).

Carmichael numbers are relatively infrequent. It has been shown for example that there are only 2163 Carmichael numbers among the first 25 000 000 000 integers. However it has been proved by Alford, Granville and Pomerance that there do exist infinitely many Carmichael numbers. There is a list of Carmichael numbers up to 10^{16} (see [CP]).

   We note that if $n > 3$ is a Carmichael number, then n must be odd. To see this suppose that n is even. We have $(n - 1, n) = 1$ and since n is a Carmichael number we have $(n - 1)^{n-1} \equiv 1 \pmod{n}$. On the other side $(n - 1)^{n-1} \equiv -1 \pmod{n}$ since n is even. Hence $n \mid 2$ so $n \leq 2$ which contradicts $n > 3$. Therefore n must be odd.

   For what follows we need the fact that the multiplicative group $\mathbb{Z}_{p^\alpha}^*$ is cyclic for $p \geq 3$ and $\alpha \geq 1$ (see [FR]).

Theorem 6.6.13 (The First Probabilistic Primality Test). *Let $n \geq 3$ be an odd integer which is not a Carmichael number. Choose $k > 0$ so that $\left(\frac{1}{2}\right)^k$ is smaller than the desired primality testing error. Consider the following algorithm*
(1) Choose a random number $b \in \{1, \ldots, n - 1\}$.
(2) Calculate (b, n). If $(b, n) > 1$ return composite number and stop.
(3) Calculate b^{n-1} modulo n. If b^{n-1} is not congruent to 1 modulo n return composite number and stop.
(4) Repeat steps (1), (2), (3) up to k times.
If at each of the k passes $b^{n-1} \equiv 1 \pmod{n}$ then return n is prime with probability $1 - \left(\frac{1}{2}\right)^k$.
   This is a probabilistic algorithm which correctly determines up to a probability $\geq 1 - \left(\frac{1}{2}\right)^k$ that n is a prime number.

For the proof of Theorem 6.6.13 we need the following.

Theorem 6.6.14. *Let $n \geq 3$ be a composite integer.*
(a) Let $b \in \mathbb{N}$ with $(b, n) = 1$. Then n is a pseudoprime to the base b if and only if the order of $\overline{b}$ in $\mathbb{Z}_n^*$ divides $n - 1$.
(b) Let $b_1, b_2 \in \mathbb{N}$ with $(b_1, n) = (b_2, n) = 1$. If n is a pseudoprime to the base b_1 then n is a pseudoprime to the base $b_1 b_2$ and also to each base $b_1 a_2$ with $\overline{a_2 b_2} = \overline{1}$ in $\mathbb{Z}_n^*$.
(c) If n is not a pseudoprime to at least one base than n is not a pseudoprime to at least 50 % of the bases $b \in \{1, \ldots, n - 1\}$ with $(b, n) = 1$.

Proof (of Theorem 6.6.14). (a) Let $i = \text{ord}(\bar{b}) = |\bar{b}|$ the order of $\bar{b}$ in $\mathbb{Z}_n^*$. From $\bar{b}^{n-1} = \bar{1}$ we get that $i|(n-1)$.

Conversely from $|\bar{b}|\,|(n-1)$ we get $\bar{b}^{n-1} = (\bar{b}^{|\bar{b}|})^k = \bar{1}$.

(b) If $\bar{b}_1^{n-1} = \bar{1}$ and $\bar{b}_2^{n-1} = \bar{1}$, then $(\bar{b}_1\bar{b}_2)^{n-1} = \bar{1}$. If $\bar{a}_2\bar{b}_2 = \bar{1}$ then

$$\bar{a}_2^{n-1} = \bar{a}_2^{n-1}\bar{b}_2^{n-1} = (\bar{a}_2\bar{b}_2)^{n-1} = \bar{1}$$

and hence

$$(\bar{b}_1\bar{a}_2)^{n-1} = \bar{b}_1^{n-1}\bar{a}_2^{n-1} = \bar{1}.$$

(c) Let $b_1, \ldots, b_s \in \{1, \ldots, n-1\}$ with $(b_1, n) = 1$ and for $i = 1, \ldots, s$ those bases for which n is a pseudoprime. Let $\bar{a} \in \mathbb{Z}_n^*$ with $\bar{a}|^{n-1} \neq \bar{1}$. Then

$$(\overline{ab}_i)^{n-1} = \bar{a}^{n-1}\bar{b}_i^{n-1} = \bar{a}^{n-1} \neq 1 \text{ for } i = 1, \ldots, s.$$

Since the $\overline{ab}_i$ are pairwise different, there are at least s bases for which n is not a pseudoprime. Further by the pairwise distinctness s is at least 50 % of the bases $b \in \{1, \ldots, n-1\}$ with $(b, n) = 1$. □

We can now give the proof of the first probabilistic primality test, Theorem 6.6.13.

Proof (of Theorem 6.6.13). The finiteness of the algorithm is obvious. The correctness of the algorithm is given by part (c) of the above theorem as follows. If n is composite then n is not a pseudoprime to at least one base $b \in \{1, \ldots, n-1\}$ with $(b, n) = 1$ because n is not a Carmichael number. Hence by part (c) of Theorem 6.6.6 the integer n is a pseudoprime to at most 50 % of the bases, that is for $\leq \frac{1}{2}$ of the bases $b \in \{1, \ldots, n-1\}, (b, n) = 1$.

Let $b_1, b_2 \in \{1, \ldots, n-1\}$ with $(b_1, n) = (b_2, n) = 1$. Then the probability that n is a pseudoprime to the basis b_1 and b_2 is $\leq \frac{1}{2}\frac{1}{2} = \frac{1}{4}$ because the calculations for b_1 and b_2 are independent.

This proves Theorem 6.6.13. □

6.6.4 Miller-Rabin Primality Testing

Miller and Rabin determined an even stronger test than the primality test given in the previous section.

Definition 6.6.15. Let n be an odd composite integer with $n - 1 = 2^s t$ with t odd. If $b > 1$ and $(n, b) = 1$ Then n is a **strong pseudoprime** to the base b if either
(1) $b^t \equiv 1 \pmod{n}$ or
(2) there exists r with $0 \leq r < s$ such that $b^{2^r t} \equiv -1 \pmod{n}$.

The Miller-Rabin Test is based on the following theorem, that was proved independently by Monier and Rabin.

Theorem 6.6.16. *For each composite integer $n > 9$ the number of bases b with $0 < b < n$ for which n is a strong pseudoprime is less than $\frac{1}{4}$.*

The proof of this is long and involved and can be found in [FR].

If n is not a strong pseudoprime to the base b we say that b is a **witness** for n (a witness that n is composite). Then Theorem 6.6.16 says that at least $\frac{3}{4}$ of all the integers in $(1, n-1)$ are witnesses for n.

Theorem 6.6.17 (Miller-Rabin Primality Test). *Let $n \geq 3$ be an odd integer. Choose $k > 0$ so that $\left(\frac{1}{2}\right)^k$ is smaller than the desired primality testing error. Consider the following algorithm.*

(1) *Write $n - 1 = 2^s t$ with $s \geq 1$ and t odd.*

(2) *Consider a random number $b \in \{1, \ldots, n-1\}$ and calculate (b, n). If $(b, n) > 1$ return composite number and stop.*

(3) *Calculate b^{n-1} modulo n. If b^{n-1} is not congruent to 1 modulo n return composite number and stop.*

(4) *Calculate b^t modulo n. If either $b^t \equiv 1 \pmod{n}$ or $b^t \equiv -1 \pmod{n}$ continue with step (6).*

(5) *Calculate $(b^t)^2, (b^t)^4, \ldots, (b^t)^{2^s}$ modulo n.*
If $b^{s^i t} \equiv -1 \pmod{n}$ for some $i \in \{1, \ldots, s-1\}$ continue with step (6).
If at some time $b^{2^i t} \equiv 1 \pmod{n}$ but $b^{2^{i-1} t}$ is not congruent to 1 modulo n and $b^{2^{i-1} t}$ is not congruent to -1 modulo n return composite number and stop.

(6) *Repeat steps (2) to (5) up to k times.*
If we always have no break off, then give "test passed" and stop.

This is a probabilistic algorithm which correctly determines up to a probability $\geq 1 - \left(\frac{1}{4}\right)^k$ that n is a prime number.

Based on the previous theorem, the proof proceeds analogously to the proof in the last section, except that now the probability is $\geq 1 - \left(\frac{1}{4}\right)^k$ based on Theorem 6.6.5.

In practice, it is sufficient to test for only a few bases. If $n < 2 \cdot 5 \cdot 10^{10}$ there exists only one strong pseudoprime n to the bases $2, 3, 5, 7$.

The Miller-Rabin Test can be made deterministic under the assumption that the Extended Riemann Hypothesis holds (see [FR]). In particular, Bach proved the following.

Theorem 6.6.18. *Assuming that the Extended Riemann Hypothesis holds, then for any odd composite integer n there is a witness less than $2(\log n)^2$.*

Hence based on the theorem, we would only have to test for witnesses less than $2(\log n)^2$. If there are none, then n is prime. This is then a deterministic polynomial time algorithm. However it depends on the unproved Extended Riemann Hypothesis.

6.6.5 Mersenne Primes and the Lucas-Lehmer Test

A large portion of primality testing, especially relative to cryptography, has centered on the Mersenne primes. In fact most of the prime *records*, that is the determination of a largest known prime, involves finding larger and larger Mersenne primes.

A Mersenne number is a positive integer of the form $M_n = 2^n - 1$, with $n = 1, 2, \ldots$. If M_n is prime, then M_n is a **Mersenne prime**. It is not known whether or not there are infinitely many Mersenne primes, however it is conjectured, and believed, that there are infinitely many. For more information on this interesting class of numbers and their properties see [FR].

Testing Mersenne numbers for primality has been particularly fruitful because of the **Lucas-Lehmer test**. This is a straightforward deterministic primality test specific to the Mersenne numbers. It is relatively easy to implement on a computer and has been quite successful in finding larger and larger Mersenne primes. For the most part historically, the largest known Mersenne prime, is also the largest known prime or current prime record. For a list of prime records see [FR].

Theorem 6.6.19 (Lucas-Lehmer Test). *Let p be an odd prime and define the sequence* (S_n) *inductively by*

$$S_1 = 4 \text{ and } S_n = S_{n-1}^2 - 2.$$

Then the Mersenne number $M_p = 2^p - 1$ *is a Mersenne prime if and only if* M_p *divides* S_{p-1}.

A proof of this can be found in [FR]. We note that if M_p is prime we must have p prime.

6.7 Exercises

6.1 Use trial division to determine if any of the following integers are prime.
(a) 10 387   (b) 269   (c) 46 411

6.2 Use the Sieve of Eratosthenes to develop a list of primes less than 300. (Note this list could be used for problem 6.1).

Given positive integers m, x, by a slight modification, the Sieve of Eratosthenes can be used to determine all the positive integers relatively prime to m and less than or equal to x.

Here suppose we are given m and x. Let $p_1, \ldots, p_k$ be the distinct prime factors of m arranged in ascending order, that is $p_1 < p_2 < \cdots < p_k$. Next list all the positive integers less than or equal to x as we did for the ordinary sieve. Start with p_1 and eliminate all multiples of p_1 on the list. Then successively do the same for p_2 through p_k. The numbers remaining on the list are precisely those relatively prime to m that are also less than or equal to x. If $p_i > x$ ignore this prime and all higher primes.

6.3 Use the modified Sieve of Eratosthenes described above to find the integers less than 100 and relatively prime to 891.

Using the Sieve of Eratosthenes, Legendre developed a formula to compute $N_m(x)$, the number of integers less than or equal to x and relatively prime to m. This is called **Legendre's Formula for the Sieve of Eratosthenes.** Let $m \in \mathbb{N}$, $x \geq 0$ then

$$N_m(x) = \sum_{d|m} \mu(d)\left[\frac{x}{d}\right]$$

where $\mu(d)$ is the Moebius function and $[\ ]$ is the greatest integer function.

6.4 Apply Legendre's formula to evaluate
(a) $N_{655}(200)$; (b) $N_{891}(100)$.

6.5 Let $P(x)$ denote the number of primes $p \leq x$ for which $p + 2$ is prime. Then by Lemma 6.2.1.4 for $x \geq 3$ we have

$$P(x) < c\frac{x}{(\ln x)^2}(\ln \ln x)^2$$

where c is a constant. Show that this implies that for $x \geq 3$

$$P(x) \leq k\frac{x}{(\ln x)^{\frac{3}{2}}}$$

where k is a constant.

6.6 Use the integral test for infinite series to show that

$$\sum_{r=1}^{\infty} \frac{1}{r(\ln(r+1))^{\frac{3}{2}}}$$

converges.

6.7 Prove that

$$(-1)^{m+1}\binom{n}{m+1} + (-1)^m\binom{n-1}{m} = (-1)^{m+1}\binom{n-1}{m+1}.$$

6.8 Use the Fermat probable prime test to determine if 42671 is prime or not.

6.9 Use the Lucas Test to establish that 271 is prime

6.10 Show that if n is prime and $k \neq 0, 1$ then the binomial coefficient $\binom{n}{k}$ is congruent to 0 modulo n.

6.11 Use problem 6.10 to show that if p is prime then

$$(x - a)^p = x^p - a \text{ in } \mathbb{Z}_p.$$

6.12 Determine the bases b (if any) $0 < b < 14$ for which 14 is a pseudoprime to the base b.

6.13 Prove Lemma 6.3.1.1: If n is a pseudoprime to the base b_1 and also a pseudoprime to the base b_2 then it is a pseudoprime to the base $b_1 b_2$.

6.14 Show that $561 = 3 \cdot 11 \cdot 17$ is the smallest Carmichael number. (Use the Korselt criterion together with Corollary 6.3.1.)

6.15 Define the sequence (S_n) inductively by

$$S_1 = 4 \quad \text{and} \quad S_n = S_{n-1}^2 - 2.$$

Let $u = 2 - \sqrt{3}, v = 2 + \sqrt{3}$. Show that $u + v = 4 = S_1$ and $uv = 1$. Then use induction to show that

$$S_n = u^{2^{n-1}} + v^{2^{n-1}}.$$

6.16 Show that if p, q are primes and e, d are positive integers with $(e, (p - 1)(q - 1)) = 1$ and $ed \equiv 1 \ (\mathrm{mod}(p - 1)(q - 1))$ then $a^{ed} \equiv a \ (\mathrm{mod}\, pq)$ for any integer a. (This is the basis if the decryption function used in the RSA algorithm.

6.17 Show that for the polynomial $p(x) = x^2 - x + 41$ the values $p(a)$ for $a = 0, 1, \ldots, 40$ are all prime numbers.

Hint: Assume that $n^2 - n + 41$ is composite for some n with $0 \le n \le 40$. Let p be the smallest prime factor of this n and show that if n is composite then 163 is a quadratic residue modulo p which gives a contradiction.

7 Public Key Cryptography

7.1 Public Key Cryptography

There are many instances where secure information must be sent over open communication lines. These include, for example, banking and financial transactions, purchasing items via credit cards over the internet, and similar things. This type of cryptographic communication is handled by **public key cryptography** in which both the encryption technique and the encrypted ciphertext are open to everyone but the encrypted transmission is considered secure.

Roughly, in classical cryptography only the sender and receiver know the encoding and decoding methods. Further it is a feature of such cryptosystems, such as the ones that we've already looked at, that if the encrypting method is known, then the decryption can be carried out. In public key cryptography, the encryption method is public knowledge, but only the receiver has the secret key and hence knows how to decode. More precisely in a classical cryptosystem, once the encrypting algorithm is known, the decryption algorithm can be implemented in approximately the same order of magnitude of time. In a public key cryptosystem, developed first by Diffie and Hellman, the decryption algorithm is much more difficult to implement. This difficulty depends on the type of computing machinery utilized, and as computers get better, new and more secure public key cryptosystems become necessary.

Recall from Chapter 1 that a **basic cryptosystem** was defined as a tuple $\{\mathcal{P}, \mathcal{C}, f, g\}$ where $\mathcal{P}$ is the set of plaintext messages, $\mathcal{C}$ is the set of ciphertext messages, $f : \mathcal{P} \to \mathcal{C}$ and $g : \mathcal{C} \to \mathcal{P}$ are functions such that $g \circ f = I$ on $\mathcal{P}$, that is f is injective and hence invertible on $f(\mathcal{P})$ and g is its inverse on $f(\mathcal{P})$. The function f is the **encryption map** or **encryption algorithm** while its inverse g is the **decryption map** or **decryption algorithm**. We placed this in a wider mathematical context called a **cryptosystem**. A (general) cryptosystem is a collection of basic cryptosystems indexed by a set $\mathcal{K}$ called the **key space**. Each $k \in \mathcal{K}$ is called a **key** and defines a basic cryptosystem.

A **cryptosystem** is then a tuple $\{\mathcal{P}, \mathcal{C}, \mathcal{K}, \mathcal{E}, \mathcal{D}\}$ where

$$\mathcal{P} = \{ \text{the plaintext space} \}$$
$$\mathcal{C} = \{ \text{the ciphertext space} \}$$
$$\mathcal{K} = \text{an index set called the } \textbf{key space}$$

The elements $k \in \mathcal{K}$ are called **keys**.

$$\mathcal{E} = \text{a set of injective functions}, \quad f : \mathcal{P} \to \mathcal{C}$$

indexed by the key space. These are called the set of **encryption functions**. Hence for each $k \in K$ there is an injective function f_k from plaintext to ciphertext.

$$\mathcal{D} = \text{a set of functions}, \quad g : \mathcal{C} \to \mathcal{P}$$

also indexed by the key space. These are called the set of **decryption functions**. Hence for each $k \in K$ there is an injective function f_k from plaintext to ciphertext with a corresponding key k_1 and function $g_{k_1} : \mathcal{C} \to \mathcal{P}$ such that g_{k_1} is the left inverse of f_k.

In our previous language this means that for each $k \in \mathcal{K}$ we have a basic cryptosystem $\{\mathcal{P}, \mathcal{C}, f_k, g_{k_1}\}$. To place further emphasis on the key space we consider the set $\mathcal{E}$ consisting of those functions $f_k; k \in \mathcal{K}$ that go from $\mathcal{P} \to \mathcal{C}$. These are the set of **encryption functions** and the index k is called the **encryption key**. The set $\mathcal{D}$ consists of the left inverse functions and we also index these by $k \in K$. These are called the decryption functions and the corresponding index is called the **decryption key**. Hence given an encryption key $e \in \mathcal{K}$ there is another key $d \in \mathcal{D}$ where g_d is the (left) inverse of f_e.

Classical cryptography is called **symmetric key cryptography** while public key cryptography is called **asymmetric cryptography**. Within this model, we can easily distinguish between the two. In a symmetric key cryptosystem if the encryption key k_1 is given then it is easy to find the corresponding decryption key k_2. In an asymmetric or public key cryptosystem, even if the encryption key is known, it is infeasible to find the decryption key. This is of course equivalent to how it was phrased in the short description in Section 1.5; if the encryption function is known it is infeasible to find the left inverse. In most public key cryptosystems the encryption key is made public while the decryption key is secret and known only to the communicating parties.

Notice that the key space must be fairly extensive to prevent a potential attacker from simply doing an exhaustive search to find the decryption key.

The basic idea in a public key cryptosystem is to have a **one-way function** or **trapdoor function**. That is a function which is easy to implement but very hard to invert. The **trapdoor** is what needs to be known in order to invert the function. Hence it becomes simple to encrypt a message, but very hard, unless you know the trapdoor, to decrypt. In the standard view of a public key system, each user has a **public key** and a **private key**. Assume that f is a one way function depending on a key k so that f_k is a specific one way function. Knowing f_k, one can easily encrypt with this function. However, it is a one-way function, so knowing f_k it is not feasible to find g_k, the left inverse of f_k. Each user then, say Alice, has two keys (f_A, g_A). f_A is her **public key** known to all users but g_A, the decryption function for her f_A, is her **private key** known only to her. Hence anyone can send her messages using f_A but only she can decrypt them. We will show how this is done in the next section.

Usually public key encryption systems are less efficient than symmetric key systems. Therefore, what is used is called a **hybrid cryptosystem**. Using the trapdoor function, a key is exchanged between the communicating parties which signals the use of some symmetric cryptosystem such as AES. Exchanging the key to be used in a communicating session is called a **key exchange protocol**. The key used or exchanged during a session is called a **session key**. Usually included in the exchange is some form of **verification**, that is, that the message actually comes from the supposed sender.

A mail slot with a locked mailbox is analogous to a public key system. Suppose Alice's address is known. That address is her public key. If Bob (or any other user) wants to communicate with her, he drops his message into the mail slot. However, Alice alone has the key to the locked mailbox, so only she can access the message. The mailbox key is her private key.

Notice that to prevent an exhaustive search, the key space must be quite large. Hence an important consideration in devising a public key cryptosystem is the size and storage requirements of the key space.

7.2 Standard Model for Public Key Encryption

The standard model for public key systems is the following. We suppose that Alice's public key is f_A and her private key is g_A. Here the function f indexed for A is a one-way function and knowing the trapdoor she knows g_A. Similarly Bob's public and private keys are (f_B, g_B).

Alice wants to send a message to Bob. The encrypting map f_A for Alice is public knowledge as well as the encrypting map f_B for Bob. On the other hand, the decryption algorithms g_A and g_B are secret and known only to Alice and Bob respectively. Let $\mathcal{M}$ be the message Alice wants to send to Bob. She sends $f_B(\mathcal{M})$. f_B is public knowledge and it is easy to encrypt. To decrypt Bob now applies g_B to what he has received. Recall, that presumably only he knows his private key g_B. Doing this he obtains

$$g_B(f_B(\mathcal{M})) = (g_B f_B)(\mathcal{M}) = \mathcal{M}$$

and hence recovers the message.

In practice what is usually sent by this public key method is the session key to be used with some symmetric cryptosystem such as AES.

If Bob receives a message how can he be certain that it actually comes from Alice. Within this standard model it is relatively easy to build in **verification** or **authentication**. As above, let $\mathcal{M}$ be the message Alice wants to send to Bob. Now instead of sending $f_B(\mathcal{M})$ she sends

$$f_B g_A(\mathcal{M}).$$

Now to decode Bob applies first g_B, which is his private key so that only he knows this. This gives him

$$g_B(f_B g_A(\mathcal{M})) = g_A(\mathcal{M}).$$

He then looks up f_A which is publically available and applies this

$$f_A(g_A(\mathcal{M})) = \mathcal{M}$$

to obtain the message. This supplies **verification** and **authentication** in the following manner. Suppose $\mathcal{M}$ is Alice's verification; signature, social security number etc. If Bob receives $f_B(\mathcal{M})$, it could be sent by anyone, since f_B is public. On the other hand, since only Alice supposedly knows g_A getting a reasonable message from $f_A(g_B f_B g_A(\mathcal{M}))$ would verify that it is from Alice. Applying g_B alone should result in nonsense.

This type of verification provides a **digital signature**. Recall that in Chapter 4 we looked at the general ideas of digital signatures. At the conclusion of this chapter we will look at a digital signature scheme using RSA, one of the public key methods that will e introduced.

7.3 The Diffie-Hellman Key Exchange and Protocol

Getting a reasonable one way function can be a formidable task. The most widely used (at present) public key systems are based on difficult to invert number theoretic functions.

Diffie and Hellman, in 1976, developed the original public key idea using the **discrete log problem**. We review the material on the DLP that was introduced in Chapter 6.

Definition 7.3.1. If G is a finite group, such as the cyclic multiplicative group of $\mathbb{Z}_p$ where p is a prime, and $h = g^k$ for some k then a **discrete logarithm** or **discrete log** of h to the base g is any integer t with $h = g^t$. We denote this by

$$k = \log_g^G(h) \quad \text{or simply} \quad k = \log_g(h) \quad \text{if } G \text{ is fixed.}$$

Example 7.3.2. If $G = \mathbb{Z}_{19}^*$ then $6 = \log_2(7)$ since $2^6 = 64 \equiv 7 \pmod{19}$.

A discrete log need not be unique as the next example shows.

Example 7.3.3. Let $G = \mathbb{Z}_8^*$. Then $2 = \log_3(1)$ since $3^2 \equiv 1 \pmod 8$. But also $4 = \log_3(1)$ since $3^4 \equiv 1 \pmod 8$ and 2 and 4 are incongruent modulo 8.

Notice that if $G = \langle g \rangle$ of order n and $h = g^i$ with $1 \leq i \leq n$ then the discrete log of h is uniquely defined for the generator g, that is i is unique modulo n for g.

The **discrete logarithm problem** denoted also by DLP is then the following, Let $G = \langle g \rangle$ and $h \in G$. Find $x \in \mathbb{Z}$ such that $h = g^x$.

The smallest non-negative x with this property is often called the **discrete logarithm** of h relative to the generator g.

Other than an exhaustive search there is no known algorithm to solve the discrete log problem. This makes it ideal as a one-way function. In Chapter 6 we examined several algorithms that attempt to solve the DLP in special situations.

In modular arithmetic it is easy to raise an element to a power but difficult to determine, given an element, if it is a power of another element. Hence in the multiplicative group of $\mathbb{Z}_p$ it is assumed that the discrete log problem is hard to solve.

The rough form of the **Diffie-Helman public key exchange protocol** is as follows. Bob and Alice will use a classical cryptosystem based on a key k with $1 < k < q-1$

where q is a prime. It is the key k that Alice and Bob must share. Let g be a multiplicative generator of $\mathbb{Z}_q^*$ the multiplicative group of $\mathbb{Z}_q$. The generator g is public. It is known that this group is cyclic if q is a prime.

Alice chooses an $a \in \mathbb{Z}_q$ with $1 < a < q - 1$. She makes public g^a. Bob chooses $b \in \mathbb{Z}_q^*$ and makes public g^b. The secret shared key is g^{ab}. Both Bob and Alice, but presumably no one else, can discover this key. Alice knows her secret power a and the value g^b is public from Bob. Hence she can compute the key $g^{ab} = (g^b)^a$. The analogous situation holds for Bob. An attacker however, only knows g^a and g^b and g. Supposedly, unless the attacker can solve the discrete log problem, the key exchange is secure. We make all this a bit more precise.

The Diffie-Hellman Key Exchange Protocol

Goal: To exchange a secret key k that indicates the encryption technique to be used in some classicial cryptosystem.

Setup:

(1) Bob and Alice choose a large prime q and a generator g of the cyclic multiplicative group $\mathbb{Z}_q^*$. The element g is public to all.

(2) Alice chooses an a with $1 < a < q - 1$. Her public information or public key is g^a given modulo q. This is open to all. Her private information or private key or secret key is a.

(3) Bob chooses a b with $1 < b < q - 1$. His public information or public key is g^b given modulo q. This is open to all. His private information or private key or secret key is g^b.

Communication: The secret shared key is g^{ab}. This can be computed easily by both Bob and Alice using their secret keys. However, an attacker must determine a or b given g^a or g^b. This determination is essentially dependent on the discrete log problem.

Given q, g, g^a, g^b the problem of determining the secret key g^{ab} is called the **Diffie-Hellman Problem**. At present, the only known solution is to solve the discrete log problem, which appears to be very hard. In choosing the prime q and the generator g, it is assumed that the prime q is very large so that the order of g is very large. There are algorithms to solve the discrete log problem if q is too small.

Clearly, solving the discrete log problem breaks the Diffie-Hellman protocol, as does solving the Diffie-Hellman problem. It is not known whether the Diffie-Hellman problem can be solved without solving the discrete log problem. The **Diffie-Hellman assumption** is that both the discrete log problem in $\mathbb{Z}_p^*$ and the Diffie-Hellman-Hellman problem in $\mathbb{Z}_p^*$ are hard to solve.

One attack on the Diffie-Hellman key exchange is a **man in the middle attack**. Since the basic protocol involves no authentication an attacker can pretend to be Bob and get information from Alice and then pretend to be Alice and get information from

Bob. In this way the attacker could get the secret shared key. To prevent this, digital signatures could be used (see the section on hash functions).

The **decision Diffie-Hellman problem** is the following: given a prime q, a generator g of $\mathbb{Z}_q^*$, and g^a modulo q, g^b modulo q and g^c modulo q then to determine if $g^c \equiv g^{ab} \pmod{q}$.

The Diffie-Hellman protocol is a **key exchange protocol**, that is a protocol that exchanges a key to be used in some other encryption system. In general, as previously mentioned, public key encryption is not as efficient as classical encryption, so some form of mixed protocol is used. In the next section we will examine a method called **ElGamal Encryption** for turning the Diffie-Hellman protocol into an encryption protocol.

Notice that the ideas of the Diffie-Hellman protocol and the discrete log problem can be used within any finite abelian group, or in any group for that matter, as long as the generator g and the overgroup G are specified. We will see in Chapter 8 that this is the basic idea in elliptic curve cryptography, where the group is taken as the group of points of an elliptic curve.

In 1997 it became known that the ideas of public key cryptography were developed by British Intelligence Services prior to Diffie and Hellman.

7.4 ElGamal Encryption

In 1984, T. ElGamal devised a method to turn the Diffie-Hellman key exchange protocol into a public key encryption protocol. This is now known as **ElGamal encryption**. Although ElGamal proposed using the cyclic groups $\mathbb{Z}_p^*$ for a large primes p this type of encryption can be used in any cyclic group, where the discrete log problem is assumed hard. If the group is a cyclic group within the group of an elliptic curve, ElGamal encryption becomes the basis for **elliptic curve cryptography**. We will discuss elliptic curves and elliptic curve cryptography in Chapter 8. Further variations of the ElGamal method can be utilized to transform any public key exchange protocol into a public key encryption system.

The basic scheme for an ElGamal encryption system is the following. Each user chooses a large prime p and a generator g for the cyclic group $\mathbb{Z}_p^*$. Given a large prime p there is a fixed efficiently invertible procedure to encrypt plaintext into residue classes within $\mathbb{Z}_p^*$, the unit group within $\mathbb{Z}_p$. Although each user may choose different primes this procedure is known once the prime p is fixed.

For each message transmission the user's public key is (p, g, A) where $A = g^a$ for some integer a.

The encryption and decryption works as follows. Suppose that Bob wants to send a message to Alice. Alice's public key, which is public knowledge, is (p, g, A) as above. The message is m, and as above, is encrypted in some workable efficient manner within $\mathbb{Z}_p^*$, that is the message is encrypted in a manner known to all users (once p is given) as an integer in $0, 1, \ldots, p - 1$. Bob now randomly chooses an integer b and computes $B = g^b$. He now sends to Alice (B, mC) where $C = g^{ab}$. Notice that C is the common shared key in the Diffie-Hellman key exchange and in the encryption this is multiplied by the message m.

To decrypt, Alice first uses B to determine the common shared key C. Since $B = g^b$ and she knows $A = g^a$, she knows $C = g^{ab}$ for the same reasons as the Diffie-Hellman key exchange works. Since she knows $C = g^{ab}$ and she knows the modulus p she can compute the inverse $g^{-(ab)}$. This is efficient since it only requires one exponentiation modulo p. She then multiplies $mC = mg^{ab}$ by g^{-ab} to obtain the message m. We have just shown the following.

Theorem 7.4.1. *ElGamal encryption is a public key cryptosystem assuming the difficulty of the Diffie-Hellman key exchange.*

Example 7.4.2. Suppose that Alice chooses $p = 23, g = 7$ and $a = 6$. Then $\mathbb{Z}_p^* = \langle \overline{7} \rangle$ the residue class modulo 23 of 7. Then $g^a = 7^6 \equiv 4 \pmod{23}$ and therefore her public key is $(23, 7, 4)$. Suppose that Bob wants to send a message m to Alice and suppose that the message is encrypted within $\mathbb{Z}_{23}$ as $\overline{7}$. Then Bob randomly chooses $b = 3$. Then with the notation above $B \equiv 7^3 \pmod{23}$ so $B = 21$ and $g^{ab} \equiv 7^{18} \equiv 4^3 \equiv 18 \pmod{23}$. Then $C = g^{ab}m \equiv (18)(7) \equiv 11 \pmod{23}$. Therefore Bob sends to Alice $(21, 11)$.

Alice knows that $21 \equiv g^b \pmod{23}$ and hence, since she knows $a = 6$ (and only she knows this presumably) she can compute $g^{ab} \equiv 18 \pmod{23}$. Hence she can compute 18^{-1} modulo 23 multiply it by $C = 11$ and recover the message 7 modulo 23.

Both encryption and decryption involve exponentiation within $\mathbb{Z}_p$ so, as explained earlier, it can be done efficiently. We consider the security. The security is based on the difficulty of the Diffie-Hellman problem. We have the following.

Theorem 7.4.3. *Breaking the ElGamal encryption scheme and breaking the Diffie-Hellman key exchange protocol are equally difficult.*

Proof. Suppose that the Diffie-Hellman key exchange protocol can be broken, that is one can determine the secret key g^{ab} from p, g, g^a, g^b. Then since all this information is transferred in the ElGamal system via $C = g^{ab}m$ one can recover m and the ElGamal encryption is broken.

Conversely suppose that any message m can be computed given the ElGamal transfers. Suppose that we are given p, g, g^a, g^b and we want to determine the secret key g^{ab}. We apply the ElGamal decryption algorithm to the message with $m = 1$ to obtain g^{ab}. $\square$

Although the difficulty of breaking ElGamal is equivalent to the Diffie-Hellman problem it is not known whether breaking ElGamal implies breaking the discrete log problem.

We now make some remarks about the ElGamal system and its security.

The ElGamal cryptosystem is based on the choice of very large prime numbers. We want to generate randomly, integers $n \in \mathbb{N}$, that are probably prime numbers and which have a certain large bit length k, that is

$$n = \sum_{i=1}^{k} b_i 2^{k-i}.$$

To do this we first generate randomly an odd integer $n \geq 1$ with bit length k as follows. We define the first and last bit to be 1 and choose the remaining $k - 2$ bits independently and randomly according to a uniform distribution. We then check if n is a prime number applying the Miller-Rabin test for n with k repetitions. If we do not obtain from this test that n is composite then we consider n as a prime number.

Experiments show that for numbers with more than 1000 bits we quickly get a prime number with a probability of $\geq 1 - \left(\frac{1}{2}\right)^{80}$.

Let p be a large prime number. We know that the unit group $\mathbb{Z}_p^*$ is cyclic and hence there exists a primitive root g modulo p so that $\mathbb{Z}_p^* = \langle \overline{g} \rangle$. In general though it is not easy to find a primitive root modulo p. Then following criterion could be helpful.

Let G be a finite group with $|G| = p_1^{e_1} \cdots p_r^{e_r}, 1 \leq e_i$ and the p_i pairwise distinct prime numbers $i = 1, \ldots, r$. Let $g \in G$ and f_i the greatest non-negative integer with $g^{d_i} = 1$ where

$$d_i = \frac{|G|}{p_i^{f_i}}.$$

Then

$$o(g) = |g| = p_1^{e_1 - f_1} \cdots p_r^{e_r - f_r}.$$

This criterion is a direct consequence of Lagrange's Theorem (see Chapter 5).

Example 7.4.4. Let $p = 101$ and $G = \mathbb{Z}_p^*$. That is $e_1 = e_2 = 2$ if we define $p_1 = 2$ and $p_2 = 5$. Then $|G| = 100 = 2^2 5^2$. Let $\overline{2} \in G$. Then $2^{2 \cdot 5^2} \equiv -1 \pmod{101}$. Hence $f_1 = 0$. Further $2^{2^2 5} \equiv -6 \pmod{101}$ and therefore $f_2 = 0$ also. Therefore $o(\overline{2}) = 100$ and 2 is a primitive root modulo 101.

The above criterion has an easy consequence. Let G be a finite group with $g \in G$. Let $n \in \mathbb{N}$ with $g^n = 1$ and $g^{\frac{n}{p}} \neq 1$ for each prime divisor p of n. Then n is the order of g.

For a prime p there is no general procedure for finding a primitive root modulo p. One can try 2, and then if 2 fails, try 3 and so on.

For a prime p we have $|\mathbb{Z}_p^*| = p - 1$. Suppose that the prime factorization of $p - 1$ is $p_1^{v_1} \cdots p_k^{v_k}$. Then as explained in Chapter 5 the discrete log problem for $\mathbb{Z}_p^*$ can be reduced to the discrete log problem in cyclic groups of orders $p_1, \ldots, p_k$.

We have seen that breaking the ElGamal encryption system is equivalent to solving the Diffie-Hellman problem. Paired with the comment in the preceding paragraph, this impacts the cryptanalysis of the ElGamal system, and the choice of primes in the system. The discrete log problem for $\mathbb{Z}_p^*$ is easy to solve if $p - 1$ has only small prime divisors. Hence we must choose p, so that $p - 1$ has at least one large prime divisor. This holds if $\frac{p-1}{2}$ is also a prime, or at least if $\frac{p-1}{4}$ is prime in the case that $p \equiv 1 \pmod 4$.

The choice of b in the ElGamal encryption must be randomized. Hence they cannot be found for example by a linear congruence generator. If the number b is chosen randomly then the pair (B, C) is uniformly distributed among the residue classes in $\mathbb{Z}_p^*$ and hence ciphertext messages using ElGamal encryption are **semantically secure** (see Chapter 3) if the Diffie-Hellman problem is hard. For each new ElGamal encryption, a new exponent b must be chosen, to maintain this semantic security.

The ElGamal encryption system then has the strength of this semantic security. However, it does have the disadvantage of sending more information than the message since we must also send the key. This is known as **message expansion**.

7.4.1 Generalizations of ElGamal

In an analogous manner, ElGamal encryption can be done over any finite cyclic group. In particular this can be done over a cyclic subgroup of the group of an elliptic curve and is the basis for elliptic curve cryptography. This has the advantage of requiring a smaller key space. We will discuss this in Chapter 8.

Here we note that ElGamal encryption can also be done over the multiplicative group of any finite field and the multiplicative group for many modular rings $\mathbb{Z}_n$. Here we recall the following two results. The first is mentioned in Chapter 5 while the second can be found in [FR].

Theorem 7.4.5. *The multiplicative group F^* of any finite field F is cyclic.*

Theorem 7.4.6. *The unit group $\mathbb{Z}_n^*$ is cyclic if and only if $n = 2, 4, 2p, p^k$ for p an odd prime and $k \geq 1$, that is there is a primitive element for these moduli.*

ElGamal encryption then can be done over any finite field and over $\mathbb{Z}_n$ for the moduli $n = 2, 4, 2p, p^k$, with p an odd prime, using a primitive root.

7.5 The RSA Algorithm and Protocol

In 1977, Rivest, Adelman and Shamir developed the **RSA Algorithm**, which is presently one of the most widely used public key protocols. It is based on the difficulty of factoring large integers and in particular on the fact that it is easier to test for primality than to factor very large integers. In this first section we look at the basic algorithm and then look a bit more deeply in subsequent sections.

In basic form, the RSA algorithm works as follows. Each user U randomly chooses two large primes p_U, q_U and computes $n_U = p_U q_U$. These integers are chosen randomly to minimize attack. The number $n_U = p_U q_U$ is called the **RSA modulus**. Let

$\phi(n_U) = \phi(p_U q_U) = (p_U - 1)(q_U - 1)$ be the value of the Euler phi function (see Chapter 5) on the RSA modulus. The user then randomly chooses an integer $e_U \in \mathbb{Z}_{n_U}$ with $(e_U, \phi(n_U)) = 1$. The integer e_U is called the **RSA encryption exponent** for the user U. Since $(e_U, \phi(n_U)) = 1$ it follows that e_U is a unit modulo $\phi(n_U)$ and therefore there is a d_U with

$$e_U d_U \equiv 1 \pmod{\phi(n_U)}.$$

The integer d_U can be found by the Euclidean algorithm (see Chapter 5). d_U is the user's **RSA decryption exponent**.

Assume that a plaintext message is given by an integer $M \in \mathbb{Z}_{n_U}$ so that $M \in \{0, 1, \ldots, n_U - 1\}$. If we let f_U stand for the encryption function then the ciphertext message is give by

$$f_U(M) \equiv M^{e_U} \pmod{n_U}.$$

That this is a valid encryption function, and can be decrypted, is based on the following number theoretic theorem.

Theorem 7.5.1. *Suppose that p, q are distinct odd primes and $(e, \phi(pq)) = 1$ and $ed \equiv 1 \pmod{\phi(pq)}$. Then for any $M \in \mathbb{Z}_{pq}$ we have*

$$M^{ed} = M \text{ in } \mathbb{Z}_{pq}$$

Proof. Let $n = pq$ be a product of two distinct odd primes, $(e, \phi(n)) = 1$ and $ed \equiv 1 \pmod{\phi(n)}$. Then since $\phi(n) = (p-1)(q-1)$ we have

$$ed = 1 + t(p-1)(q-1) \quad \text{for some integer } t.$$

Let $M \in \mathbb{Z}_n$. Then

$$M^{ed} = M^{1+t(p-1)(q-1)} = M \cdot M^{t(p-1)(q-1)} = M \cdot (M^{p-1})^{t(q-1)}.$$

If $m = \overline{0}$ in $\mathbb{Z}_p$ then certainly $M^{ed} = M$ in $\mathbb{Z}_p$, and if $M \neq \overline{0}$ in $\mathbb{Z}_p$ then $(M^{p-1})^{t(q-1)} = \overline{1}$ in $\mathbb{Z}_p$ by Fermat's Theorem. Therefore

$$M^{ed} = M \text{ in } \mathbb{Z}_p.$$

Identically since q is also a prime

$$M^{ed} = M \text{ in } \mathbb{Z}_q.$$

Since p and q are distinct primes by the Chinese Remainder Theorem, it follows that $M^{ed} = M$ in $\mathbb{Z}_{pq}$, completing the theorem. $\qquad\square$

The theorem then allows RSA decryption. Suppose the user U receives the encrypted message M^{e_U}. The user then applies the decryption exponent d_U to obtain

$$(M^{e_U})^{d_U} = M^{e_U d_U} \equiv M \pmod{n}$$

and recovers the message.

In the next section we show how to make this into a public key cryptosystem.

7.5.1 The RSA Cryptosystem

We now show how to make the brief encryption protocol described in the last section into a workable cryptosystem. Important first is the preparation

Each user must randomly choose two distinct large primes. Primality tests arise in this process. The user first randomly chooses a large odd integer m and tests it for primality. If it is prime it is then used in the protocol. If not, the user tests $m + 2, m + 4, \ldots$, and so on until the first prime p_U is obtained. The user then repeats the process to get q_U. Now the user must randomly choose the encryption exponent e_U. To do this the user chooses another odd integer m and tests until getting an e_A relatively prime to $\phi(p_A q_A)$. The primes chosen should be quite large. Originally RSA used primes of approximately 100 decimal digits, but as computing and attack have become more sophisticated, larger primes have had to be utilized. Presently keys with 400 decimal digits are not uncommon. Once the user has obtained p_U, q_U, e_U, the values $n_U = p_U q_U$ and d_U, the multiplicative inverse of e_U modulo $\phi(n_U)$, are computed. The user makes public the enciphering key $k_U = (n_U, e_U)$ and the encryption algorithm known to all is

$$f_U(M) = M^{e_U} = C \text{ in } \mathbb{Z}_{n_U}$$

where $M \in \mathbb{Z}_{n_U}$ is a message unit. If $e_U d_U \equiv 1 \pmod{(p_U - 1)(q_U - 1)}$ then $M^{e_U d_U} = M$ in $\mathbb{Z}_{n_U}$. Therefore the decryption algorithm is

$$g_U(C) = C^{d_U} \text{ in } \mathbb{Z}_{n_U}.$$

Notice then that $g_U(f_U(M)) = M^{e_U d_U} = M$ in $\mathbb{Z}_{n_U}$ so it is the left inverse.

It follows that in the RSA protocol each user's public key is (n_U, e_U) while the private or secret key is (p_U, q_U, d_U).

Encryption and decryption proceed as follows: Suppose that Alice wants to send a message to Bob. Alice's has made all the required choices and has her public key (n_A, e_A) and her private key (p_A, q_A, d_A), Bob makes the same type of choices to obtain his public key (n_B, e_B) and his private key (p_B, q_B, d_B).

If Alice wants to send a message to Bob, she first formulates the message as an integer M in $\mathbb{Z}_{n_B}$. We will say more about how to formulate a message in Section 7.6, Alice knows n_B and e_B since they have been made public by Bob. She now sends Bob the encrypted message

$$f_B(M) = M^{e_B} \text{ in } \mathbb{Z}_{n_B}.$$

To decrypt Bob uses his secret key d_B to find

$$M^{e_B d_B} = M \text{ in } \mathbb{Z}_{n_B}$$

recovering the message.

Here we have assumed that the message M is a single letter $M \in \mathbb{Z}_{n_B}$. In the next section we will show how to make RSA a block cipher and describe a method to formulate the messages from some natural alphabet.

Authentication can be done as in the general procedure by sending to Bob

$$f_B\left(g_A(M)\right).$$

However, in RSA care must be taken. The domain of g_A is $\mathbb{Z}_{n_A}$ and n_B might be larger than n_A so that the message M might not be in the domain of g_A. In this case, Alice could send

$$g_A\left(f_B(M)\right).$$

Now Bob first applies f_A which is public to obtain $f_A(g_A f_B(M)) = f_B(M)$. Now he applies his secret key g_B to recover the message. Summarizing this, in the RSA algorithm for authentication, if $n_B > n_A$ Alice sends $g_A(f_B(M))$. If $n_A \geq n_B$ she sends $f_B(g_A(M))$. Both n_A and n_B are public, so Bob knows which way to do the encryption.

We summarize the basic RSA setup, encryption and decryption.

RSA Encryption Preparation

(1) Each user U must randomly choose two large distinct odd primes p_U, q_U and calculate $n_U = p_U q_U$. This is called the **RSA modulus.**
(2) All users must have a fixed procedure to encrypt plaintext blocks into residue classes modulo n_U.
(3) Each user calculates $\phi(n_U) = (p_U - 1)(q_U - 1)$ and must randomly choose a random integer $e_U \in \{1, \ldots, \phi(n_U) - 1\}$ with $(e_U, \phi(n_U)) = 1$. This integer e_U is called the **encryption exponent.**
(4) Using the extended Euclidean algorithm determine the integer $d_U \in \{0, 1, \ldots, \phi(n_U) - 1\}$ such that

$$d_U e_U \equiv 1 \pmod{\phi(n_U)}.$$

This integer d_U is called the **decryption exponent.**

We now describe the essential encryption and decryption procedure. From the preparatory steps we have for each user

$$p_U, q_U, n_U = p_U q_U, e_U, d_U.$$

Then:

RSA Encryption and Decryption Procedure

(1) The **public key** for each user is the pair (n_U, e_U) while the **secret key** is the triple (p_U, q_U, d_U).
(2) To encrypt a message $M \in \mathbb{Z}_{n_U}$ the encryption algorithm is $f_U(M) \equiv M^{e_U} \pmod{n_U}$.
(3) To decrypt, given $C \in \mathbb{Z}_{n_U}$ the decryption algorithm is then $g_U(C) \equiv C^{d_U} \pmod{n_U}$.

7.5.2 RSA as a Block Cipher

We now show how to make RSA into a block cipher and how to encrypt with RSA from a natural language such as English or German.

Suppose there is an N letter alphabet which is to be used for both plaintext and ciphertext. The plaintext message is to consist of k vectors of letters and the ciphertext message of l vectors of letters with $k < l$. To ensure that the range of plaintext messages and ciphertext messages are the same, integers $k < l$ are chosen so that

$$N^k < n_U < N^l$$

for each user U, that is $n_U = p_U q_U$. In this case any plaintext message M is an integer less than N^k considered as an element of $\mathbb{Z}_{n_U}$.

Each of the k plaintext letters in a message unit M are then considered as integers modulo N and the whole plaintext message is considered as a k digit integer written to the base N (see example below). The transformed message is then written as an l digit integer modulo N and then the digits are considered integers modulo N from which encrypted letters are found. Since $n_U < N^l$ the image under the power transformation corresponds to an l digit integer written to the base N and hence to an l letter block. We give an example with relatively small primes. In real world applications the primes would be chosen to have over a hundred digits and the computations and choices must be done using good computing machinery.

Example 7.5.2. Suppose $N = 26, k = 2$ and $l = 3$. Suppose further that Alice chooses $p_A = 29, q_A = 41, e_A = 13$. Here $n_A = 29 \cdot 41 = 1189$, so she makes public the key $k_A = (1189, 13)$. She then computes the multiplicative inverse d_A of 13 modulo $1120 = 28 \cdot 40$. Now suppose we want to send her the message TABU. Since $k = 2$ the message units in plaintext are 2 vectors of letters so we separate the message into TA BU. We show how to send TA.

We make the usual assignment of numbers to letters;

$$A \to 0, \quad B \to 1, \ldots, \quad T \to 19, \ldots, \quad Z \to 26.$$

The numerical sequence for the letters TA modulo 26 is then $(19, 0)$. We then use these as the digits of a 2-digit number to the base 26. Hence

$$TA = 19 \cdot 26 + 0 \cdot 1 = 494.$$

We now compute the power transformation using her $e_A = 13$ to evaluate

$$f(19, 0) = 494^{13} \pmod{1189}.$$

This is evaluated as 320. Now we write 320 to the base 26. By our choices of k, l this can be written with a maximum of 3 digits to this base. Then

$$320 = 0 \cdot 26^2 + 12 \cdot 26 + 8.$$

The letters in the encoded message then correspond to $(0, 12, 8)$ and therefore the encryption of TA is AMI since 0 corresponds to A, 12 to M and 8 to I.

To decode the message Alice knows d_A and applies the inverse transformation.

Since we have assumed that $k < l$ this seems to restrict the direction in which messages can be sent. In practice to allow messages to go between any two users the following is done. Suppose Alice is sending an authenticated message to Bob. The keys $k_A = (n_A, e_A)$, $k_B = (n_B, e_B)$ are public. If $n_B \leq n_A$ Alice sends $f_B g_A(M)$. On the other hand, if $n_B > n_A$ she sends $g_A f_B(M)$.

There have been attacks on RSA for special types of primes so care must be chosen in choosing the primes.

The computations and choices used in real world implementations of the RSA algorithm must be done with computers. Similarly, attacks on RSA are done via computers. As computing machinery gets stronger and factoring algorithms get faster, RSA becomes less secure and larger and larger primes must be used. In order to combat this, other public key methods are in various stages of ongoing development. RSA and Diffie-Hellman and many related public key cryptosystems use properties in abelian groups. In recent years a great deal of work has been done to encrypt and decrypt using certain non-abelian groups such as linear groups or braid groups (see Chapters 9 and 10).

To formally describe the ideas for changing RSA into a block cipher we have the following: We assume that RSA encryption and decryption has been set up so that each user has a public key $(n_U, e_U) = (n, e)$ and a private key $(p_U, q_U, d_U) = (p, q, d)$. Let N be the number of pairwise distinct plaintext units. These are the letters of our alphabet $A = \{0, 1, \ldots, N - 1\}$ which we represent as $\mathbb{Z}_N$. We code these units as elements of the modular ring $\mathbb{Z}_N$ and we should have $N < n$ where $n = pq$ is the RSA modulus.

We now choose integers $k, l \geq 1$ such that

$$N^k < n < N^l.$$

As a rule we may take $k = [\log_N n]$ where $[\alpha]$ is the greatest integer $\leq \alpha$ and $l = k + 1$.

To encrypt a message block m we break it up into k-letter blocks that is elements $(m_1, m_2, \ldots, m_k) \in A^k$. We write $m_1 \cdots m_k$ instead of $(m_1, \ldots, m_k)$.

A block $m_1 \cdots m_k$, $m_i \in A$, $1 \leq i \leq k$ can be transformed into the integer

$$x = \sum_{i=1}^{k} m_i N^{k-i}.$$

Since $k = [\log_N n]$ we get that

$$0 \leq x \leq (N - 1) \sum_{i=1}^{k} N^{k-i} = N^k - 1 < n.$$

Let e with $(e, \phi(n)) = 1$ be an encryption exponent. We encrypt x by determining

$$c \equiv x^e \pmod{n}, \ 0 \leq c \leq n.$$

The integer c can now be written as

$$c = \sum_{i=0}^{k} c_i N^{k-i}, \quad c_i \in \mathcal{A}, \quad 0 \le i \le k$$

in its N-adic representation. Recall that the N-adic representation can have a length $k + 1$ (this we considered when we chose l with $n < N^l$).

The cipher block then is

$$c = c_0 c_1 \cdots c_k.$$

The gives the RSA-encryption maps from plaintext blocks of length k injectively onto ciphertext blocks of length $k + 1$.

Example 7.5.3. We present a numerical example. Let $n = 253 = 11 \cdot 23$. Then $\phi(n) = 220 = 4 \cdot 5 \cdot 11$. We take $e = 3$ as the encryption exponent. Since e must be relatively prime to $\phi(n)$ this is the smallest possible encryption exponent. Using the Euclidean algorithm we find the decryption exponent as $d = 147$.

Let $\mathcal{B} = \{o, a, b, c\}$ be a four letter alphabet. We encode this by $o \to 0, a \to 1$, $b \to 2, c \to 3$ and so as a numerical alphabet $\mathcal{A} = \{0, 1, 2, 3\}$. If $n = 253$ then $k = [\log_4 253] = 3$. This then becomes the length of the plaintext blocks and thus the ciphertext blocks have length 4.

Suppose that we want to encrypt the word *abb* which is represented by the string 122. Since $N = 4$ we then get

$$x = 1 \cdot 4^2 + 2 \cdot 4^1 + 2 \cdot 4^0 = 26.$$

With the encryption exponent $e = 3$ the integer x will be encrypted into

$$26^3 \equiv 119 \ (\mathrm{mod}\ 253)$$

and therefore $c = 119$. Expressing the 4-adic representation of this we get

$$c = 1 \cdot 4^3 + 3 \cdot 4^2 + 1 \cdot 4^1 + 3 \cdot 4^0$$

and hence the ciphertext block is 1313 and so the encrypted word is *acac*.

If $d = 147$ then $119^{147} \equiv 26 \ (\mathrm{mod}\ 253)$.

Example 7.5.4. We take the following example from [BD]. We choose $p = 29, q = 53$ and $n = (29)(53) = 1537$. Then $\phi(n) = (28)(52) = 1456$.

Suppose that $N = 10$ so that we have $\mathcal{A} = \{0, 1, 2, \ldots, 9\}$ as the plaintext units which we represent as elements of the modular ring $\mathbb{Z}_{10}$. We then define $\mathcal{P} = \mathcal{C} = \mathbb{Z}_n$.

We take the letters of the Latin alphabet as

$$A = 01, \quad B = 02, \ldots, \quad Z = 26$$

and punctuation given by

$$, = 27, \quad . = 28, \quad ? = 29, \quad 0 = 30, \quad 1 = 31, \ldots, \quad 9 = 39, \quad ! = 40.$$

Further we take 00 as the designation for the space character between words.

We want to encrypt the plaintext

LISA KAM IM JANUAR

Using the allocations above we get the following plaintext number, which becomes the message:

$$m = \underset{=\text{LISA}}{12091901}\underset{=\text{KAM}}{00110110009}\underset{=\text{IM}}{1300100}\underset{=\text{JANUAR}}{114210118}$$

We choose $k = 3$ and $l = k + 1 = 4$ since $10^3 < n = 1537 < 10^4$. Hence the block length for a plaintext block is 3 and the block length for a ciphertext block is 4.

Now we define $\tilde{\mathcal{P}} = (\mathbb{Z}_{10})^3$ and $\tilde{\mathcal{C}} = (\mathbb{Z}_{10})^4$. We decompose the plaintext number M into blocks of length 3.

$$120|919|010|111|300|091|300|100|114|210|118.$$

This defines the maps $\rho : \tilde{\mathcal{M}} \to \mathcal{M} = \mathbb{Z}_n$ and $\gamma : \mathcal{C} = \mathbb{Z}_n \to \tilde{\mathcal{C}}$. For example, if

$$(\overline{1}, \overline{2}, \overline{0}) \in (\mathbb{Z}_{10})^3 \quad \text{and} \quad (\overline{1}, \overline{3}, \overline{1}, \overline{0}) \in (\mathbb{Z}_{10})^4$$

then

$$\rho((\overline{1}, \overline{2}, \overline{0})) = 120 + 1537\mathbb{Z}$$

and

$$\gamma(1130 + 1537\mathbb{Z}) = (\overline{1}, \overline{3}, \overline{1}, \overline{0}).$$

Both ρ and γ are injective and invertible on their range since

$$10^3 < n = 1537 < 10^4.$$

We choose the random integer $e = 47$ and $(47, 1456) = 1$ so we use e as the encryption exponent. We then calculate the decryption exponent as $d = 31$, so that $ed \equiv 1 \pmod{1456}$.

To find the cipher number of the first plaintext block 120 of the plaintext message m we take this block to the exponent e modulo 1537. Hence

$$120^{47} \equiv 734 \pmod{1537}.$$

Therefore the ciphertext number for the block 120 is then 734. Doing this for the whole plaintext message we get the complete ciphertext number, put in blocks of length ≤ 4:

$$734|720|978|1262|1262|340|589|340|470|496|1002|1366.$$

We recover the message m if we decrypt blockwise with $d = 31$, for example

$$734^{31} \equiv 120 \pmod{1537}.$$

7.5.3 Practical Implementation of RSA

We formalize the ideas in the last section. First we look at the preparatory steps necessary for RSA encryption.

RSA Encryption Preparation

(1) Choose two large distinct odd primes p, q and calculate $n = pq$. This is called the **RSA modulus**.

(2) Define encrypting maps from plaintext blocks into residue classes modulo N and from residue classes modulo n to ciphertext blocks as follows:

   (a) Let N be the number of pairwise distinct plaintext units. Let $A = \{0, 1, \ldots, N - 1\}$ be the set of these units. We represent A as the modular ring $\mathbb{Z}_N$ and we should have $N < n$.

   (b) Choose integers $k, l \geq 1$ such that

$$N^k < n < N^l.$$

As a rule we may take $k = [\log_N n]$ where $[\alpha]$ is the greates integer $\leq \alpha$ and $l = k + 1$.

   (c) Let $\tilde{\mathcal{M}} = \mathbb{Z}_N^k$ and $\tilde{\mathcal{C}} = \mathbb{Z}_N^l$

   (d) Choose injective, efficient invertible maps

$$\rho : \tilde{\mathcal{M}} \to \mathcal{P} = \mathbb{Z}_n$$

and

$$\gamma : \mathcal{C} = \mathbb{Z}_n \to \tilde{\mathcal{C}}$$

(invertible on the respective images).

(3) Calculate $\phi(n) = (p - 1)(q - 1)$ and choose a random integer $e \in \{1, \ldots, \phi(n) - 1\}$ with $(e, \phi(n)) = 1$. This integer e is called the **encryption exponent**.

(4) Using the extended Euclidean algorithm, determine $d \in \{0, 1, \ldots, \phi(n) - 1\}$ such that

$$de \equiv 1 \ (\mathrm{mod} \ \phi(n)).$$

This integer d is called the **decryption exponent**.

We now describe the essential encryption and decryption procedure. From the preparatory steps we have

$$p, q, n = pq, \tilde{\mathcal{M}}, \tilde{\mathcal{C}}, \rho, \gamma, e, d.$$

Then:

RSA Encryption and Decryption Procedure

(1) The **public key** is (n, e) while the **secret key** is d.

(2) To encrypt a message block $m \in \tilde{M}$ calculate $\bar{x} = \rho(m) \in \mathbb{Z}_n$. Then calculate $\bar{y} = (\bar{x})^e \in \mathbb{Z}_n$. Finally calculate $c = \gamma(\bar{y}) \in \tilde{C}$.

(3) To decrypt, given $c \in \tilde{C}$ find a $\bar{y} \in \mathbb{Z}_n$ with $\gamma(\bar{y}) = c$. Then calculate

$$\bar{z} = (\bar{y})^d \in \mathbb{Z}_n$$

and then find $\tilde{m} \in \tilde{M}$ with $\rho(\tilde{m}) = \bar{z}$.

The key point of the decryption is the map $\delta : \mathbb{Z}_n \to \mathbb{Z}_n$ given by $\bar{y} \to (\bar{y})^d$.

Let $k = [\log_N n]$ with $n = pq$ the product of two distinct odd primes. We write $m_1 \cdots m_k$ instead of $(m_1, \ldots, m_k)$ with $m_i \in A, 1 \le i \le k$.

A block $m_1 \cdots m_k, m_i \in A, 1 \le i \le k$ can be transformed into the integer

$$x = \sum_{i=1}^{k} m_i N^{k-i}.$$

Since $k = [\log_N n]$ we get that

$$0 \le x \le (N-1) \sum_{i=1}^{k} N^{k-i} = N^k - 1 < n.$$

This represents the application of a map $\rho : \tilde{M} \to P$. Let e with $(e, \phi(n)) = 1$ be an encryption exponent. We encrypt x by determining

$$c \equiv x^e \pmod{n}, \quad 0 \le c \le n.$$

The integer c can now be written as

$$c = \sum_{i=0}^{k} c_i N^{k-i}, \quad c_i \in A, \quad 0 \le i \le k$$

in its N-adic representation. Recall that the N-adic representation can have a length $k + 1$ (this we considered when we chose l with $n < N^l$).

The cipher block then is

$$c = c_0 c_1 \cdots c_l.$$

The gives the RSA-encryption maps from plaintext blocks of length k injectively onto ciphertext blocks of length $k + 1$.

Suppose a user receives $(c_0, c_1, \ldots, c_l)$. He computes this as the N-adic representation of an integer as in the encryption to form

$$c = \sum_{i=0}^{k} c_i N^{k-i}, \quad c_i \in A, \quad 0 \le i \le k.$$

He then applies the decryption exponent d to obtain

$$m \equiv c^d \pmod{n}.$$

Writing the N-adic representation of m the user recovers $(m_1, \ldots, m_k)$.

7.5.4 Feasibility of the RSA Algorithm

The main work in using RSA is the setup, especially in choosing appropriate large primes. We may generate integers p, q with quick probabilistic primality tests, like the Miller-Rabin test (see Chapter 6).

Another simple possibility to generate p and q with p and q eventually prime numbers is as follows. Choose an interval $I = [a, b] \subset \mathbb{N}$ which should contain a prime number. For example let a be a large odd integer and $b = a + 10^5$. Start with a and check $a, a + 2, a + 4, \ldots$ for primality as long as we find hopefully a prime number.

Note that in general there is no definite such interval. Recall that there exist arbitrarily long sequences of natural numbers consiting of all composites (see [FR]). However, if $a \in \mathbb{N}$ there always exists a prime p with $a \leq p \leq 2a$. This is called Bertrand's Theorem (see [FR]).

Once appropriate p and q are chosen we can randomly choose an odd integer $m \in \{1, \ldots, \phi(n)\}$ and then test $m, m + 2, m + 4, \ldots$ and so on to find the encryption exponent e with $(e, \phi(n)) = 1$. Having e we can then find the decryption exponent d.

The exponentiation maps $\bar{x} \rightarrow \bar{x}^e$ and $\bar{y} \rightarrow \bar{y}^d$ can be efficiently calculated by iterated exponentiation based on the following recursion for powers:

$$a^n = (a^{\frac{n}{2}})^2 \quad \text{if } n \text{ is even}$$
$$a^n = (a^{\frac{n-1}{2}})^2 a \quad \text{if } n \text{ is odd}.$$

7.5.5 Security of RSA

The security certificate of the RSA cryptosystem is based on the assumption that the factorization into prime factors is difficult for large integers.

It is not really known how difficult the factorization problem really is. It is possible that there exists an easy solution to the factorization problem that is not yet known. At the present time we can say that the factorization problem is in the complexity class NP. Recall that a mathematical problem Π belongs to NP if there exists a polynomial time algorithm (see Chapter Two) which can prove if a general solution is correct or not.

The factorization problem for an integer $n \geq 1$ is in **NP** because it can be checked with the division algorithm if a general divisor really is or is not a divisor of n.

Each problem in **NP** can be solved by testing for each candidate. This approach to find the solution is, in general, exponential in time. If the input value is of size n, then in general, the set to consider has $\mathcal{O}(2^n)$ elements, that is we have to make $\mathcal{O}(2^n)$ tests. Since each individual test is polynomial in time the naive algorithm of going through everything is, in general, exponential in time.

In contrast is the complexity class **P**. A mathematical problem belongs to **P** if there exists a polynomial time algorithm for the solution to the problem.

The problem **Prim(n)** given by *check if a given n ∈ ℕ is a prime number or not* is in **P**. The AKS-algorithm for this problem is deterministic with polynomial running time (see [FR] or [KKi]).

We mention here again the main open problem of whether $P \equiv NP$.

Let $n = pq$ be the product of two distinct odd primes. Being able to factor this product would break the RSA cryptosystem. We make some remarks about this.

(a) We can efficiently calculate the pair $\{p, q\}$ if and only if we can calculate $\phi(n)$. (See Chapter 6).

(b) The following conditions are equivalent:

   [1] If we know an integer e with $(e, \phi(n)) = 1$ then we can efficiently find an integer d with $ed \equiv 1 \pmod{\phi(n)}$.

   [2] If we can determine square roots modulo n, then we can can efficiently find the prime factors p and q of n (see the material on square roots and products of two primes in Chapter 6).

(c) The two problems; to factorize $n = pq$ and to determine square roots modulo n are probabilistically algorithmically equivalent. Again see the material in Chapter 5.

7.5.6 Cryptanalysis of RSA

Based on the security ideas mentioned in the previous subsection, we can see that the primes p, q must be chosen carefully so that it is not easy to factor $n = pq$. From the material on factorizing products of two primes (see Chapter 6) we see that the primes should have the following properties:

(a) $|p - q|$ should be large

(b) $(p - 1, q - 1)$ should be small

(c) $p - 1$ and $q - 1$ should have at least one large prime factor.

(b) and (c) would be satisfied for instance if $\frac{p-1}{2}$ and $\frac{q-1}{2}$ are also prime numbers. Such primes p, q are called **secure prime numbers**.

Some further ideas involving the cryptanalysis of the RSA system are:

(1) The same encrypted message should not be sent to several receivers with different n and the same e.

(2) e and d should be sufficiently large.

(3) The prime numbers p and q should be chosen randomly (modulo the appropriate conditions from above). The list of possible p and q should be large.

7.6 Rabin Encryption

The RSA cryptosystem is based on the supposed difficulty of factorization of $n = pq$ with p, q large odd primes. However, as explained, there is no security certificate saying that breaking the RSA cryptosystem requires an algorithm to factor pq. It might be possible that there is a method to break RSA without solving the factorization problem.

As pointed out in the last section, and proved among several results in Chapter 6, the two problems; to factorize $n = pq$ and to determine square roots modulo n; are probabilistically algorithmically equivalent. This is an advantage of the **Rabin Cryptosystem** that we will discuss. This cryptosystem is based on the determination of square roots in $\mathbb{Z}_n^*$ with $n = pq$ and p, q two distinct primes. First we recall some necessary information from Chapters 5 and 6. The proofs can be found there.

7.6.1 Quadratic residues and Rabin Encryption

Rabin encryption is based on the difficulty of finding square roots modulo n where n is the product of two primes. We review some necessary information on modular square roots that was presented in Chapter 6.

First recall that if $(a, m) = 1$ and and $x^2 \equiv a \pmod{m}$ has a solution, then a is called a **quadratic residue** modulo m. If $x^2 \equiv a \pmod{m}$ has no solution, then a is a **quadratic non-residue**. Further if $x^2 \equiv a \pmod{m}$ then x is a **squareroot** of a modulo m.

The basic idea in using quadratic residues in cryptography stems from the following. If n is a prime, then we can decide effectively if a is a quadratic residue modulo n or not. However, if $n = pq$, the product of two distinct odd primes p and q, then in general there is no known probabilistic algorithm to decide if a is a quadratic residue modulo n unless p and q are known. We first need the following.

Theorem 7.6.1. *Let $p \geq 3$ be an odd prime.*
(1) *The set $\{1, 2, \ldots, p - 1\}$ contains exactly $\frac{p-1}{2}$ quadratic residues and hence exactly $\frac{p-1}{2}$ quadratic non-residues.*
(2) *If $a \in \mathbb{Z}$ and $(a, p) = 1$ then a has 0 or 2 square roots modulo p.*
(3) *If p is an odd prime and $(a, p) = 1$ then a is a quadratic residue modulo p if and only if $a^{\frac{p-1}{2}} \equiv 1 \pmod{p}$. If a is a quadratic non-residue then $a^{\frac{p-1}{2}} \equiv -1 \pmod{p}$.*
(4) *If $a, b \in \mathbb{Z}$ with $(a, p) = (b, p) = 1$ then ab is a quadratic residue modulo p if both a and b are quadratic residues modulo p or if both a and b are quadratic non-residues modulo p.*

Determining if a is a quadratic residue is helped by the use of the **Legendre symbol**. If p is an odd prime and $(a, p) = 1$ then the Legendre symbol $\left(\frac{a}{p}\right)$ is defined by
(1) $\left(\frac{a}{p}\right) = 1$ if a is a quadratic residue modulo p,
(2) $\left(\frac{a}{p}\right) = -1$ if a is a quadratic non-residue modulo p.

The determination of $\left(\frac{a}{p}\right)$ is then accomplished with the utilization of quadratic reciprocity.

Theorem 7.6.2 (Law of Quadratic Reciprocity). *If p, q are distinct odd primes then*

$$\left(\frac{p}{q}\right)\left(\frac{q}{p}\right) = (-1)^{\left(\frac{p-1}{2}\right)\left(\frac{q-1}{2}\right)}.$$

Alternatively, if p, q are distinct odd primes then
(1) *If at least one of p, q is congruent to 1 modulo 4 then*

$$x^2 \equiv q \pmod{p} \quad and \quad x^2 \equiv p \pmod{q}$$

are either both solvable or both unsolvable.
(2) *If both p and q are congruent to 3 modulo 4 then one of*

$$x^2 \equiv q \pmod{p} \quad and \quad x^2 \equiv p \pmod{q}$$

is solvable and the other is unsolvable.

We now describe a probabilistic algorithm to calculate square roots modulo an odd prime p. By a **probabilistic algorithm** we mean an algorithm that will return either that an inputted integer a with $(a, p) = 1$ does not have a square root modulo p or that a has a given degree of probability of having a square root.

Suppose first that $p \equiv 3 \pmod 4$. Then $\frac{p+1}{4} \in \mathbb{N}$. If $\overline{a} = \overline{b}^2$ in $\mathbb{Z}_p^*$ then

$$\overline{a}^{\frac{p+1}{4}} = \overline{b}^{\frac{p+1}{2}} = \overline{b}^{\frac{p-1}{2}+1} = \overline{b}^{\frac{p-1}{2}}\overline{b}.$$

Since $\overline{b}^{\frac{p-1}{2}} = \pm\overline{1}$ it follows that the whole expression equals $\pm\overline{b}$. Therefore

$$\overline{a}_p = \overline{a}^{\frac{p+1}{4}} \in \mathbb{Z}_p^*$$

is a square root of $\overline{a}$ and $\pm\overline{a}_p$ are the only squareroots of $\overline{a}$. Therefore in the case $p \equiv 3 \pmod 4$ we get the square roots of modulo p by simple exponentiation. We mention again the following theorem which was also given as Theorem 6.3.2. The proof can be found in Section 6.3.

Theorem 7.6.3. *Let p be an odd prime and $a \in \{1, \ldots, p-1\}$ a quadratic residue modulo p. Then the following algorithm produces with arbitrary probability $\geq 1 - \left(\frac{1}{2}\right)^k$ a square root of a modulo p.*
(1) *If $p \equiv 3 \pmod 4$ then give $a^{\frac{p+1}{4}}$ modulo p and stop.*
(2) *If $p \equiv 1 \pmod 4$ write $p - 1 = 2^s t$ with $s \geq 2$ and t odd.*
    (a) *Choose a random $\overline{u} \in \mathbb{Z}_p^*$ and compute $\overline{u}^{\frac{p-1}{2}}$. Repeat this as long as $\overline{u}^{\frac{p-1}{2}} = -1$ up to a maximum of k times. Now calculate $\overline{v} = \overline{u}^t$.*
    (b) *Determine $\overline{y} = \overline{a}^{\frac{t+1}{2}}$.*

(c) *Calculate $\bar{a}^{2^{s-i}t}$ for $i = 2, 3, \ldots, s$ and then define $b_0, b_1, \ldots, b_{s-2} \in \{0, 1\}$ recursively by*

$$\bar{a}^{2^{s-i}t} = \bar{v}^{b_0 2^{s-i+1} + b_1 2^{s-i+2} + \cdots + b_{s-2} 2^{s-1}}. \tag{$\star$}$$

Define

$$b = b_0 + 2b_1 + \cdots + 2^{s-2} b_{s-2}$$

and return $\bar{x} = \overline{yv}^{-b}$.

If $p \equiv 1 \pmod 4$ then we obtain the $b_0, b_1, \ldots, b_{s-2}$ by taking square roots in each step.

Finally $\bar{a}^t = \bar{v}^{2b}$ and hence

$$\bar{x}^2 = \bar{y}^2 \bar{v}^{-2b} = \bar{a}^{t+1} \bar{a}^{-t} = \bar{a}.$$

We now consider the situation $n = pq$ with p, q two distinct primes.

Theorem 7.6.4. *Let $p, q \geq 2$ be two distinct primes. Let $a \in \mathbb{Z}$ with $(a, p) = (a, q) = 1$. Then a is a quadratic residue modulo pq if and only if a is a quadratic residue modulo both p and q.*

Theorem 7.6.5. *Let $n = pq$ with p, q two distinct odd primes. Let a be an integer with $(a, p) = (b, p) = 1$. Then the congruence $x^2 \equiv a \pmod n$ has either 0 or 4 solutions modulo n.*

If the prime factorization $n = pq$ is known and if a is a quadratic residue modulo n then there exists a probabilistic algorithm to efficiently calculate the four solutions of $x^2 \equiv a \pmod n$.

If $n = pq$ with p, q distinct large odd primes then there is no known general probabilistic algorithm which can be applied to check efficiently if an integer a is a quadratic residue modulo n or not unless the prime factors p and q are known. The following theorem will provide a security equivalence for the Rabin cryptosystem.

Theorem 7.6.6. *Let p and q be two distinct primes and $n = pq$. If there exists a probabilistic algorithm to efficiently determine a square root b modulo n for every quadratic residue a modulo n, then there exists a probabilistic algorithm to factorize n.*

7.6.2 The Rabin Cryptosystem

We describe first the preparation necessary for Rabin encryption.

Rabin Cryptosystem Preparation
(1) Choose two large odd distinct primes p and q and calculate $n = pq$
(2) We take the plaintext message units $\mathcal{P}$ and the ciphertext $\mathcal{C}$ as $\mathbb{Z}_n{}^*$. Hence

$$\mathcal{P} = \mathcal{C} = \mathbb{Z}_n^*.$$

We next describe the encryption method.

Rabin Encryption

We assume that we have made the choices for p, q and have computed $n = pq$.

(1) The public key is n while the secret key is (p, q)
(2) The encryption works as follows:
   If $m \in \mathcal{P}$ then we calculate $\bar{c} = m^2$ in $\mathbb{Z}_n^*$.

Rabin Decryption

(1) We assume that we have made the choices for p, q and have computed $n = pq$.
(2) Given an encrypted message $\bar{c}$ the decryption proceeds as follows:
   Calculate with the help of the algorithm described in Theorem 7.6.3 the four square roots $\bar{x}_1, \bar{x}_2, \bar{x}_3, \bar{x}_4$ of $\bar{c}$ in $\mathbb{Z}_n^*$. Then determine those $\bar{x}_i$ for which the plaintext message makes sense.

We give a straightforward simple example.

Example 7.6.7. Let $p = 3, q = 7$ so that $n = 21$. Suppose that the message is $\overline{10} \in \mathbb{Z}_{21}^*$.
Calculate $\bar{c} = \overline{10}^2 = \overline{16}$ in $\mathbb{Z}_{21}^*$. The ciphertext message is then $\overline{16}$.
   Receiving $\overline{16}$ to decrypt we calculate with the help of the secret primes p and q (in this case we can calculate directly but in practice this would be infeasible) the four square roots of 16 modulo 21.
   The square roots of 16 modulo 3 are $\overline{a_3} = \overline{1}$ and $-\overline{a_3} = \overline{2}$. The square roots of 16 modulo 7 are $\overline{a_7} = \overline{4}$ and $-\overline{a_7} = \overline{3}$.
   Therefore we must consider the four combinations

$$w_1 = (\overline{1}, \overline{4}), \quad w_2 = (\overline{1}, \overline{3}), \quad w_3 = (\overline{2}, \overline{4}), \quad w_4 = (\overline{2}, \overline{3}).$$

Using the Chinese remainder theorem we then get the four square roots of 16 modulo 21:

$$\overline{x_1} = \overline{4}, \quad \overline{x_2} = \overline{10}, \quad \overline{x_3} = \overline{11}, \quad \overline{x_4} = \overline{17}.$$

The plaintext message m is then one of these and indeed is $\overline{10}$.

The Rabin Cryptosystem is clearly a valid cryptosystem. One of the four squareroots of a ciphertext must be the plaintext.

7.6.3 Security Equivalence of the Rabin Cryptosystem

Both the RSA algorithm and Rabin encryption use the difficulty of factoring $n = pq$ where p, q are large primes. In the RSA algorithm if the factorization problem is solvable then RSA is broken. That is if there is an algorithm to find p, q given $n = pq$, then the secret key can be determined. However, there is no proof that breaking RSA requires a solution of the factorization problem. Formally, let $n = pq, p, q, e, d$ be the parameters in the RSA protocol for a given user. The **factorization problem** asks given $n = pq$ determine p and q. The **RSA Problem** asks whether given n and e determine d. The solution to the factorization problem implies the solution to the RSA problem. However, it is not known whether a solution to the RSA problem must require a solution to the factorization problem.

However, Theorems 7.6.4, 7.6.5 and 7.6.6 show that finding square roots modulo $n = pq$ and factoring a product of two odd primes are probabilistically algorithmically equivalent. Therefore the security of the Rabin cryptosystem is equivalent to the difficulty of the factorization problem for two primes. Hence in distinction to RSA there is a security equivalence for the Rabin protocol. The following is really a restatement of the combined results in Theorems 7.6.4, 7.6.5 and 7.6.6.

Theorem 7.6.8. *Breaking the Rabin cryptosystem is probabilistically algorithmically equivalent to the factorization problem.*

We mention that the choice of the primes p and q in Rabin encryption should follow the same guidelines as for the RSA cryptosystem.

7.7 Session Keys and Mixed Encryption

In general, symmetric key encryption methods, such as AES and simple block and stream cipher methods, are much more efficient in terms of time and storage requirements, than public key methods. Therefore in real practice, what is used is **mixed encryption**. By this we mean that the message m is encrypted by some symmetric key method that is dependent on a key k called the **session key**. It is assumed that this encrypted message is **semantically secure**, that is secure against ciphertext only attacks. The session key k is transmitted, or shared, by some public key protocol. Notice that there are several different keys in this protocol. The users' public and private keys which are then used to share or transport the session key. The session key is then used to encrypt and decrypt the message. There are key spaces for the public and private keys used in the key transfer, and a key space used for the session key. These key spaces may or may not be the same. A mixed encryption system is often called a **digital envelope**.

In outline we have the following for real world mixed encryption.

(1) The communicating parties have agreed upon a symmetric key method, such as AES, that will be used for message encryption and decryption. It is assumed that encryption and decryption in this method is known once a given key is known. Each transmission session will use a different key called the **session key** so that once the session key is known the communicating parties can encrypt and de-

crypt messages based on it. It is further assumed that the encryption protocol is semantically secure.

(2) The users agree upon a public key method, such as Diffie-Hellman or RSA, and choose necessary parameters and public and private keys.

(3) The communicating parties then use the public key protocol to share or transport the session key k.

(4) Once the session key is known to the communicating parties, the message m can be encrypted and decrypted using the agreed upon symmetric key method.

(5) The security of the mixed encryption protocol is then dependent on both the security of the private key method as well as the security of the symmetric key method.

7.8 The RSA Signature Method

As explained in Section 7.2, the standard model for a public key cryptosystem provides a method for authentication and verification. This was also described in Chapter 4 when authentication protocols and digital signature protocols were introduced.

This type of verification can be done with any public key cryptosystem such as RSA. In practice, as explained in the last section, what is usually sent by this public key method is the session key to be used with some symmetric cryptosystem such as AES. Transmitting messages is usually done in some symmetric key fashion since message transmission becomes lengthy in a public key method (see Chapter 6). Here cryptographic hash functions play a role. Suppose that m is the true message and h is a hash function. What is done is to verify, using a public key system, $h(m)$, the hash value of the message. From the collision and preimage resistance of a hash function, verification of the hash function of the message provides verification of the message itself. Since the hash function value has a fixed length, this handles the problem of public key time transmission of messages.

We show how this can be done with an RSA cryptosystem. This is called the **RSA Signature Method**. Each party A and B has its own RSA parameters (n_A, d_A, e_A) and (n_B, d_B, e_B) respectively. Let h be a chosen hash function that acts on elements of $\mathbb{Z}_{n_A}$. This takes modular integers in $\mathbb{Z}_{n_A}$ and returns fixed length bit strings that can be considered as elements in $\mathbb{Z}_{n_A}$. Let m be the original message. This can also be a message unit for size requirements. We assume that the communicating parties each have m which has be transmitted by some other method and we are interested in the verification aspects. Since each of A and B have m, each knows the hash value $h(m)$. We assume that A has transmitted this message to B.

A calculates $x = h(m)^{d_A}$ (recall that d_A is A's secret key). A now sends x via RSA encryption to B using B's RSA parameters. That is A sends B the values x^{e_B} modulo n_B. B now decrypts using his d_B to find x and now computes x^{e_A} to get a value x_1 which is presumably the hash value of the transmitted message. B checks if $x_1 = h(m)$. If it is the message, m is verified from A. This verification procedure has the same security as

RSA. Further since x and m are both sent encrypted, an attacker has no knowledge of who sent the message, only knowledge of the ciphertext.

This RSA signature method handles all the verification requirements;

(1) B can be certain that the message came from A. Since they have determined the correct hash value $h(m)$ this must have come from A since d_A was used.

(2) B is verified to A since B, if he accepts the hash value, must have the original message,

(3) B cannot alter the original message since this will change the hash value which has been verified.

(4) The method provides a verification of undeniability. A cannot subsequently claim that she did not send the message. Only A can encrypt x with d_A.

7.9 Exercises

7.1 Given prime numbers p, q with $q < p$ and $n = pq$. For an RSA cryptosystem assume that $p-q$ is very small. Show that n can be factorized using the following procedure:

(a) Let $t \in \mathbb{N}$ be the smallest natural number with $t \geq \sqrt{n}$.

(b) If $t^2 - n$ is a square, that is, $t^2 - n = s^2$ for some $s \in \mathbb{N}$, then $p = t + s$ and $q = t - s$ provides the factorization.

(c) Otherwise take the next integer $t \geq \sqrt{n}$ and go back to (b).

7.2 Use the procedure of exercise 7.1 to factorize $n = 9898\,828\,507$.

7.3 A user has the public RSA key (n, e). By a security gap, the number $\phi(n)$ becomes known. Show that the user has to reject the key.

(a) Explain how the secret key d can be calculated, that is, the number d such that $ed \equiv 1 \pmod{\phi(n)}$.

(b) Explain how n can be factorized.

7.4 The colleagues of a company use an RSA-cryptosystem for their internal communication. To simplify the encryption and decryption they only use keys of the form $(n, 3)$.

Let $(n_1, 3), (n_2, 3), (n_3, 3)$ be three of these public keys with pairwise coprime $n_1, n_2, n_3 \in \mathbb{N}$ with $n_i \neq 1, i = 1, 2, 3$. One colleague uses these three keys to encrypt a coded message $m \in \{1, \ldots, \min\{n_1, n_2, n_3\} - 1\}$. He then sends the secret messages c_1, c_2, c_3 to the respective key holders. Here the secret message c_i was calculated with the respective public key $(n_i, 3)$ for $i = 1, 2, 3$.

Explain why in such a manner the security of the communication is endangered. Take as an example $n_1 = 391, n_2 = 319, n_3 = 204$ and $c_1 = 236, c_2 = 44$, $c_3 = 203$.

7.5 Let $(n, e) = (2047, 179)$ be the public RSA-key. A plaintext alphabet has the 26 letters $A, B, \ldots, Z$ and the empty sign $\cup$ between words. The plaintext mes-

sage c with ⊔ between words will be subdivided into double blocks with the empty sign at the end, if necessary, to get double blocks.
By the assignment

$$A \rightarrow 00, \quad B \rightarrow 01, \ldots, \quad Z \rightarrow 25, \quad ⊔ \rightarrow 26$$

each double block gives a block with 4 digits. We consider the 4 digit numbers as residue classes modulo 2047. Encryption with the public key $(2047, 179)$ gives the ciphertext message

$$\overline{1054}, \overline{92}, \overline{1141}, \overline{1571}, \overline{92}, \overline{832}$$

in the form of residue classes modulo 2047.
(a) Break the encryption by factoring 2047 and give the plaintext message.
(b) Why is the number 2047, besides the small size, a particularly unfavorable choice? Is it possible to break the encryption without factoring 2047?

7.6 This problem shows that if the plaintext number is too small in the RSA algorithm, the system can be broken
(a) Let $e \in \mathbb{N}, n_1, \ldots, n_e \in N$ be pairwise coprime and $x \in \mathbb{N}$ with $0 \le x \le n_i, 1 \le i \le e$. Let $c \in N$ with $c \equiv x^e \pmod{n_i}$ for $0 \le i \le e$ and $0 \le c \le n_1 n_2 \cdots n_e$. Then $c = x^e$.
(b) Use part (a) for a low-exponent attack on an RSA cryptosystem. A bank sends the same plaintext block as a plaintext number x to three different customers. Here the bank uses the three different public keys $n_1 = 143$, $n_2 = 391$ and $n_3 = 899$ but always uses the same encryption exponent 3. The attacker knows the ciphertext numbers $c_1 = 60, c_2 = 203$ and $c_3 = 711$. Show how the attacker can determine x by calculating roots.

7.7 Let $n \in \mathbb{N}, x, y \in \mathbb{Z}$ with $x^2 \equiv y^2 \pmod{n}$, and x not congruent to either y or $-y$ modulo n. Show that $gcd(x + y, n)$ and $gcd(x - y, n)$ are proper divisors of n.

7.8 Let $n = 7 \cdot 11 = 77$ and $e = 43$
(a) What is the order of the multiplicative group $\mathbb{Z}_{77}^*$.
(b) Calculate $d \in \mathbb{N}$ with $ed \equiv 1 \pmod{\phi(n)}$.
(c) The message $y = \overline{5}$ is encrypted with the RSA cryptosystem and the public key is $(77, 43)$. Calculate the plaintext x.

7.9 The plaintext message x is encrypted with the two RSA keys $(551, 5)$ and $(551, 11)$. The resulting ciphertexts are respectively $\overline{277}$ and $\overline{429}$. From this calculate x.

7.10 Let $n = 11 \cdot 23 = 253$.
(a) Encrypt the message $\overline{36}$ using the Rabin cryptosystem with public key n.
(b) Decrypt the ciphertext $\overline{36}$ using the Rabin encryption with the secret key $(11, 23)$.

7.11 Let $n = 124573$ be the public key of the Rabin cryptosystem. An attacker knows an algorithm to calculate squareroots modulo n. For $c = \overline{113}^2 = \overline{12769}$, using his algorithm he obtains the squareroot $\overline{110\,459}$. Use this information to determine the prime factorization of n.

7.12 Let $n = 2 \cdot 23 + 1 = 47$ and $g = 5$.

(a) What is the order of $\overline{g}$ in the multiplicative group $\mathbb{Z}_n^*$?

(b) Let $a = \overline{16}$ and $b = \overline{9}$ be the secret keys of Alice and Bob. What are the public keys A and B and the common public key k by the Diffie-Hellman public key exchange protocol?

(c) Bob will encrypt the message $x = \overline{33}$ with (n, g, A) and the secret key b using ElGamal encryption. What is the secret message (y, B)?

7.13 Justify that for an RSA cryptosystem the choice of $n = pq$ with $|p - q| < 1000$ is not suitable.

7.14 Alice and Bob agree on the following public key cryptosystem:

(1) Alice chooses $a, b \in \mathbb{Z}$ and calculates $M = ab - 1$. Then Alice chooses two integers $a', b' \in \mathbb{Z}$ and calculates $e = a'M + a$ and $d = b'M + b$. She then calculates $n = \frac{ed-1}{M}$.

(2) Alice publishes the pair (n, e). The secret key is d.

(3) Bob wants to send a message $m \in \{0, 1, \ldots, n - 1\}$ to Alice. He calculates $c \equiv em \pmod{n}$ and sends c to Alice.

(4) She decrypts the message by calculating cd modulo n.

(a) Show that this is a valid cryptosystem, that is, Alice gets the message.

(b) How can this cryptosystem be used for digital signatures?

(c) Break this cryptosystem.

8 Elliptic Curve Cryptography

8.1 The ElGamal and Elliptic Curve Encryption System

The standard public key systems that we have described so far, Diffie-Hellman, El-Gamal, RSA and Rabin, require very large key spaces. In an attempt to use the same ideas but reduce the key space size it was suggested that Diffie-Hellman be applied to other abelian groups. To accomplish this, algebraic geometry was introduced into cryptography. In 1985, Neil Koblitz, and independently Victor Miller, suggested the use of elliptic curves over finite fields, and their corresponding groups, as possible cryptographic platforms. These methods have been quite successful and result, in many cases, in faster encryption and smaller key spaces than standard RSA methods. First, let us recall the basic ElGamal system and then we must introduce elliptic curves.

In 1984, T. ElGamal devised a method to turn the Diffie-Hellman key exchange protocol into a public key encryption protocol. This is now known as **ElGamal encryption**. We discussed this in detail in Section 6.4. The basic scheme for an ElGamal encryption system is the following. Each user chooses a large prime p and a generator g for the cyclic group $\mathbb{Z}_p^*$. Given a large prime p there is a fixed efficiently invertible procedure to encrypt a plaintext into residue classes within $\mathbb{Z}_p^*$, the unit group within $\mathbb{Z}_p$. Although each user may choose different primes this procedure is known once p is known.

For each message transmission the user's public key is (p, g, A) where $A = g^a$ for some integer a.

The encryption and decryption works as follows. Suppose that Bob wants to send a message to Alice. Alice's public key, which is public knowledge, is (p, g, A) as above. The message is m and, as above, is encrypted in some workable efficient manner within $\mathbb{Z}_p^*$, that is the message is encrypted in a manner known to all users (once p is given) as an integer in $0, 1, \ldots, p - 1$. Bob now randomly chooses an integer b and computes $B = g^b$. He now sends (B, mC) to Alice where $C = g^{ab}$. Notice that C is the common shared key in the Diffie-Hellman key exchange and in the encryption this is multiplied by the message m.

To decrypt Alice first uses B to determine the common shared key C. Since $B = g^b$ and she knows $A = g^a$ she knows $C = g^{ab}$ for the same reasons as the Diffie-Hellman key exchange works. Since she knows $C = g^{ab}$ and she knows the modulus p she can compute the inverse $g^{-(ab)}$. This is efficient since it only requires one exponentiation modulo p. She then multiplies $mC = mg^{ab}$ by g^{-ab} to obtain the message m.

Although ElGamal proposed using the cyclic groups $\mathbb{Z}_p^*$ for large primes p, this type of encryption can be used in any cyclic group where the discrete log problem is assumed hard. If the group is a cyclic group within the group of an elliptic curve, ElGamal encryption becomes the basis for **elliptic curve cryptography**. Before discussing elliptic curve cryptography in detail, we must introduce the necessary algebraic geometric material concerning elliptic curves.

8.2 Elliptic Curves

We start with a fixed finite field F of characteristic not equal to 2 or 3. For cryptographic purposes an **elliptic curve** over F consists of the points $(x, y) \in F^2$ satisfying the equation

$$C : y^2 = x^3 + ax + b; \quad a, b \in F$$

together with a distinguished point at infinity denoted by $\mathcal{O}$. A group operation can be placed on the points on C to form an abelian group, usually denoted $E(C)$, called the **elliptic curve group over** C. There must be additional conditions met by the elliptic curve C to be useful in cryptography and we will discuss these later.

The standard public key cryptographic protocols such as Diffie-Hellman, ElGamal and RSA use as platforms the cyclic groups in modular rings and, as explained in Chapter 6, are based on supposedly hard number theoretic problems. Miller and Koblitz noted that these problems are also difficult in elliptic curve groups. Hence the methods of public key encryption and key exchange, embodied in the above protocols, can be mimicked with some great advantages within the groups of elliptic curves. The use of elliptic curve RSA encryption is basically of only theoretical interest since the difficulty of the factoring problem turns out to be equally difficult in number theory and elliptic curves. However, the discrete log problem, and hence Diffie-Hellman and ElGamal, provides a distinct advantage for elliptic curve groups.

Before introducing elliptic curve cryptography, we discuss the algebraic preliminaries necessary for the study of elliptic curves.

8.2.1 Fields and Field Extensions

Elliptic curve cryptography involves elliptic curves over finite fields. We first look at the necessary material from fields and field extensions. For more details and proofs see the book [CFR1].

Recall that a **field** F is a commutative ring with an identity such that every non-zero element has a multiplicative inverse (see [CFR1]). The rationals $\mathbb{Q}$, the reals $\mathbb{R}$ and the modular rings $\mathbb{Z}_p$, where p is a prime, all form fields.

The characteristic of a field F denoted char(F) is the smallest $n \in \mathbb{N}$ such that $n \cdot 1 = 0$. If no such n exists then the characteristic is zero. It is clear that the characteristic of $\mathbb{Q}$ is 0. Further if p is a prime the characteristic of the finite field $\mathbb{Z}_p$ is p. Since there are no zero divisors (non-zero elements whose product is zero) in a field we get the following.

Theorem 8.2.1. *The characteristic of a field is zero or a prime.*

If a field F has characteristic zero then it must be infinite. Hence if F is a finite field its characteristic must be a prime p.

Lemma 8.2.2. *If F is a finite field then* char$(F) = p$ *for some prime p.*

A field F is a **prime field** if its only non-trivial subfield is the field itself. It is easy to show that the rationals $\mathbb{Q}$ and the finite fields $\mathbb{Z}_p$ are prime fields. However, the real number field $\mathbb{R}$ is not a prime field.

   If F and F' are fields with F a subfield of F', then F' is an **extension field**, or **field extension**, or simply an **extension**, of F. F' is then a vector space over F and the **degree of the extension** is the dimension of F' as a vector space over F. We denote the degree by $|F' : F|$. If the degree is finite, that is, $|F' : F| < \infty$, so that F' is a finite-dimensional vector space over F, then F' is called a **finite extension** of F.

   From vector space theory we easily obtain that the degrees are multiplicative. Specifically:

Lemma 8.2.3. *If $F \subset F' \subset F''$ are fields with F'' a finite extension of F then $|F' : F|$ and $|F'' : F'|$ are also finite, and $|F'' : F| = |F'' : F'||F' : F|$.*

In the case of the lemma we say that F' is an **intermediate field** (when F and F'' are understood) and F is the **ground field**.

Lemma 8.2.4. *The field $\mathbb{C}$ is a finite extension of $\mathbb{R}$, but the field $\mathbb{R}$ is an infinite extension of $\mathbb{Q}$.*

The elements $1, i$ form a basis for $\mathbb{C}$ over $\mathbb{R}$. Hence the dimension $\mathbb{C}$ over $\mathbb{R}$ is 2 and $\mathbb{C}$ is a finite extension of $\mathbb{R}$. The fact that $|\mathbb{R} : \mathbb{Q}|$ is infinite can be shown using the existence of transcendental numbers (see [CFR1]).

   Let F be a field. Denote by $F[x]$ the set of all polynomials with coefficients from F. A polynomial $f(x) \in F[x]$ of degree ≥ 1 is **irreducible** if it cannot be factored into two polynomials each of smaller degree. A **root** or **zero** of a polynomial $f(x) \in F[x]$ is an element c with $f(c) = 0$. If c is a root of $f(x)$ then the linear factor $x - c$ divides $f(x)$.

Lemma 8.2.5. *If $f(x) \in F[x]$ and $f(c) = 0$ for $c \in F$ then $f(x) = (x - c)h(x)$ for some $h(x) \in F[x]$.*

Suppose F' is an extension field of F and $\alpha \in F'$. Then α is **algebraic over F** if there exists a polynomial $0 \neq p(x) \in F[x]$ with $p(\alpha) = 0$. This means that the element α is a root of a polynomial with coefficients in F. If every element of F' is algebraic over F, then F' is an **algebraic extension** of F. If $\alpha \in F'$ is non-algebraic over F, then α is called **transcendental** over F. A non-algebraic extension is called a **transcendental extension**.

Lemma 8.2.6. *Every element of F is algebraic over F.*

Proof. If $f \in F$ then $p(x) = x - f \in F[x]$ and $p(f) = 0$. □

The tie-in to finite extensions is via the following theorem (see [CFR1] for a proof).

Theorem 8.2.7. *If F' is a finite extension of a field F, then F' is an algebraic extension.*

If a polynomial $f(x) \in F[x]$ of degree 2 or higher has a root in F then it is reducible. Hence if $f(x) \in F[x]$ of degree greater than one is irreducible it cannot have root in F. The next result, due to Kronecker, is fundamental because it says that given any irreducible polynomial $f(x) \in F[x]$ we can construct an extension field F' of F in which $f(x)$ has a root.

Theorem 8.2.8 (Kronecker's Theorem). *Let F be a field and $f(x) \in F[x]$ an irreducible polynomial over F. Then there exists a finite extension F' of F where $f(x)$ has a root.*

The proof of Kronecker's Theorem follows from the following construction. There is a **division algorithm** in $F[x]$. Given $f(x), g(x) \in F[x]$ then there exist $q(x), r(x) \in F[x]$ such that

$$f(x) = q(x)g(x) + r(x)$$

where $r(x) = 0$ or $\deg r(x) < \deg g(x)$. From this we can define the greatest common divisor or gcd of $g(x), f(x)$ as a monic polynomial $d(x)$ such that $d(x) \mid g(x)$, $d(x) \mid f(x)$ and if $d_1(x)$ is another common divisor of $f(x)$ and $g(x)$ then $d_1(x) \mid d(x)$. Then just as in the integers we get the following results (see [CFR1] for proofs).

Theorem 8.2.9. *Given $f(x), g(x) \in F[x]$ then their gcd exists, is unique and can be written as a linear combination, that is*

$$d(x) = k(x)f(x) + h(x)g(x)$$

for some $h(x), k(x) \in F[x]$.

From this we also get, using the division algorithm, that the Euclidean algorithm works, just as in the integers to find the gcd of two polynomials. We also get Euclid's Lemma

Lemma 8.2.10. *If $p(x)$ is an irreducible polynomial and $p(x) \mid f(x)g(x)$ then $p(x) \mid f(x)$ or $p(x) \mid g(x)$.*

As a consequence we obtain that $F[x]$ has unique factorization into irreducible polynomials, that is $F[x]$ is a unique factorization domain. The proof follows the same general outline as the proof of the Fundamental Theorem of Arithmetic.

Theorem 8.2.11. *$F[x]$ is a unique factorization domain. That is given $f(x) \in F[x]$ then $f(x)$ can be written as a product of irreducible polynomials and this factorization is unqiue up to ordering and unit factors.*

Kronecker's Theorem follows from this construction in the following manner. If $f(x)$ is an irreducible polynomial then

$$\langle f(x)\rangle = \{g(x)f(x) \mid g(x) \in F[x]\}$$

forms a maximal ideal in $F[x]$ and hence $K = F[x]/\langle f(x)\rangle$ forms a field. $F \subset K$ and $x + \langle f(x)\rangle \in K$ is a root of $f(x)$ (see [CFR1]). If we denote this root by α then K has a basis $1, \alpha, \ldots, \alpha^{n-1}$ as a vector space over F. Hence the degree of this extension is precisely the degree of the irreducible polynomial.

If $F = \mathbb{Z}_p$ then for each $n \geq 1$ there exists an irreducible polynomial of degree n, and hence for each n there is a finite field of size p^n. This can be shown to be unique up to isomorphism and is denoted $\mathbb{F}_{p^n}$ and called the **Galois field** of order p^n, that is with p^n elements.

From Kronecker's Theorem, we have seen that given an irreducible polynomial over a field F, we can always find a field extension in which this polynomial has a root. By repeating this procedure we can push this further to obtain field extensions in which a given polynomial has all its roots.

Definition 8.2.12. If $0 \neq f(x) \in F[x]$ and F' is an extension field of F, then $f(x)$ **splits** in F' (F' may be F) if $f(x)$ factors into linear factors in $F'[x]$. Equivalently, this means that all the roots of $f(x)$ are in F'. F' is a **splitting field** for $f(x)$ over F if F' is the smallest extension field of F in which $f(x)$ splits. (A splitting field for $f(x)$ is the smallest extension field in which $f(x)$ has all its possible roots.) F' is a **splitting field** over F if it is the splitting field for some finite set of polynomials over F.

Theorem 8.2.13. *If $0 \neq f(x) \in F[x]$, then there exists a splitting field for $f(x)$ over F.*

Pushing this further, we can construct a field, called the **algebraic closure** of F, which we will denote by $\overline{F}$, which is the smallest field where every $f(x) \in F[x]$ splits. We say that such a field F' is **algebraically closed**. Note that in this language the famous **Fundamental Theorem of Algebra** says that the complex number field $\mathbb{C}$ is algebraically closed. The next result gives several clearly equivalent formulations of being algebraically closed.

Theorem 8.2.14. *Let F be a field. Then the following are equivalent:*
(1) *F is algebraically closed.*
(2) *Every non-constant polynomial $f(x) \in F[x]$ splits in $F[x]$.*
(3) *F has no proper algebraic extensions, that is, there is no algebraic field extension E with $F \subset E$ and $F \neq E$.*

Example 8.2.15. For this example we assume the Fundamental Theorem of Algebra, that is, that $\mathbb{C}$ is algebraically closed. Note that $\mathbb{C}$ is not the algebraic closure of $\mathbb{Q}$ since $\mathbb{C}$ is not algebraic over $\mathbb{Q}$. However, the field of complex algebraic numbers $A_{\mathbb{C}}$, that is, the set of complex numbers which are algebraic over $\mathbb{Q}$, is the algebraic closure of $\mathbb{Q}$. To see this, notice that $A_{\mathbb{C}}$ is algebraic over $\mathbb{Q}$ by definition. Now we show that it is algebraically closed. Let $f(x) \in A_{\mathbb{C}}[x]$. If α is a root of $f(x)$, then $\alpha \in \mathbb{C}$, and then α is also algebraic over $\mathbb{Q}$ since each element of $A_{\mathbb{C}}$ is algebraic over $\mathbb{Q}$. Therefore, $\alpha \in A_{\mathbb{C}}$ and $A_{\mathbb{C}}$ is algebraically closed.

More generally, if K is an extension field of F and K is algebraically closed, then the algebraic closure of F in K is the algebraic closure of F.

We need the following concept.

Definition 8.2.16. Let F', F'' be extension fields of F. An **F-isomorphism** is an isomorphism $\sigma : F' \to F''$ such that $\sigma(f) = f$ for all $f \in F$. That is, an F-isomorphism is an isomorphism of the extension fields that **fixes each element of the ground field.** If F', F'' are F-isomorphic, we denote this relationship by $F' \underset{F}{\cong} F''$.

Theorem 8.2.17. *Every field F has an algebraic closure, and any two algebraic closures of F are F-isomorphic.*

To study elliptic curves we will need the following.

Theorem 8.2.18. *Let F be a field with $char(F) \neq 2, 3$ and let $f(x) = x^3 + ax + b \in F[x]$. Then $f(x)$ has three pairwise distinct zeros in the algebraic closure of F if and only if $4a^3 + 27b^2 \neq 0$ in F. The element*

$$\Delta = -4a^3 - 27b^2$$

*is called the **discriminant** of $f(x)$.*

The result follows from Cardano's formula (see [CFR1]).

8.2.2 Elliptic Curves

We now introduce **elliptic curves** over a field F. Elliptic curves play an important role in many areas of mathematics and they were crucial in the proof, given by A.Wiles, of Fermat's Last Theorem. In the next two sections we look at the material we need for elliptic curve cryptography. We refer to [Sil] for further information on elliptic curves in general. We consider only the case where the field F has characteristic $\neq 2, 3$. In the case of characteristic 2, the considered Weierstrass form (normal form) that we will introduce shortly, is different. For information about elliptic curves over fields of characteristic 2 or 3, see [Sil] or [Was].

Definition 8.2.19. Let F be a field with $char(F) \neq 2, 3$. Given an equation of the form

$$C : y^2 + a_1 xy + a_3 y = x^3 + a_2 x^2 + a_4 x + a_6 \tag{$\star\star$}$$

with $a_1, a_2, a_3, a_4, a_6 \in F$ then the set

$$C(F) = \{(x, y) \in F^2 \mid (x, y) \text{ satisfies } (\star\star)\}$$

is called a **(planar affine) cubic curve** over F.

We note that the numeration of the coefficients may look a bit strange but there are historical reasons for it and it is used in the literature.

Under an appropriate coordinate transformation a cubic curve can be put into a standard form.

Theorem 8.2.20. *Let F be a field with* $\mathrm{char}(F) \neq 2, 3$. *Then there exists a coordinate transformation*

$$x \mapsto \alpha_1 x + \alpha_2 y + \alpha_3, \quad y \mapsto \beta_1 x + \beta_2 y + \beta_3$$

with $\alpha_i, \beta_i \in F$ *which carries the equation* (**) *into an equation of the form*

$$y^2 = x^3 + ax + b \quad \text{with} \quad a, b \in F.$$

The equation above is called the **Weierstrass form** *of the cubic curve.*

The proof is computational and uses only transformation methods from linear algebra. For the remainder of this chapter we always assume that a cubic curve is given in its Weierstrass form.

Let F be a field and $\overline{F}$ be its algebraic closure. Let $C(F) : y^2 = x^3 + ax + b$ be a cubic curve over F. Consider the polynomial $g(x, y) = y^2 - x^3 - ax - b$ in two variables over $\overline{F}$ and the corresponding extended curve

$$C(\overline{F}) : y^2 = x^3 + ax + b.$$

A point $(r, s) \in \overline{F}^2$ is called a **singular point** if

$$g(r, s) = 0, \quad \frac{\partial g}{\partial x}(r, s) = 0, \quad \frac{\partial g}{\partial y}(r, s) = 0.$$

If $g(x, y) = y^2 - x^3 - ax - b$ and $(r, s) \in \overline{F}^2$ is a singular point then

$$s^2 - r^3 - ar - b = 0, \quad -3r^2 - a = 0 \quad \text{and} \quad 2s = 0.$$

It follows that $a = -3r^2$ and $b = 2r^3$ and hence

$$\Delta = -4a^3 - 27b^2 = 4 \cdot 27 \cdot r^6 - 4 \cdot 27 \cdot r^6 = 0,$$

that is the discriminant Δ of $x^3 + ax + b$ is 0. We say that the cubic curve $C(F)$ is **non-singular** or **smooth** if the extended curve $C(\overline{F})$ has no singular point. The above calculations easily show that $C(F)$ is smooth if and only if $\Delta \neq 0$.

If $\Delta = 0$ then the above equation defines a point $(r, s) \in \overline{F}^2$ that is singular. Combined with Theorem 8.2.2.1 we get the following.

Theorem 8.2.21. *Let F be a field with* $\mathrm{char}(F) \neq 2, 3$ *and let* $C(F) : y^2 = x^3 + ax + b$ *be a cubic curve in Weierstrass form over F. The the following are equivalent:*
(1) *The polynomial* $f(x) = x^3 + ax + b$ *has three pairwise distinct zeros in* $\overline{F}$, *the algebraic closure of F.*

(2) *The discriminant $\Delta = -4a^3 - 27b^2 \neq 0$ in F,*

(3) *C(F) is smooth.*

Definition 8.2.22. Let F be a field with char(F) $\neq 2, 3$. Then an **elliptic curve** is a smooth curve $C(F) : y^2 = x^3 + ax + b$ over F.

An elliptic curve $C(F)$ is smooth and therefore from Theorem 8.3.2 it follows that the discriminant $\Delta = -4a^3 - 27b^2 \neq 0$ and that $x^3 + ax + b$ has three distinct zeros in $\overline{F}$.

To provide an explanation of the name ellipitc curve, we briefly describe the connection with elliptic functions.

The Weierstrass p-function is defined as

$$p(z) = \frac{1}{z^2} + \sum_{w \neq 0} \left(\frac{1}{(z - w)^2} - \frac{1}{w^2} \right),$$

where $z \in \hat{\mathbb{C}} = \mathbb{C} \cup \{i\infty\}$, $w_1, w_2 \in \mathbb{C} \setminus \{0\}$, $\frac{w_1}{w_2} \notin \mathbb{R}$ and $w = mw_1 + nw_2$, $m, n \in \mathbb{Z}$.

The function $p(z)$ is an elliptic or doubly periodic function, that is, we have $p(z + w) = p(z)$ for all $w \in \{mw_1 + nw_2 | m, n \in \mathbb{Z}\}$. The function $p(z)$ satisfies the differential equation

$$(p'(z))^2 = 4(p(z))^3 - g_2 p(z) - g_3$$

where g_2, g_3 are constants depending only on w_1, w_2.

Now let $F = \mathbb{C}$ and $C(F) : y^2 = x^3 + ax + b$ with $-4a^2 - 27b^2 \neq 0$ be an elliptic curve over $\mathbb{C}$. If we replace x by $\tilde{x} = \frac{x}{4^{1/3}}$, a by $g_2 = -4^{1/3}a$ and b by $-g_3$ then we get

$$y^2 = 4(\tilde{x})^3 + 4^{1/3}a\tilde{x} + b = 4(\tilde{x})^3 - g_2\tilde{x} - g_3.$$

From $\Delta = -4a^3 - 27b^2 \neq 0$ we get that $g_2^3 - 27g_3^2 \neq 0$.

There exist $w_1, w_2 \in \mathbb{C} \setminus \{0\}$ with $\frac{w_1}{w_2} \notin \mathbb{R}$ and a parametrization $y = p'(z)$, $\tilde{x} = p(z)$ where $p(z)$ is the Weierstrass p-function corresponding to w_1, w_2. For more details see [Sch].

8.2.3 Elliptic Curve Groups

We will now write $E(F)$ for an elliptic curve over a field F. This implies that the curve is smooth and the discriminant is non-zero. We will add on a point at infinity with we will denote by $\mho$ and denote by $\overline{E}(F)$ the extended curve $E(F) \cup \mho$. On the points of the extended elliptic curve $\overline{E}(F)$ we will define an addition that will make the points form an abelian group, called the **elliptic curve group**. It is on this group that we will construct an elliptic curve cryptosystem. If the underlying field is the reals $\mathbb{R}$ this operation is very easily motivated by geometry so we will do this first. From the equations that arise in $\mathbb{R}$ we will define the addition over any field F of characteristic $\neq 2, 3$. Elliptic curve cryptography is usually done over a finite field.

Consider now an elliptic curve

$$E(F) : y^2 = x^3 + ax + b$$

over the real numbers $\mathbb{R}$ and let $\overline{E}(F) = E(F) \cup \mathcal{O}$. We will geometrically define a binary operation on $\overline{E}(F)$. Now the elliptic curve is given by a curve in the complex plane. Suppose that P, Q are two points on the curve. Then we define the following:

(1) If $P = (x, y)$ then $-P = (x, -y)$, and we define $P + (-P) = \mathcal{O}$.
(2) If $P \neq Q \neq -P$ then the line $\overline{PQ}$ will hit the curve at a third point $R = (x, y)$. Let $R^* = (x, -y)$ be the reflection of R through the x-axis. Then we define $P + Q = R^*$.
(3) $P + P = R^*$ where we do the same procedure as when $P \neq Q$ but use the tangent line at P.
(4) $P + \mathcal{O} = \mathcal{O} + \mathcal{P} = \mathcal{P}$.
(5) $\mathcal{O} + \mathcal{O} = \mathcal{O}$ and $-\mathcal{O} = \mathcal{O}$.

These operations over $\mathbb{R}$ can easily be put into coordinate form. Suppose $P = (x_1, y_1)$, $Q = (x_2, y_2)$ and $P + Q = (x_3, y_3)$. Then if $P \neq Q$, $x_2 \neq x_1$ then

$$x_3 = \left(\frac{y_2 - y_1}{x_2 - x_1}\right)^2 - x_1 - x_2 \quad \text{and} \quad y_3 = -y_1 + \left(\frac{y_2 - y_1}{x_2 - x_1}\right)(x_1 - x_3). \tag{$*$}$$

If $P \neq Q$ and $x_2 = x_1, y_2 \neq y_1$ then

$$P + Q = \mathcal{O}. \tag{$**$}$$

If $P = Q$ so $(x_3, y_3) = 2P$, $y_1 \neq 0$, then

$$x_3 = \left(\frac{3x_1^2 + a}{2y_1}\right)^2 - 2x_1 \quad \text{and} \quad y_3 = -y_1 + \left(\frac{3x_1^2 + a}{2y_1}\right)^3 (x_1 - x_3). \tag{$***$}$$

If $P = Q$ so $(x_3, y_3) = 2P$, $y_1 = 0$ then

$$P + Q = \mathcal{O}. \tag{$****$}$$

Now we consider an elliptic curve $E(F) : y^2 = x^3 + ax + b$ over a finite field F with discriminant $\Delta = -4a^3 - 27b^2 \neq 0$ in F.

Theorem 8.2.23. *Let F be a field with characteristic not equal 2 or 3 and let $E(F)$ be an elliptic curve over F. Let $\mathcal{O}$ denote the point at infinity and let $\overline{E}(F) = E(F) \cup \mathcal{O}$. If $P_1 = (x_1, y_1)$ and $P_2 = (x_2, y_2)$ with $P_1, P_2 \neq \mathcal{O}$ are two points in $\overline{E}(F)$ then we define the point $P_1 + P_2$ and $-P_1$ by the coordinate equations given in $(*)$, $(**)$, $(***)$ and $(****)$. Addition with $\mathcal{O}$ is defined to make $\mathcal{O}$ an additive identity. Then:*

(a) *This addition is a valid binary operation of $\overline{E}(F)$, that is, the addition of two points in $\overline{E}(F)$ is again in $\overline{E}(F)$.*
(b) *The set $\overline{E}(F)$ together with this operation forms an abelian group. If F is a finite field this is then a finite abelian group.*

Proof. (a) is purely computational and can be done with a suitable computer algebra system like CoCoA (see [KR1] and [KR2]).

(b) The existence of an additive identity and inverses are built into the definition of the operation. Similarly the commutativity of the operation follows easily from the definition. The associativity of the operation can be done computationally with CoCoA. A formal theoretical proof can be found in the book by Knapp on elliptic curves [Kna]. $\quad\square$

Note that if $n \geq 3$ is an integer with $(6, n) = 1$ then we may consider cubic curves

$$C(\mathbb{Z}_n) = \{(x, y) \in \mathbb{Z}_n^2 \mid y^2 = x^3 + ax + b, \ a, b \in \mathbb{Z}_n\}.$$

We call $C(\mathbb{Z}_n)$ an elliptic curve over $\mathbb{Z}_n$ if $\overline{4a}^3 + \overline{27b}^2 \neq 0$ in $\mathbb{Z}_n$. With this condition, we get that for each prime divisor p of n, the cubic curve

$$E_p : y^2 = x^3 + ax + b$$

is a cubic curve over $\mathbb{Z}_p$, if we just reduce modulo p. We may try to define an operation $+$ as in the theorem above. Let $P_1 = (x_1, y_1), P_2 = (x_2, y_2)$ with $x_1 \neq x_2$ or $x_1 = x_2, y_1 = y_2 \neq \overline{0}$ be elements of $C(\mathbb{Z}_n)$. If $x_1 - x_2 \neq \overline{0}$ or $y_1 \neq \overline{0}$ respectively are not invertible in $\mathbb{Z}_n$, then the operation is not well-defined and n is composite. Then $d = (x_1 - x_2, n)$ and $d = (y, n)$ respectively are divisors of n. This is roughly the idea for the elliptic curve factorization algorithm for integers $n \geq 1$ by H. W. Lenstra [Len]. One has to start with an elliptic curve $C(\mathbb{Z}_n) : y^2 = x^3 + ax + b$ which contains elements $(x, y) \in \mathbb{Z}_n^2$; for instance with such an elliptic curve with $b = 1$ then $(0, 1) \in C(\mathbb{Z}_n)$.

Now again let F be a finite field with $\mathrm{char}(F) \neq 2, 3$ and $\overline{E}(F)$ the elliptic curve group for the elliptic curve $E(F)$. As usual we write

$$nP = P + P + \cdots + P, \quad n \text{ times for } n \in \mathbb{N},$$
$$P_1 - P_2 = P_1 + (-P_2),$$

and

$$(-n)P = n(-P) \text{ for } n \in \mathbb{N}.$$

We want to use the abelian group $\overline{E}(F)$ as a platform group for a public key cryptosystem based on the additive version of the discrete log problem in $\overline{E}(F)$. In additive notation this is, given P and nP determine n. Constructing this cryptosystem we will do later in this chapter. Here we consider the following questions that are important for cryptographic purposes.

(1) How can we in general find an $\overline{E}(F)$ over a finite field F with enough elements?
(2) How can we calculate the order $|P|$ of an element $P \in \overline{E}(F)$ with F a finite field?
(3) How can we calculate the overall order $|\overline{E}(F)|$ with F a finite field?

Before we consider these questions we give some examples.

Example 8.2.24. Let $F = \mathbb{Z}_{11}$ and $y^2 = x^3 + x + \overline{6}$. This equation defines an elliptic curve over $\mathbb{Z}_{11}$ since

$$\Delta = -(\overline{4} \cdot \overline{1}^3 + \overline{27} \cdot \overline{6}^2) = \overline{3} \neq \overline{0} \text{ in } \mathbb{Z}_{11}.$$

In a systematic manner we find that

$$\overline{E}(\mathbb{Z}_{11}) = \{\mathcal{O}, (\overline{2}, \overline{4}), (\overline{2}, \overline{7}), (\overline{3}, \overline{5}), (\overline{3}, \overline{6}), (\overline{5}, \overline{2}), (\overline{5}, \overline{9}),$$
$$(\overline{7}, \overline{2}), (\overline{7}, \overline{8}), (\overline{8}, \overline{3}), (\overline{8}, \overline{8}), (\overline{10}, \overline{2}), (\overline{10}, \overline{9})\}.$$

Therefore $|\overline{E}(\mathbb{Z}_{11})| = 13$. Since 13 is a prime, it follows that $\overline{E}(\mathbb{Z}_{11})$ is a cyclic group generated by any non-zero element, for instance by $(\overline{2}, \overline{4})$.

Example 8.2.25. Let $F = \mathbb{Z}_5$ and $y^2 = x^3 + x$. This equation defines an elliptic curve over $\mathbb{Z}_5$. Again in a systematic manner we find that

$$\overline{E}(\mathbb{Z}_5) = \{\mathcal{O}, (\overline{0}, \overline{0}), (\overline{2}, \overline{0}), (\overline{3}, \overline{0})\}.$$

Hence

$$\overline{E}(\mathbb{Z}_5) = \mathbb{Z}_2 \times \mathbb{Z}_2 = V_4$$

the Klein four group.

If we take the same equation over $F = \mathbb{Z}_7$ instead of $\mathbb{Z}_5$, then the elliptic curve group is cyclic of order 8.

These types of groups, cyclic groups and direct products of two cyclic groups, are typical for elliptic curve groups over finite fields.

8.2.4 The Order of an Elliptic Curve Group

We now present some theorems about the structure and order of elliptic curve groups over finite fields. For this section, F is a finite field, so F is the Galois field $F = \mathbb{F}_q$ where $q = p^n$ for some prime p and $n \geq 1$. The structure and order are relevant to the answers to the questions we presented in the last section, and to the development of elliptic curve cryptosystems. For proofs we need a great deal more informations about the theory of elliptic curves and for the proofs we solely provide references.

For this section we assume that $F = \mathbb{F}_q, q = p^n$ for some prime $p \geq 5$ and $n \geq 1$. Since the multiplicative group F^* is cyclic we get

$$|\{x^2 \mid x \in F^*\}| = \frac{q-1}{2}.$$

Therefore we expect about q elements in $\overline{E}(F)$.

It is easy to provide an upper bound for $|\overline{E}(F)|$. Consider $f(x) = x^3 + ax + b$. If $f(x)$ is a square in F, that is $f(x) = y^2$ in F, then we get the two points (x, y) and $(x, -y)$ on $E(F)$. Together with the element $\mathcal{O}$ we then get the upper bound $|\overline{E}(F)| \leq 2q + 1$. A much stronger result is given by the following.

Theorem 8.2.26 (Hasse's Theorem). *Let $F = \mathbb{F}_q$ with $q = p^n$, p prime and $n \geq 1$ and $E(F) : y^2 = x^3 + ax + b$ be an elliptic curve. Then*

$$q + 1 - 2\sqrt{q} \leq |\overline{E}(F)| \leq q + 1 + 2\sqrt{q}.$$

A proof of Hasse's Theorem can be found in the book by J.H. Silverman [Sil].

The number $t = q + 1 - |\overline{E}(F)|$ is called the **trace of the Frobenius map**. We discuss this later for the case $q = p \geq 5$. The elliptic curve $E(F)$ is called **super singular** if $p \mid t$. We will see later that super singular elliptic curves are not suitable for cryptographic purposes.

Theorem 8.2.27. *Let $F = \mathbb{F}_q$ with $q = p^n$, p prime and $n \geq 1$. Then there exists an elliptic curve $E(F)$ with $|\overline{E}(F)| = q + 1 - t$ if and only if one of the following conditions hold:*
(1) *t is non-congruent to 0 modulo p and $|t| \leq 2\sqrt{q}$. In this case $E(F)$ is not super singular.*
(2) *2 does not divide n and $t = 0$.*
(3) *2 divides n and $t = 0$ and p is not congruent to 1 modulo 4.*
(4) *2 divides n and $t^2 = q$ and p is not congruent to 1 modulo 3.*
(5) *2 divides n and $t^2 = 4q$.*

A proof is given in the paper by Waterhouse [Wat].

Corollary 8.2.28. *Let $F = \mathbb{Z}_p$ with p prime and $p \geq 5$. Let*

$$I = [p + 1 - 2\sqrt{p}, p + 1 + 2\sqrt{p}] \cap \mathbb{N}.$$

Then there exists for each $k \in I$ at least one elliptic curve $E(\mathbb{Z}_p)$ with $|\overline{E}(\mathbb{Z}_p)| = k$.

Proof. Let $k = p + 1 - t \in I$. Since $q = p \geq 5$ we have $|t| < p$ and we have to consider the cases (1) and (2) of the theorem. If $t \neq 0$ then t is non-congruent to 0 modulo p for case (1) which shows the existence for $t \neq 0$. But 2 does not divide n because $n = 1$ which shows the existence for $t = 0$. $\qquad\square$

Theorem 8.2.29. *Let $F = \mathbb{F}_q$ with $q = p^n$, $p \geq 5$ prime and $n \geq 1$ and $E(F)$ be an elliptic curve over F. Then $\overline{E}(F)$ is either cyclic or the direct product $\overline{E}(F) = C_1 \times C_2$ where $C_1 \cong (\mathbb{Z}_{n_1}, +)$ and $C_2 \cong (\mathbb{Z}_{n_2}, +)$ where $n_1, n_2 \in \mathbb{N}$ with $n_2 | n_1$ and $n_2 | (q - 1)$.*

For a proof of this result see [Sil]. A proof is also available in [Sc1] where there is also a complete description of $\overline{E}(F)$ in the case of a super singular curve.

If $E(F)$ is not super singular we have the following.

Theorem 8.2.30. *Let $F = \mathbb{F}_q$ with $q = p^n$, $p \geq 5$ prime and $n \geq 1$. Let $k = q + 1 - t$ with t not divisible by p and $|t| \leq 2\sqrt{q}$. Suppose $n_1, n_2 \in \mathbb{N}$ with $k = n_1 n_2, n_2 | n_1$ and $n_2 | (q - 1)$. Then there exists an elliptic curve $E(F)$ with*

$$|\overline{E}(F)| = k \quad and \quad \overline{E}(F) \cong (\mathbb{Z}_{n_1}, +) \times (\mathbb{Z}_{n_2}, +).$$

Theorem 8.2.31 (Lang-Trotter Procedure). *Let* $F = \mathbb{Z}_p$, $p \geq 5$ *and* p *prime, and let* $E(\mathbb{Z}_p) : y^2 = x^3 + ax + b$ *be an elliptic curve over* $\mathbb{Z}_p$. *Then*

$$|\bar{E}(\mathbb{Z}_p)| = p + 1 + \sum_{x=0}^{p-1} \left(\frac{x^3 + ax + b}{p} \right)$$

where $\left(\frac{x^3+ax+b}{p} \right)$ *is the Legendre symbol.*

Proof. If $f(x) = x^3 + ax + b$ we have p values for x in $\mathbb{Z}_p$. If $f(x) = 0$ then there is exactly one point $(x, \bar{0})$ on $E(\mathbb{Z}_p)$. Here $\left(\frac{x^3+ax+b}{p} \right) = 0$.

Now let $f(x) \neq 0$. Then $f(x)^{\frac{p-1}{2}} \in \{\bar{1}, -\bar{1}\}$. If $f(x)^{\frac{p-1}{2}} = \bar{1}$, then $f(x)$ is a quadratic residue modulo p so for x there are two points (x, y) and $(x, -y)$ on $E(\mathbb{Z}_p)$. Hence the number of points is $\left(\frac{x^3+ax+b}{p} \right) + 1 = 2$ since $\left(\frac{x^3+ax+b}{p} \right) = 1$ if $f(x)$ is a quadratic residue modulo p.

Now let $f(x)^{\frac{p-1}{2}} = -\bar{1}$. Then $f(x)$ is a quadratic non-residue modulo p so for x there is no point on $E(\mathbb{Z}_p)$. Hence the number of points is $\left(\frac{x^3+ax+b}{p} \right) + 1 = 0$ since $\left(\frac{x^3+ax+b}{p} \right) = -1$ if $f(x)$ is a quadratic non-residue modulo p. Now the result follows since we must add a 1 for the element $\mathcal{O}$. $\square$

8.2.5 Calculating Points in Elliptic Curve Groups

In this section we describe efficient probabilistic algorithms to calculate points on an elliptic curve and to calculate their orders. The first algorithm is equivalent to the algorithm given in Chapter 6 to calculate square roots.

Theorem 8.2.32. *Let* $F = \mathbb{Z}_p$ *with* $p \geq 5$ *a prime and* $E(\mathbb{Z}_p) : y^2 = x^3 + ax + b$ *be an elliptic curve over* $\mathbb{Z}_p$. *Then a probabilistic algorithm for calculating a point* $(x, y) \in E(\mathbb{Z}_p)$ *is given as follows.*

(1) *Choose an arbitrary* $x \in \mathbb{Z}_p$ *and calculate* $f(x) = x^3 + ax + b$.
(2) *If* $f(x) = \bar{0}$ *then* $(x, \bar{0}) \in E(\mathbb{Z}_p)$.
(3) *Repeat Step* (1) *as long as* $f(x) \neq \bar{0}$. *Such an* x *must exists since* $p \geq 5$ *and* $f(x)$ *has at most 3 zeros in* $\mathbb{Z}_p$.
(4) *Calculate* $f(x)^{\frac{p-1}{2}} \in \{\bar{1}, -\bar{1}\} \subset \mathbb{Z}_p$. *Check if the result is* $\bar{1}$, *that is, if* $f(x)$ *is a square in* $\mathbb{Z}_p^*$.
(5) *If the result is* $-\bar{1}$ *then repeat steps* (1) *through* (4) *as long as we get the result* $\bar{1}$
(6) *If* $f(x) \neq \bar{0}$ *and* $f(x)^{\frac{p-1}{2}} = \bar{1}$ *then calculate* y *with* $y^2 = f(x)$ *as follows:*
  (a) *If* $p \equiv 3 \pmod 4$ *then calculate* $y = f(x)^{\frac{p+1}{4}}$. *Return the pair* (x, y) *and stop.*
  (b) *If* $p \equiv 1 \pmod 4$ *then write* $p - 1 = 2^s t$ *with* $s \geq 2$ *and* t *odd.*
Choose a random element $u \in \mathbb{Z}_p$ *and calculate* $u^{\frac{p-1}{2}} \in \{\bar{1}, -\bar{1}\}$. *Repeat this as long as* $u^{\frac{p-1}{2}} = -\bar{1}$.

Calculate $v = u^t$ *and* $y_1 = f(x)^{\frac{t+1}{2}}$. *For* $i = 2, 3, \ldots, s$ *calculate* $f(x)^{2^{s-i}t}$ *and define recursively* $c_0, c_1, \ldots, c_{s-2} \in \{0, 1\}$ *with*

$$f(x)^{2^{s-i}t} = v^{c_0 2^{s-i+1} + \cdots + c_{s-2} 2^{s-1}}.$$

Define $c = c_0 + 2c_1 + \cdots + 2^{s-2}c_{s-2}$ *and* $y = y_1 v^{-c}$. *Then give the pair* (x, y) *and stop.*

Proof. Since $f(x) = x^3 + ax + b$ has at most three zeros in $\mathbb{Z}_p$ and $p \geq 5$ it follows that there exists an $x \in \mathbb{Z}_p$ with $f(x) \neq \bar{0}$. The rest of the proof is done analogously to the proof of the algorithm for finding square roots modulo p given in Chapter 6. □

If we only want an elliptic curve $E(\mathbb{Z}_p)$ with a pair $P = (x_0, y_0) \in E(\mathbb{Z}_p)$ we may proceed as follows:
(1) Choose $x_0, y_0, a \in \mathbb{Z}_p$ randomly. Then define $b = y_0^2 - x_0^3 - ax_0$.
(2) Calculate $\Delta = -4a^3 - 27b^2$.
(3) Repeat (1) and (2) as long as $\Delta \neq \bar{0}$.

Then $E(\mathbb{Z}_p) : y^2 = x^3 + ax + b$ is an elliptic curve with $(x_0, y_0) \in E(\mathbb{Z}_p)$.

We now consider the question of determining the order of a point P on an elliptic curve $E(\mathbb{Z}_p)$.

Theorem 8.2.33 (Shank's Baby Step-Giant-Step (BSGS) Algorithm for the order of elements in $\bar{E}(\mathbb{Z}_p)$). *Let* $F = \mathbb{Z}_p, p \geq 5$ *and* $E(\mathbb{Z}_p) : y^2 = x^3 + ax + b$ *an elliptic curve over* $\mathbb{Z}_p$. *Let* $P \in E(\mathbb{Z}_p)$ *and* $B \in \mathbb{N}$ *with* $B > p + 2 + 2\sqrt{p}$. *Then the following procedure calculates the order* $|P| = ord_{\bar{E}(\mathbb{Z}_p)}(P)$ *of* P *in* $\bar{E}(\mathbb{Z}_p)$.
(1) *Let* $\beta = \min\{k \in \mathbb{N} | k^2 \geq B\}$. *(Then we have* $\beta^2 \geq B$.) *Calculate the points* $P, 2P, \ldots, \beta P$ *and define* $P_1 = \beta P$.
(2) *Calculate* αP_1 *for* $\alpha = 1, \ldots, \beta$ *and check after each calculation if* $\alpha P_1 \in \{P, 2P, \ldots, \beta P\}$. *If this is the case go to step* (3).
(3) *If* $\alpha P_1 = \gamma P$ *with* $\alpha \in \{1, \ldots, \beta\}$ *and* $\gamma \in \{1, \ldots, \beta\}$ *then factor* $\alpha\beta - \gamma$ *and find the smallest positive divisor of* δ *with* $\delta P = 0$. *The value of* δ *provides the order of* P.

Proof. Let $n = |P|$. Write $n = \tilde{\alpha}\beta + \tilde{\gamma}$ with $\tilde{\alpha} \in \{0, 1, \ldots, \beta - 1\}$ and $\tilde{\gamma} \in \{0, 1, \ldots, \beta - 1\}$. Since

$$n = |P| \leq p + 1 + 2\sqrt{p} < B \leq \beta^2$$

it follows that

$$0 = nP = \tilde{\alpha}\beta P + \tilde{\gamma}P = \tilde{\alpha}P_1 + \tilde{\gamma}P = \alpha P_1 - \gamma P$$

with $\alpha = \tilde{\alpha} + 1$ and $\gamma = \beta - \tilde{\gamma}$ since $P_1 = \beta P$. Here $\alpha, \gamma \in \{1, \ldots, \beta\}$. Hence in (2) we must have a match $\alpha P_1 = \gamma P$. □

We mention that there is a strong optimization of the BSGS algorithm given in the book by H. Cohn [Coh].

We now describe how to determine the order of the whole group $\bar{E}(\mathbb{Z}_p)$ if we have the order of an element P. Recall from Hasses' Theorem that if $p \geq 5$ then

$$p + 1 - 2\sqrt{p} \leq |\bar{E}(\mathbb{Z}_p)| \leq p + 1 + 2\sqrt{p}.$$

Theorem 8.2.34 (Shank's Method)**.** *Let* $F = \mathbb{Z}_p, p \geq 5$ *and* $E(\mathbb{Z}_p)$ *an elliptic curve over* $\mathbb{Z}_p$*. Suppose there is a point* $P \in E(\mathbb{Z}_p)$ *with order* $|P| = ord_{\bar{E}(\mathbb{Z}_p)}(P) > 4\sqrt{p}$*. Then there is exactly one multiple* $n|P|$ *with* $n \in \mathbb{N}$ *of* P *in the interval* $[p + 1 - 2\sqrt{p}, p + 1 + 2\sqrt{p}]$*. Then* $|\bar{E}(\mathbb{Z}_p)| = n|P|$*.*

If there exists such a point P *with* $|P| > 4\sqrt{p}$ *then following is a probabilistic algorithm to determine* $|\bar{E}(\mathbb{Z}_p)|$*.*

(1) *Choose a random* $P \in E(\mathbb{Z}_p)$*.*
(2) *Calculate the order* $|P|$ *of* P *with the help of the BSGS algorithm*
(3) *Repeat* (1) *and* (2) *as long as* $|P| > 4\sqrt{p}$*.*
(4) *Calculate the unique* $n \in \mathbb{N}$ *with*

$$n|P| \in [p + 1 - 2\sqrt{p}, p + 1 + 2\sqrt{p}].$$

Proof. We must only recall that $4\sqrt{p}$ is the length of the involved interval. We get $|\bar{E}(\mathbb{Z}_p)|$ from Hasse's Theorem and Lagrange's Theorem. $\square$

There is one problem in the Shank's method; when there does not exist such a $P \in E(\mathbb{Z}_p)$ with $|P| > 4\sqrt{p}$ in $\bar{E}(\mathbb{Z}_p)$ for each elliptic curve $E(\mathbb{Z}_p)$.

The idea is to use Shank's method universally by constructing an elliptic curve $\tilde{E}(\mathbb{Z}_p)$ which is closely related to $E(\mathbb{Z}_p)$ and which contains a point $\tilde{P}$ with $|\tilde{P}| > 4\sqrt{p}$ in $\bar{\tilde{E}}(\mathbb{Z}_p)$.

To do this we need some deep results from the theory of elliptic curves. Let $F = \mathbb{Z}_p$ with $p \geq 5$ and $E(\mathbb{Z}_p) : y^2 = x^3 + ax + b$ an elliptic curve. Let $d \in \mathbb{Z}_p^*$ be a quadratic non-residue modulo p so that $d^{\frac{p-1}{2}} = -\bar{1}$ in $\mathbb{Z}_p$. Now construct

$$\tilde{E}(\mathbb{Z}_p) : y^2 = x^3 + d^2 ax + d^3 b.$$

The curve $\tilde{E}(\mathbb{Z}_p)$ is an elliptic curve because $\tilde{\Delta} = d^6\Delta \neq \bar{0}$ in $\mathbb{Z}_p$. The curve $\tilde{E}(\mathbb{Z}_p)$ is called the **twist** of $E(\mathbb{Z}_p)$.

Theorem 8.2.35. *Let* $F = \mathbb{Z}_p, p \geq 5, E(\mathbb{Z}_p)$ *an elliptic curve over* $\mathbb{Z}_p$ *and* $\tilde{E}(\mathbb{Z}_p)$ *be a twist of* $E(\mathbb{Z}_p)$*. Then*

(1) $|\bar{E}(\mathbb{Z}_p)| + |\bar{\tilde{E}}(\mathbb{Z}_p)| = 2p + 2$*.*
(2) *If* $p > 457$ *then* $E(\mathbb{Z}_p)$ *or* $\tilde{E}(\mathbb{Z}_p)$ *contains a point* P *or* $\tilde{P}$ *respectively with order* $|P| > 4\sqrt{p}$ *in* $\bar{E}(\mathbb{Z}_p)$ *or* $|\tilde{P}| > 4\sqrt{p}|$ *in* $\bar{\tilde{E}}(\mathbb{Z}_p)$ *respectively.*
(3) *The number of suitable points is about* $\frac{Cp}{\log\log p}$ *where* C *is a constant.*

Proofs are available in [Coh] and [Sc1].

Theorem 8.2.36 (The Shank's-Mestre Algorithm). *Let $F = \mathbb{Z}_p$ with $p > 457$ and $E(\mathbb{Z}_p)$: $y^2 = x^3 + ax + b$ an elliptic curve over $\mathbb{Z}_p$. Then the following probabilistic algorithm determines the order $|\overline{E}(\mathbb{Z}_p)|$.*

(1) *Find a d with $d^{\frac{p-1}{2}} = -\overline{1}$ and define the twist*

$$\tilde{E}(\mathbb{Z}_p) : y^2 = x^3 + d^2ax + d^3b.$$

Let $f(x) = x^3 + ax + b$ and $\tilde{f}(x) = x^3 + d^2ax + d^3b$.

(2) *Choose randomly $x \in \{0, 1, \ldots, p-1\}$ and calculate $f(x)^{\frac{p-1}{2}}$ in $\mathbb{Z}_p$. If $f(x) = \overline{0}$ then choose another x and repeat this until $f(x) \neq \overline{0}$. If $f(x)^{\frac{p-1}{2}} = \overline{1}$ then calculate $y \in \mathbb{Z}_p$ with $y^2 = f(x)$ and set $P = (x, y)$. Otherwise go to step (4).*

(3) *Determine the order $|P|$ of P in $\overline{E}(\mathbb{Z}_p)$ with the BSGS algorithm. If $|P| > 4\sqrt{p}$ then continue to Step (6). Otherwise go to Step (2).*

(4) *Define $\tilde{x} = dx$. If $f(x)^{\frac{p-1}{2}} = -\overline{1}$ then*

$$\tilde{f}(x)^{\frac{p-1}{2}} = ((dx)^3 + d^2a(dx) + d^3b)^{\frac{p-1}{2}} = (d^3)^{\frac{p-1}{2}}(f(x))^{\frac{p-1}{2}} = \overline{1}.$$

Calculate $\tilde{y} \in \mathbb{Z}_p$ with $(\tilde{y})^2 = \tilde{f}(\tilde{x})$ and set $\tilde{P} = (\tilde{x}, \tilde{y}) \in \tilde{E}(\mathbb{Z}_p)$.

(5) *Determine the order $|\tilde{P}|$ of $\tilde{P}$ in $\tilde{E}(\mathbb{Z}_p)$ with the BSGS algorithm. If $|\tilde{P}| > 4\sqrt{p}$ then go to Step (6). Otherwise go to Step (2).*

(6) *Determine $n \in \mathbb{N}$ such that $n|P|$ or $n|\tilde{P}|$ respectively is in the interval $[p + 1 - 2\sqrt{p}, p + 1 + 2\sqrt{p}]$. Then return $n|P|$ or $2p + 2 - n|\tilde{P}|$ respectively and stop.*

With practical improvements to the Shank's-Mestre algorithm it is possible to calculate $|\overline{E}(\mathbb{Z}_p)|$ for prime numbers with 500 digits.

Another method to determine $|\overline{E}(\mathbb{Z}_p)|$ is to calculate the trace t of the Frobenius map, the Frobenius trace for short. Recall that $t = p + 1 - |\overline{E}(\mathbb{Z}_p)|$. We briefly describe this method.

Let $F = \mathbb{Z}_p$ with $p \geq 5$ and let $\overline{\mathbb{Z}_p}$ be the algebraic closure of $\mathbb{Z}_p$. The Frobenius map $\phi : \overline{\mathbb{Z}_p} \to \overline{\mathbb{Z}_p}$ given by $x \mapsto x^p$ is an injective field homomorphism by Fermat's theorem and $\mathbb{Z}_p$ is the set of fixed points of ϕ, that is, the set $\{x \mid \phi(x) = x\}$ (see [CFR1] or [KKi]). Let $E(\mathbb{Z}_p) : y^2 = x^3 + ax + b$ be an elliptic curve over $\mathbb{Z}_p$ and

$$\overline{E}(\overline{\mathbb{Z}_p}) = \{(u, v) \in \overline{\mathbb{Z}_p}^2 \mid v^2 = u^3 + au + b\} \cup \{\mathcal{O}\}\}$$

the elliptic curve group over $\overline{\mathbb{Z}_p}$. Then $\overline{E}(\mathbb{Z}_p) \subset \overline{E}(\overline{\mathbb{Z}_p})$ is a subgroup of $\overline{E}(\overline{\mathbb{Z}_p})$.

If $n \in \mathbb{N}$ then we define

$$\overline{E}[n] = \{P \in \overline{E}(\overline{\mathbb{Z}_p}) \mid nP = \mathcal{O}\}.$$

For $n \in \mathbb{N}$ the set $\overline{E}[n]$ is a subgroup of $\overline{E}(\overline{\mathbb{Z}_p})$. If $k \in \mathbb{Z}$ and $k \neq 0$ then $\overline{E}[|k|]$ is the kernel of the group endomorphism

$$[k] : \overline{E}(\overline{\mathbb{Z}_p}) \to \overline{E}(\overline{\mathbb{Z}_p}) \quad \text{given by} \quad P \to kP.$$

The Frobenious map ϕ induces a group endomorphism of $\overline{E}(\mathbb{Z}_p)$, denoted also by ϕ, and is called the **Frobenius map** on $\overline{E}(\mathbb{Z}_p)$ via $\phi : \overline{E}(\mathbb{Z}_p) \to \overline{E}(\mathbb{Z}_p)$ with

$$(x,y) \to (\phi(x), \phi(y)) = (x^p, y^p) \quad \text{if } (x,y) \in E(\mathbb{Z}_p) \text{ and } \mathbb{O} \to \mathbb{O}.$$

By the definition of the group operation in $\overline{E}(\mathbb{Z}_p)$ we get that ϕ is a group endomorphism. Also

$$E(\mathbb{Z}_p) = \{P \in \overline{E}(\mathbb{Z}_p) \mid \phi(P) = P\}$$

and hence $\phi(\overline{E}(\mathbb{Z}_p)) = \overline{E}(\mathbb{Z}_p)$ (recall that $\phi(a) = a, \phi(b) = b$).

Theorem 8.2.37. *Let $F = \mathbb{Z}_p, p \geq 5$ and $E(\mathbb{Z}_p) : y^2 = x^3 + ax + b$ be an elliptic curve over $\mathbb{Z}_p$. Let ϕ be the Frobenius map on $\overline{E}(\mathbb{Z}_p)$. Let $t = p + 1 - |\overline{E}(\mathbb{Z}_p)|$. Then the following hold:*

(1) $\overline{E}([n]) \cong (\mathbb{Z}_{n_1}, +) \times (\mathbb{Z}_{n_2}, +)$ *for some $n_1, n_2 \in \mathbb{N}$ if $n \in \mathbb{N}$ and p does not divide n*

$$\overline{E}([p^r]) \cong \{\mathbb{O}\} \text{ if } p|t, \text{ that is, } E(\mathbb{Z}_p) \text{ is super singular and}$$

$$\overline{E}([p^r]) \cong (\mathbb{Z}_{p^r}, +) \text{ if } p \text{ does not divide } t.$$

(2) *The map $\mathbb{Z} \to End(\overline{E}(\mathbb{Z}_p))$ given by $k \to [k]$ is an injective ring homomorphism. We call this the **Tate pairing**.*

(3) $\phi^2 - [t]\phi + [p] = [0]$ *in $End(\overline{E}(\mathbb{Z}_p))$.*

Proof. Here $End(\overline{E}(\mathbb{Z}_p))$ is the ring of endomorphisms of $(\overline{E}(\mathbb{Z}_p))$ via $k + l \to [k + l]$ where $[k + l](P) = [k](P) + [l](P)$ for $P \in (\overline{E}(\mathbb{Z}_p))$ and $kl \to k \circ l$ where $[k \circ l](P) = [k]([l](P))$ for $P \in (\overline{E}(\mathbb{Z}_p))$. It is easy to check that $End(\overline{E}(\mathbb{Z}_p))$ is a ring with two binary operations.

A proof of (1) and (3) can be found in [Sil]. Here we prove (2). Certainly the map is a ring homomorphism. We have $[k] = 0$ if and only if $kP = \mathbb{O}$ for all $P \in \overline{E}(\mathbb{Z}_p))$. By (1) there exist points in $E(\mathbb{Z}_p))$ with arbitrarily large orders in $(\overline{E}(\mathbb{Z}_p))$. Hence $k = 0$ and the kernel of the ring homomorphism is trivial. □

From the above theorem we have the reason for calling the number $t = p + 1 - |(\overline{E}(\mathbb{Z}_p)|$ the trace of the Frobenius map.

Based on the above theorem, Schoof (see [Sc1] and [Sc2]) developed a probabilistic algorithm to determine the trace t of the Frobenius map. If we know t then we have the order $|\overline{E}(\mathbb{Z}_p)|$ of $\overline{E}(\mathbb{Z}_p)$ via $|\overline{E}(\mathbb{Z}_p)| = p + 1 - t$. For more details see also [KKi]. Schoof's algorithm works especially well if we find a point $P \in \overline{E}(\mathbb{Z}_p)$ with $|P| = p$. The algorithm deals with $[t]$ and uses that $k \to [k]$ is injective.

The theorem is also the reason that super singular elliptic curves $E(\mathbb{Z}_p)$ are not suitable for cryptographic purposes.

8.3 Elliptic Curve Cryptography

We now apply the ElGamal method to the group of an elliptic curve to obtain the **elliptic curve cryptosystem**. We restrict ourselves to odd prime numbers $p \geq 5$ and the corresponding finite fields $\mathbb{Z}_p$.

Consider the elliptic curve (in Weierstrass form) over $\mathbb{Z}_p$ given by

$$E(\mathbb{Z}_p) : y^2 = x^3 + ax + b, \quad a, b \in \mathbb{Z}_p$$

with $\Delta = -\overline{4}a^3 - \overline{27}b^2 \neq 0$ in $\mathbb{Z}_p$.

Now let

$$\overline{E}(\mathbb{Z}_p) = E(\mathbb{Z}_p) \cup \{\mathcal{O}\}$$

be the elliptic curve group of $E(\mathbb{Z}_p)$. The basic idea is to use the ElGamal method, and its dependence on the corresponding discrete log problem, in $\overline{E}(\mathbb{Z}_p)$, this is, given $P \in E(\mathbb{Z}_p)$ and $nP \in \overline{E}(\mathbb{Z}_p)$ find n.

We now define the **elliptic curve encryption scheme** which we will abbreviate by ECES. This is also known as the **Elliptic Curve ElGamal Cryptosystem** or the **Meneses-Vanstone Cryptosystem**.

ECES Preparation

(1) Choose a large odd prime p with $p \geq 5$ and $a, b \in \mathbb{Z}_p$ such that

$$E(\mathbb{Z}_p) : y^2 = x^3 + ax + b$$

is an elliptic curve.
(2) Choose an injective efficiently invertible (on the image) map $\rho : \mathcal{M} \to E(\mathbb{Z}_p)$ from the set $\mathcal{M}$ of plain text units to $E(\mathbb{Z}_p)$. We describe such a choice below.
(3) Choose a point $P \in E(\mathbb{Z}_p)$.
(4) Choose a secret integer $d \in \mathbb{Z}$ and calculate $dP \in \overline{E}(\mathbb{Z}_p)$.

Encryption and Decryption in ECES

(1) The public key is $(P, dP) \in E(\mathbb{Z}_p) \times \overline{E}(\mathbb{Z}_p)$ and the elliptic curve $E(\mathbb{Z}_p)$ itself. The secret key is d.
(2) **Encryption:** Let $m \in \mathcal{M}$ be a plain text message unit. Calculate $Q = \rho(m)$. Choose a random integer $k \in \mathbb{Z}$ and define

$$c = (kP, Q + k(dP)) \in (\overline{E}(\mathbb{Z}_p))^2 = \mathcal{C}.$$

This is the encrypted message unit
(3) **Decryption:** Let $c = (c_1, c_2) \in \mathcal{C}$ be a ciphertext unit. Calculate $Q = c_2 - dc_1$ and $m = \rho^{-1}(Q)$ the preimage of Q.
Recall that $Q \in E(\mathbb{Z}_p)$ if $Q = \rho(m)$ and $(c_1, c_2) = (kP, Q + k(dP))$.

Theorem 8.3.1. *ECEC provides a valid cryptosystem.*

Proof. Let $(c_1, c_2) = (kP, Q + k(dP))$. Then $c_2 - dc_1 = Q = \rho(m)$. $\square$

Notice that if the discrete log problem for $\overline{E}(\mathbb{Z}_p)$ is solvable, that is if we can calculate d from (P, dP) then the ECES is broken.

We now show how to construct an injective, efficiently invertible map $\mathcal{M} \to E(\mathbb{Z}_p)$.

Let $E(\mathbb{Z}_p) : y^2 = x^3 + ax + b$ be an elliptic curve over $\mathbb{Z}_p$ with $p \geq 5$. By Hasse's theorem (see last section) we have

$$|\overline{E}(\mathbb{Z}_p)| \in [p + 1 - 2\sqrt{p}, p + 1 + 2\sqrt{p}] \cap \mathbb{N}.$$

There are efficient probabilistic algorithms to generate points of $E(\mathbb{Z}_p)$ (see Theorem 8.2.10). We need many points in $E(\mathbb{Z}_p)$.

(1) Choose $k \in \mathbb{N}$ such that the permitted probability of error is $< \frac{1}{2^k}$.
(2) Let $\mathcal{M} = \{0, 1, \ldots, M\}$. We should have $p > (M + 2)k$.
(3) Define an injective map:

$$\Psi : \mathcal{M} \times \{1, \ldots, k\} \to \mathbb{Z}_p \quad \text{by} \quad (m, j) \to \overline{mk} + \overline{j}.$$

Recall that $0 \leq mk + j < p$ because $p > (M + 2)k$.

(4) Let $x = \Psi(m, 1)$. Calculate $f(x) = x^3 + ax + b$ and check if there exists $y \in \mathbb{Z}_p$ with $y^2 = f(x)$. If this is the case then choose y so that $y \in \{\overline{0}, \overline{1}, \ldots, \frac{p-1}{2}\}$ and define $\rho(m) = (x, y)$.

We note that $f(x)$ is a quadratic residue modulo p for about half of the $f(x)$ with $f(x) \neq \overline{0}$ and $x \in \mathbb{Z}_p$ gives 0, 1 or 2 points on the elliptic curve.

(5) If $x = \Psi(m, 1)$ and there is no $y \in \mathbb{Z}_p$ with $y^2 = f(x)$ then try $x = \Psi(m, 2), x = \Psi(m, 3)$ and so on.

With probability $> 1 - \frac{1}{2^k}$ we find an element $x \in \{\Psi(m, 1), \ldots, \Psi(m, k)\}$ with $f(x) = y^2$ for some $y \in \mathbb{Z}_p$.

If j with $1 \leq j \leq k$ is the smallest integer j such that $x = \Psi(m, j)$ and $f(x) = y^2$ for some $y \in \mathbb{Z}_p$ - such a j exists with probability $> 1 - \frac{1}{2^k}$ - then choose $y \in \{\overline{0}, \ldots, \frac{p-1}{2}\}$ and define $\rho(m) = (x, y)$.

(6) If $(x, y) \in Im(\rho) \subset E(\mathbb{Z}_p)$ then we may recover m efficiently. If $x = \overline{mk} + \overline{j}$ then $m = \overline{m} = \frac{x - j}{k}$ because $k \in \mathbb{N}$ and $p > (M + 2)k$.

8.4 Cryptoanalysis of Elliptic Curve Cryptosystems

We now examine the cryptoanalysis of elliptic curve based cryptosystems. First we look at some general ideas.

Recall that $E(\mathbb{Z}_p)$ and P are public. An attacker has to calculate $|\overline{E}(\mathbb{Z}_p)|$ or $|P|$.

(1) ECES is not secure if $|\overline{E}(\mathbb{Z}_p)|$ has only small prime factors (see the section on the reduction of the DLP to groups of prime power order in Chapter 6). Hence $|\overline{E}(\mathbb{Z}_p)|$ should have at least one large prime factor.

(2) Analogously $|P|$ should have at least one large prime factor.

(3) ECES is not secure if $|P| = p$. Here we can determine $|\overline{E}(\mathbb{Z}_p)|$ effectively via the trace t of the Frobenius map using what is called Schoof's algorithm (see [Sc1] and [Sc2]).

Elliptic curves that have passed all known attacks so far can be found at the website http://www.ecc-brainpool.org/ecc-standards.htm.

8.5 The MOV-Algorithm

In this section we look at the **MOV algorithm** for attacking elliptic curve cryptosystems. This algorithm, named after Menezes, Okamoto and Vanstone, who developed it (see [MOV]), is a probabilistic algorithm to determine the discrete logarithm in the group $\overline{E}(F)$ of an elliptic curve over a finite field. The MOV algorithm reduces the elliptic curve discrete log problem (ECDSP) to the discrete log problem (DLP) in the multiplicative group of an extension field $\mathbb{F}_{q^k}$ of the base field $\mathbb{F}_q$ and then solves the DLP using a known best algorithm. It was originally believed that to solve the ECDLP required exponential time. However, using the MOV algorithm for elliptic curves it can be solved in probabilistically subexponential time. The theoretical background and a more detailed description of the MOV algorithm can be found in [Sil].

To start, let F be a field of characteristic $p \geq 5$. We note that the concepts can also be described with modifications for fields of characteristic 2 or 3 but here we only consider $p \geq 5$. Let

$$E(F) : y^2 = x^3 + ax + b$$

be an elliptic curve over F and $\overline{E}(F) = E(F) \cup \mathcal{O}$ be the corresponding elliptic curve group.

For each $P, Q \in \overline{E}(F)$ we take a formal symbol $[P]$ with $[P] \neq [Q]$ if $P \neq Q$. A **divisor** D of $\overline{E}(F)$ is defined by the formal sum

$$D = \sum_{P \in \overline{E}(F)} n_P[P]$$

with all $n_P \in \mathbb{Z}$ and at most finitely many $n_P \neq 0$. Together with the sum

$$\sum_{P \in \overline{E}(F)} n_P[P] + \sum_{P \in \overline{E}(F)} m_P[P] = \sum_{P \in \overline{E}(F)} (n_P + m_P)[P]$$

the set of divisors of $\overline{E}(P)$ is a free abelian group.

Let $\overline{F}$ be the algebraic closure of F and let $R(\overline{E}(\overline{F}))$ be the field of rational functions of $\overline{E}(\overline{F})$. Recall that $\mathcal{O}$ is the point at infinity.

Let $f \in R(\overline{E}(\overline{F}))^*$ so that $f \in R(\overline{E}(\overline{F}))$ and $f \neq \mathcal{O}$. The divisor of f is defined by

$$\mathrm{div}(f) = \sum_{P \in \overline{E}(F)} n_P[P]$$

with $n_P \in \mathbb{Z}$, $n_P \geq 0$ if $P \in \overline{E}(F)$ is a zero of f with multiplicity n_P and $n_P < 0$ if $P \in \overline{E}(F)$ is a pole of f with multiplicity $|n_P|$. A divisor D of $\overline{E}(F)$ is called a **principal divisor** if $D = \mathrm{div}(f)$ for some $f \in R(\overline{E}(\overline{F}))^*$.

The **degree** of a divisor $D = \sum_{P \in \overline{E}(F)} n_P[P]$ is defined by

$$\deg(D) = \sum_{P \in \overline{E}(F)} n_P$$

and the **sum** of the divisor D is defined by

$$\mathrm{sum}(D) = \sum_{P \in \overline{F}} n_P P.$$

Theorem 8.5.1. *Let $f \in R(\overline{E}(\overline{F}))^*$.*
(1) *We have $\mathrm{div}(f) = \mathcal{O}$ if and only if $f \in \overline{F}$ and $f \neq \mathcal{O}$.*
(2) *We have $\deg(\mathrm{div}(f)) = 0$.*
(3) *A divisor $D = \sum_{P \in \overline{E}(F)} n_P[P]$ is a principal divisor if and only if we have $\deg(D) = 0$ and $\mathrm{sum}(D) = \mathcal{O}$.*

We now describe the **Weil-pairing**. Let $m \geq 1$ be a natural number with p not a divisor of m. We call $P \in \overline{E}(F)$ an m-torsion point if $mP = \mathcal{O}$. The point at infinity $\mathcal{O}$ is an m-torsion point for all m. The m-torsion subgroup $\overline{E}(F)[m]$ of $\overline{E}(F)$ is the set of all m-torsion points of $\overline{E}(F)$, that is,

$$\overline{E}(F)[m] = \{P \in \overline{E}(F); \; mP = \mathcal{O}\}.$$

This is clearly a subgroup of $\overline{E}(F)$ since $\overline{E}(F)$ is abelian.

Definition 8.5.2. Given $P, Q \in \overline{E}(F)[m]$, we let $f_P, f_Q \in R(\overline{E}(\overline{F}))^*$ such that $\mathrm{div}(f_P) = m[P] - m[\mathcal{O}]$ and $\mathrm{div}(f_Q) = m[Q] - m[\mathcal{O}]$. The **Weil-pairing** $e_m(P, Q)$ of P and Q is defined by

$$e_m(P, Q) = \frac{f_P(Q + S)/f_P(S)}{f_Q(P - S)/f_Q(-S)}$$

for $S \in \overline{E}(F)[m]$ with $S \notin \{\mathcal{O}, \mathcal{P}, -\mathcal{Q}, \mathcal{P} - \mathcal{Q}\}$.

Notice that if $P, Q \in \overline{E}(F)[m]$ then $mP = mQ = \mathcal{O}$. Then by Theorem 8.4.1.1 there exist rational functions $f_P, f_Q \in R(\overline{E}(\overline{F}))^*$ with $\mathrm{div}(f_P) = m[P] - m[\mathcal{O}]$ and $\mathrm{div}(f_Q) = m[Q] - m[\mathcal{O}]$ which are related to the divisors $m[P] - m[\mathcal{O}]$ and $m[Q] - m[\mathcal{O}]$ respectively. P is a zero of f_P with multiplicity m and $\mathcal{O}$ is a pole of f_P with multiplicity m. The analogous statement holds for f_Q. This shows that the Weil-pairing $e_m(P, Q)$ is defined and non-zero for $S \notin \{\mathcal{O}, P, -Q, P - Q\}$.

Further notice that the Weil-pairing is well-defined, that is, independent of the choice of the rational functions f_P, f_Q and independent of the choice of the point $S \notin \{O, P, -Q, P - Q\}$.

Theorem 8.5.3 (Properties of the Weil-Pairing).
(1) $e_m(P, Q)^m = 1$ for all $P, Q \in \overline{E}(F)[m]$
(2) $e_m(P_1 + P_2, Q) = e_m(P_1, Q)e_m(P_2, Q)$ and
    $e_m(P, Q_1 + Q_2) = e_m(P, Q_1)e_m(P, Q_2)$ for all $P, P_1, P_2, Q_1, Q_2 \in \overline{E}(F)[m]$
(3) $e_m(P, P) = 1$ for all $P \in \overline{E}(F)[m]$
(4) If $e_m(P, Q) = 1$ for all $Q \in \overline{E}(F)[m]$ then $P = O$.

The Weil-pairing e_m is a map

$$e_m : \overline{E}(F)[m] \times \overline{E}(F)[m] \to \Phi_m \subset F^*$$

where Φ_m is the group of the m-th roots of unity in F^*. There is an efficient algorithm to calculate the Weil-pairing (see [Sil]).

We now describe the **embedding degree** κ. This is important for the running time of the probabilistic algorithm to determine the discrete logarithm in $\overline{E}(F)$.

Note that if $\kappa \in \mathbb{N}$ with $\kappa \geq 1$ then $\mathbb{Z}_p$ embeds into the finite field $\mathbb{F}_q$, $q = p^\kappa$. With this embedding, any elliptic curve $E(\mathbb{Z}_p)$ can be considered as an elliptic curve $E(\mathbb{F}_q)$, that is if $P = (a, b)$ is a point on $E(\mathbb{Z}_p)$ then P can be considered as a point on $E(\mathbb{F}_q)$.

Theorem 8.5.4. Let $E(\mathbb{Z}_p)$ be an elliptic curve and let $\ell > \sqrt{p} + 1$ be a prime number ($p \geq 5$ prime). If $\overline{E}(\mathbb{Z}_p)$ contains an element P of order ℓ then $\overline{E}(\mathbb{Z}_p)[\ell] \cong (\mathbb{Z}_\ell, +) = \mathbb{Z}_\ell$.

Proof. Since $\overline{E}(\mathbb{Z}_p)$ contains an element P of order ℓ the subgroup $\overline{E}(\mathbb{Z}_p)[\ell]$ contains at least ℓ elements. We now show that $\overline{E}(\mathbb{Z}_p)[\ell])$ contains at least ℓ elements and hence $\overline{E}(\mathbb{Z}_p)[\ell])$ is isomorphic to $\mathbb{Z}_\ell$.

Assume that there are more than ℓ elements in $\overline{E}(\mathbb{Z}_p)[\ell]$. The element P generates a cyclic subgroup $\langle P \rangle$ of order ℓ in $\overline{E}(\mathbb{Z}_p)[\ell]$. Hence there exists an element $Q \in \overline{E}(\mathbb{Z}_p)[\ell]$ which cannot be written as aP with $a \in \{0, 1, \ldots, \ell - 1\}$. The element Q also generates a cyclic subgroup $\langle Q \rangle$ of order ℓ in $\overline{E}(\mathbb{Z}_p)[\ell]$ and we have $\langle P \rangle \cap \langle Q \rangle = \{O\}$. Recall that ℓ is a prime number. Each element of the form $aP + bQ$ with $b \in \{0, 1, \ldots, \ell - 1\}$ is also in $\overline{E}(\mathbb{Z}_p)[\ell]$. Hence $\overline{E}(\mathbb{Z}_p)[\ell]$ has at least ℓ^2 elements. We have $\ell^2 > p + 1 + 2\sqrt{p}$. On the other hand, the number of elements of $\overline{E}(\mathbb{Z}_p)$ is by Hasse's Theorem in the interval $[p + 1 - 2\sqrt{p}, p + 1 + 2\sqrt{p}]$ which gives a contradiction and therefore there is no such Q. It follows that $\overline{E}(\mathbb{Z}_p)[\ell] = \langle P \rangle \cong \mathbb{Z}_\ell$. $\square$

Using this we get the following theorem.

Theorem 8.5.5. *There is a $\kappa \in \mathbb{N}, \kappa \geq 1$ with*

$$\overline{E}(\mathbb{F}_{p^{j\kappa}})[m] \cong \mathbb{Z}_m \times \mathbb{Z}_m$$

for all $j \geq 1$.

Definition 8.5.6. The **embedding degree** for m is the smallest natural number $\kappa \geq 1$ such that for all $j \geq 1$ we have

$$\overline{E}(\mathbb{F}_{p^{j\kappa}})[m] \cong \mathbb{Z}_m \times \mathbb{Z}_m.$$

Theorem 8.5.7. *Let $\ell \neq p$ be a prime number. If $\overline{E}(\mathbb{Z}_p)$ contains an element of order ℓ then the embedding degree κ of $\overline{E}(\mathbb{Z}_p)$ for ℓ is given by one of the following cases:*
(1) *$\kappa = 1$ and this can only happen if $\ell \leq \sqrt{p} + 1$.*
(2) *$p \equiv 1 \pmod{\ell}$ and $\kappa = \ell$.*
(3) *p is not congruent to $1 \pmod{\ell}$ and κ is the smallest natural number r with $p^r \equiv 1 \pmod{\ell}$.*

We now describe the MOV-algorithm for the discrete logarithm in $\overline{E}(\mathbb{Z}_p)$, where $p \geq 5$ is a prime number. Let $P \in \overline{E}(\mathbb{Z}_p)$ of order $\ell > \sqrt{p} + 1$ where ℓ is a prime number with $\ell \neq p$. Let κ be the embedding degree of $\overline{E}(\mathbb{Z}_p)$ for ℓ.

Let $Q = kP$ where k is the discrete logarithm of Q for the base P in $\overline{E}(\mathbb{Z}_p)$, especially $1 \leq k < \ell$. We consider the following steps to calculate k:
(1) Calculate $N = |\overline{E}(\mathbb{F}_{p^\kappa})|$
(2) Choose an element $T \in \overline{E}(\mathbb{F}_{p^\kappa})$ with $T \notin \overline{E}(\mathbb{Z}_p)$. If $T' = (\frac{N}{\ell})T = \mathcal{O}$ then repeat this step with a different T. If $T' \neq \mathcal{O}$ then continue on to setp (3).
(3) Calculate the values $g = e_\ell(P, T')$ and $b = e_\ell(Q, T')$ of the Weil-pairing.
(4) Solve the discrete log problem $b = g^k$ in $\mathbb{F}_{p^\kappa}^*$.

This is a probabilistic algorithm to solve the discrete log problem in $\overline{E}(\mathbb{Z}_p)$ in subexponential time. This probabilistic algorithm is correct as we can see from the following remarks:
(1) In Step (1) we have to calculate $N = |\overline{E}(\mathbb{F}_{p^\kappa})|$. We discussed several methods to calculate $|\overline{E}(\mathbb{Z}_p)|$ and these methods can be extended to calculate $|\overline{E}(\mathbb{F}_{p^\kappa})|$. This can be done polynomially in $\log(p^k)$. For details we refer to [HPS].
(2) In Step (2) we need an element $T \in \overline{E}(\mathbb{F}_{p^\kappa})$ with $T \notin \overline{E}(\mathbb{Z}_p)$ and $T' = (\frac{N}{\ell})T \neq \mathcal{O}$. It is easy to see that there are $T \in \overline{E}(\mathbb{F}_{p^\kappa})$ with $T \notin \overline{E}(\mathbb{Z}_p)$ by Theorems 8.4.4 and 8.4.5. We have $\kappa \geq 2$ since $\ell > \sqrt{p} + 1$. Now

$$n = |\overline{E}(\mathbb{Z}_p)| \leq p + 1 + 2\sqrt{p} \quad \text{and} \quad N \geq p^\kappa + 1 - 2\sqrt{p^\kappa}$$

by Hasse's Theorem. These give that $N > n$ since $p \geq 5$ and $\kappa \geq 2$.
To construct such a T with the additional property that $T' = (\frac{N}{\ell})T \neq \mathcal{O}$ is computationally much more complicated. However, it exists and can be constructed polynomially in $\log(p^\kappa)$ (see [SZSI]).
(3) We may calculate $g = e_\ell(P, T')$ and $b = e_\ell(Q, T')$. T' is not of the form rP. If this would be the case then we would already have $T = sP$ from $T' = (\frac{N}{\ell})T$ which contradicts $T' \neq \mathcal{O}$ because $(\frac{N}{\ell})P = \mathcal{O}$. Hence we may consider P and T' as a basis of the two-dimensional vector space $\mathbb{Z}_\ell \times \mathbb{Z}_\ell \cong \overline{E}(\mathbb{F}_{p^\kappa})[\ell]$.

Assume that $g = 1$. Then $P = \mathcal{O}$ by Theorem 8.4.2 which contradicts $P \neq \mathcal{O}$. Recall that $e_\ell(P, P) = 1$ and $e_\ell(P, aT') = 1$ for all a. Therefore we have $g \neq 1$. Then $g^\ell = 1$ and g generates the subgroup in $GF(p^\kappa)^*$ of the ℓ-th roots of unity.

(4) From $Q = kP$ we get $b = g^k$ by Theorem 8.5.2. Hence k is also the discrete logarithm for the base g in $\mathbb{F}_{p^\kappa}^*$. The discrete logarithm problem in $\mathbb{F}_{p^\kappa}$ can be solved subexponentially in $\log(p^\kappa)$, for instance with the index calculus method for $\mathbb{F}_{p^\kappa}^*$ (see [CP]). It follows that altogether the MOV-algorithm is subexponential and bounded by the running time of the index calculus method.

We end with some closing comments on the MOV-algorithm.

First of all, the MOV-algorithm for $\overline{E}(\mathbb{Z}_p)$ with $p \geq 5$ is advantageous if the running time is smaller than the running time of Pollard's ρ-method for $\overline{E}(\mathbb{Z}_p)$ (see Chapter 6). Computational experiments show that this happens if $2 \leq \kappa \leq 6$ for the embedding degree of $\overline{E}(\mathbb{Z}_p)$.

Second, the MOV-algorithm can be extended for $\overline{E}(\mathbb{F}_q), q = p^n, p \geq 5$.

Next, C. Diem has recently shown in [Di1] that the discrete logarithm problem in $\overline{E}(\mathbb{Z}_p)$ can be solved probabilistically in an expected time of $e^{\mathcal{O}(\log p)}$. His algorithm is based on an extension of the index calculus method for $\overline{E}(\mathbb{Z}_p)$. In fact Diem further has a strong extension for $\overline{E}(\mathbb{F}_q), q = p^n, p \geq 5$. Most importantly he proved that there exists an infinite sequence (F_i) of fields, $F_i = \mathbb{F}_{q_i}, q_i = p^{n_i}$ with p prime such that the discrete logarithm problem for all $\overline{E}(F_i)$ can be solved probabilistically in subexponential expected time (in the bit length of the group size).

8.6 The Elliptic Curve Digital Signature

In Chapter 4 we described the **Digitial Signature Algorithm** and then in Chapter 7 the RSA version of this. In this final section of Chapter 8, we present an elliptic curve analogue of DSA which puts the protocol within an elliptic curve group, rather than in $\mathbb{Z}_p^*$. This analogue is called the **Elliptic Curve Digital Signature Algorithm** abbreviated **ECDSA**. In 2006 this was certified by the United States National Bureau of Standards and Technology for use in the US administration. This was the third signature algorithm certified in this way, after DSA and the RSA signature Method.

In the ECDSA a signed message $m \in \mathcal{M}$ is to be sent from party A to party B. To do this we proceed as follows.

Phase I: The key generation

(1) A chooses a large prime number $p \geq 5$ and an elliptic curve $E(\mathbb{Z}_p)$.

(2) A finds a point $P \in E(\mathbb{Z}_p)$ whose order is a prime number $q \neq p$.

(3) A chooses randomly $x \in \{1, \ldots, q - 1\}$ and calculates $Q = xP$. The public key of A is then Q and the secret key of A is x.

(4) A chooses a hash function $h : \mathcal{M} \to \{1, 2, \ldots, N\}$ with N sufficiently large.

Phase II: *A* now signs the message as follows

(1) A chooses randomly $k \in \{1, \ldots, q - 1\}$ and calculates $kP = (x_1, y_1) \in E(\mathbb{Z}_p)$.

(2) A calculates the number $r \equiv x_1 \pmod{q}$ with $0 \leq r \leq q - 1$. If $r = 0$ then A repeats II (1) and II (2) until he finds an $r \neq 0$.

(3) Let $r \neq 0$. Using the extended Euclidean algorithm A finds a number ℓ with $1 \leq \ell \leq q - 1$ and $\ell k \equiv 1 \pmod{q}$.

(4) A calculates the number $s \equiv \ell(h(m) + xr) \pmod{q}$ with $0 \leq s \leq q - 1$. If $s = 0$ then A starts again with step II (1).

(5) Now let $r \neq 0$ and $s \neq 0$. A adds the signature (r, s) to the message and sends everything to B.

Phase III: *B* verifies the signature

(1) B takes an authenticated copy of the public key Q of A.

(2) B checks if $r, s \in \{1, \ldots, q - 1\}$.

(3) B calculates $w \equiv s^{-1} \pmod{q}$ and $h(m) \in \{1, \ldots, N\}$.

(4) B calculates $\mu_1 \equiv h(m)w \pmod{q}$ and $\mu_2 \equiv rw \pmod{q}$.

(5) B calculates $\mu_1 P + \mu_2 Q = (x_0, y_0) \in E(\mathbb{Z}_p)$ and $v \equiv x_0 \pmod{q}$, $0 \leq v \leq q - 1$.

(6) B accepts the signature if $v = r$.

We note that this signature is manageable. The probability for $r = 0$ in Step II (2) is about $\frac{1}{p}$. Analogously the probability for $s = 0$ in Step II (4) is about $\frac{1}{q}$ and hence very small. Therefore the loops in phase II are finite with probability 1.

Number theoretically we can show that the ECDSA is correct. First, modulo q we have

$$v \equiv x_0 \equiv \mu_1 P + \mu_2 Q \equiv h(m)wP + rwQ \equiv s^{-1}(h(m) + rx)P$$
$$\equiv k(h(m) + rx)^{-1}(h(m) + rx)P \equiv kP \equiv x_1 \equiv r.$$

Therefore B accepts the correct signature.

If $r = 0$ in Step II (2) then we would get $s \equiv \ell h(m) \pmod{q}$ in Step (8) and (r, s) would not depend on the knowledge of x.

If $s = 0$ in Step I (4) then B could not calculate w in Step III (3).

We now examine the cryptanalysis of this digital signature protocol.

(1) We have roughly $q \approx p$. To obtain similar security as in DSA q should have about 160 bits. The signature then has about 320 bits.

(2) To get (k, x) from (r, s) we must get x from $Q = xP$, that is, we must solve the discrete log problem in the subgroup $\langle P \rangle$ of $\overline{E}(\mathbb{Z}_p)$.

(3) The prime numbers p and q should not be equal. Otherwise the Froebenius trace is congruent to 1 modulo p and we would be able to calculate the discrete logarithm efficiently.

(4) The elliptic curve should not be a super singular curve.

(5) For each user we could choose $E(\mathbb{Z}_p)$ and P differently. But then (a, b, p) should be in the public key of A.

8.7 Exercises

8.1 Construct fields with 4 and 9 elements explicitly. Show their addition and multiplication tables.

8.2 Let K be a field of characteristic unequal to 2 or 3. Given an equation of the form

$$y^2 + a_1 xy + a_3 y = x^3 + a_2 x^2 + a_4 x + a_6 \qquad (\star)$$

with $a_1, a_2, a_3, a_4, a_6 \in K$.
Then there is a transformation

$$x \mapsto \alpha_1 x + \alpha_2 y + \alpha_3, \quad y \mapsto \beta_1 x + \beta_2 y + \beta_3$$

with $\alpha_1, \alpha_2, \alpha_3, \beta_1, \beta_2, \beta_3 \in K$ which transforms $(\star)$ into the Weierstrass form

$$y^2 = x^3 + ax + b$$

with $a, b \in K$.

8.3 Let K be a field and $\overline{K}$ its algebraic closure. Show that the polynomial $x^3 + ax + b$ with $a, b \in K$ has a multiple zero if and only if $4a^3 + 27b^2 = 0$.
Take care that your argument is correct if K has characteristic 2 or 3.

8.4 Let K be a finite field with $|K| = q$. Show that there are exactly $q^2 - q$ polynomials of the form $x^3 + ax + b$ with $a, b \in K$ which have no multiple zeros in any extension field of K.

8.5 Let $y^2 = x^3 + x + \overline{6}$ be a curve over $\mathbb{Z}_{11}$. Show that
(a) $y^2 = x^3 + x + \overline{6}$ is an elliptic curve over $\mathbb{Z}_{11}$.
(b) $\overline{E}(\mathbb{Z}_{11})$ is cyclic of order 13.

8.6 Let $y^2 = x^3 + x$ be a curve over $\mathbb{Z}_5$. Show that
(a) $y^2 = x^3 + x$ is an elliptic curve over $\mathbb{Z}_5$.
(b) $\overline{E}(\mathbb{Z}_5)$ is isomorphic to the Klein 4-group $\mathbb{Z}_2 \times \mathbb{Z}_2$.

8.7 Determine all possible groups $\overline{E}(\mathbb{Z}_5)$ for elliptic curves over $\mathbb{Z}_5$. Give all possible orders for a group $\overline{E}(\mathbb{Z}_5)$.

8.8 Let K be an algebraically closed field of characteristic unequal to 2 or 3. Let $E(K)$ be an elliptic curve over K. Show that

$$\{P \in E(K) \mid 3P = \mathcal{O}\} \cup \mathcal{O}$$

is isomorphic to the group $\mathbb{Z}_3 \times \mathbb{Z}_3$.

8.9 Describe in detail the Diffie-Hellman key exchange protocol for elliptic curves.

8.10 Let p be a prime number with $p \equiv 3 \pmod{4}$ and let $E(\mathbb{Z}_p)$ be an elliptic curve over $\mathbb{Z}_p$. Find a polynomial time algorithm which constructs a point (x, y) on $E(\mathbb{Z}_p)$ for a given $x \in \mathbb{Z}_p$ if such a point exists.
Use this algorithm to find a point $(2, y)$ on

$$y^2 = x^3 + x + \overline{1}$$

over $\mathbb{Z}_p$ with $p = 111119$.
Hint: Use the construction of square roots in $\mathbb{Z}_p$ for $p \equiv 3 \pmod{4}$.

9 Basic Concepts from Group Theory

9.1 Groups and Group Theory

Up to this point we have been using algebraic objects arising from number theory, such as the modular rings $\mathbb{Z}_n$ and elliptic curve groups, to do encryption. These objects are all commutative and hence we can call the type of encryption we have already done as **commutative cryptography**. In an effort to improve upon cryptographic security, **non-commutative cryptography** was introduced. Non-commutative cryptography is essentially **group based cryptography**. In group based cryptography, non-abelian groups and their properties are used in encryption and decryption. There are two primary sources for non-abelian groups: linear groups, that is, groups of matrices, and combinatorial group theory. In this chapter we describe the necessary material from group theory in general, and combinatorial group theory in particular, that is essential for non-commutative cryptographic purposes. First we define a group.

Definition 9.1.1. A **group** G is a set with one binary operation which we will denote by either multiplication $\cdot$ or just juxtaposition, such that:

(1) The operation is associative, that is, $(g_1 g_2)g_3 = g_1(g_2 g_3)$ for all $g_1, g_2, g_3 \in G$.
(2) There exists an identity for this operation, that is, an element 1 such that $1g = g$ and $g1 = g$ for each $g \in G$.
(3) Each $g \in G$ has an inverse for this operation, that is, for each g there exists a g^{-1} with the property that $gg^{-1} = 1$ and $g^{-1}g = 1$.

If in addition the operation is commutative, that is $g_1 g_2 = g_2 g_1$ for all $g_1, g_2 \in G$, the group G is called an **abelian group**.

The **order** of a group G, denoted $|G|$, is the number of elements in the group G. If $|G| < \infty$, G is a called a **finite group**, otherwise it is an **infinite group**.

It follows easily from the definition that the identity is unique and that each element has a unique inverse.

Lemma 9.1.2. *If G is a group then there is a unique identity. Further if $g \in G$ its inverse is unique. Finally if $g_1, g_2 \in G$ then $(g_1 g_2)^{-1} = g_2^{-1} g_1^{-1}$.*

Proof. Suppose that 1 and e are both identities for G. Then $1e = e$ since e is an identity and $1e = 1$ since 1 is an identity. Therefore $1 = e$ and there is only one identity.

Next suppose that $g \in G$ and g_1 and g_2 are inverses for g. Then

$$g_1 g g_2 = (g_1 g)g_2 = 1g_2 = g_2$$

since $g_1 g = 1$. On the other hand,

$$g_1 g g_2 = g_1(g g_2) = g_1 1 = g_1$$

since $g g_2 = 1$. It follows that $g_1 = g_2$ and g has a unique inverse.

Finally consider

$$(g_1g_2)(g_2^{-1}g_1^{-1}) = g_1(g_2g_2^{-1})g_1^{-1} = g_1 1 g_1^{-1} = g_1 g_1^{-1} = 1.$$

Therefore $(g_2^{-1}g_1^{-1})$ is an inverse for g_1g_2, and since inverses are unique, it is the inverse of the product. □

Groups most often arise as permutations on a set. We will see this as well as other specific examples of groups in subsequent sections.

Finite groups can be completely described by their **group tables** or multiplication tables. These are sometimes called **Cayley tables**. In general, let $G = \{g_1, \dots, g_n\}$ be a group, then the **multiplication table** of G is:

$$
\begin{array}{c|ccccccc}
 & g_1 & g_2 & \cdots & g_j & \cdots & g_n \\
\hline
 & -- & -- & \cdots & -- & \cdots & -- \\
g_1 & \cdots & & & & & \\
g_2 & \cdots & & & & & \\
\vdots & & & & & & \\
g_i & \cdots & \cdots & \cdots & g_ig_j & & \\
\vdots & & & & & & \\
g_n & \cdots & & & & & \\
\end{array}
$$

The entry in the row of $g_i \in G$ and column of $g_j \in G$ is the product (in that order) g_ig_j in G.

Groups satisfy the **cancellation law for multiplication**.

Lemma 9.1.3. *If G is a group and $a, b, c \in G$ with $ab = ac$ or $ba = ca$ then $b = c$.*

Proof. Suppose that $ab = ac$. Then a has an inverse a^{-1} so we have

$$a^{-1}(ab) = a^{-1}(ac).$$

From the associativity of the group operation we then have

$$(a^{-1}a)b = (a^{-1}a)c \implies 1 \cdot b = 1 \cdot c \implies b = c. \qquad \square$$

A consequence of the above lemma is that each row and each column in a group table is just a permutation of the group elements. That is, each group element appears exactly once in each row and each column.

A subset $H \subset G$ is a **subgroup** of G if H is also a group under the same operation as G. As for rings and fields a subset of a group is a subgroup if it is non-empty and closed under both the group operation and inverses.

Lemma 9.1.4. *A subset $H \subset G$ is a subgroup if $H \neq \emptyset$ and H is closed under the operation and inverses. That is, if $a, b \in H$ then $ab \in H$ and $a^{-1}, b^{-1} \in H$.*

Let G be a group and $g \in G$; we denote by g^n, $n \in \mathbb{N}$, as with numbers, the product of g taken n times. A negative exponent will indicate the inverse of the positive exponent. As usual, let $g^0 = 1$. Clearly group exponentiation will satisfy the standard laws of exponents. Now consider the set

$$H = \{1 = g^0, g, g^{-1}, g^2, g^{-2}, \dots\}$$

of all powers of g. We will denote this by $\langle g \rangle$.

Lemma 9.1.5. *If G is a group and $g \in G$ then $\langle g \rangle$ forms a subgroup of G called the **cyclic subgroup** generated by g. The subgroup $\langle g \rangle$ is abelian even if G is not.*

Proof. If $g \in G$ then $g \in \langle g \rangle$ and hence $\langle g \rangle$ is non-empty. Suppose then that $a = g^m$, $b = g^n$ are elements of $\langle g \rangle$. Then $ab = g^n g^m = g^{n+m} \in \langle g \rangle$ so $\langle g \rangle$ is closed under the group operation. Further $a^{-1} = (g^n)^{-1} = g^{-n} \in \langle g \rangle$ so $\langle g \rangle$ is closed under inverses. Therefore $\langle g \rangle$ is a subgroup.

Finally $ab = g^n g^m = g^{n+m} = g^{m+n} = g^m g^n = ba$ and hence $\langle g \rangle$ is abelian. $\square$

Suppose that $g \in G$ and $g^m = 1$ for some positive integer m. Then let n be the smallest positive integer such that $g^n = 1$. It follows that the set of elements $\{1, g, g^2, \dots, g^{n-1}\}$ are all distinct but for any other power g^k we have $g^k = g^t$ for some $k = 0, 1, \dots, n - 1$ (see exercises). The cyclic subgroup generated by g then has order n and we say that g has **order** n which we denote by $o(g) = n$. If no such n exists we say that g has **infinite order**.

9.2 Cosets and Normal Subgroups

In applying group theory to cryptography we must understand some things about the structure of groups. In this section we briefly discuss cosets and Lagrange's theorem, normal subgroups and factor groups and the isomorphism theorems. For more information we refer the reader to [FGR].

We first need the idea of the cosets of subgroup. Let G be a group and let H be a subgroup of G. Then for any $a \in G$, the **left coset** of a in G with respect to H, denoted by aH, is the set

$$aH = \{ah \mid h \in H\}.$$

The **right coset** of a in G with respect to H, written Ha, is defined similarly as the set

$$Ha = \{ha \mid h \in H\}.$$

While left cosets do not have to equal right cosets, the number of distinct left cosets of a subgroup H is exactly the same as the number of distinct right cosets. This number is called the **index** of H in G and is denoted $[H : G]$. It can be shown that the set of cosets of a subgroup partition the group and further each has the same size as the subgroup H. For finite groups this leads to the following result called **Lagrange's theorem** which shows that the order of any subgroup of a finite group must divide the order of the group.

Theorem 9.2.1 (Lagrange's Theorem). *Let G be a finite group and $H \subset G$ a subgroup. Then $|G| = |H|[G : H]$.*

For a finite group this implies that both the order of a subgroup and the index of a subgroup are divisors of the order of the group.

This theorem plays a crucial role in the structure theory of finite groups since it greatly restricts the size of subgroups. For example in a group of order 10 there can be proper non-trivial subgroups only of orders 2 and 5.

As an immediate corollary, we have the following result.

Corollary 9.2.2. *The order of any element $g \in G$, where G is a finite group, divides the order of the group. In particular if $|G| = n$ and $g \in G$ then $o(g)|n$ and $g^n = 1$.*

Normal subgroups are special types of subgroups which play a crucial role in determining the structure of a group. Let G be an arbitrary group. A subgroup H is a **normal subgroup** of G, which we denote by $H \triangleleft G$, if $g^{-1}Hg = H$ for all $g \in G$.

Since $g^{-1}Hg = H$ for all $g \in G$ it follows that $gH = Hg$ for all $g \in G$ and hence for normal subgroups each left coset is also a right coset and vice versa.

Suppose that H_1 and H_2 are subgroups of a group G. We say that H_2 is **conjugate** to H_1 in G if there exists an element $a \in G$ such that $H_2 = aH_1a^{-1}$. H_1, H_2 are the called **conjugate subgroups** of G. For a subgroup H being normal is equivalent to having only one conjugate.

Normal subgroups allow us to build a group structure on the set of cosets. Let G be an arbitrary group and H a normal subgroup of G. Let G/H denote the set of distinct left (and hence also right) cosets of H in G. On this set G/H define the multiplication

$$(g_1H)(g_2H) = g_1g_2H$$

for any elements g_1H, g_2H in G/H.

We then have the following theorem. For a proof see [FGR].

Theorem 9.2.3. *Let G be a group and H a normal subgroup of G. Then G/H under the operation defined above forms a group. This group is called the **factor group** or **quotient group** of G modulo H. The identity element is the coset $1H = H$ and the inverse of a coset gH is $g^{-1}H$.*

Earlier in Chapters 5 and 8 we used the idea of a group homomorphism and group isomorphism. These are closely tied to normal subgroups and factor groups. We now formally discuss them.

If G, H are groups then a **homomorphism** is a mapping $F : G \rightarrow H$ such that $f(g_1g_2) = f(g_1)f(g_2)$ for all $g_1, g_2 \in G$. We say that f preserves the group operations. A group homomorphism is an **isomorphism** if it is also a bijection. Two groups are

isomorphic if there exists an isomorphism between them. Isomorphic groups are algebraically the same. If G_1 and G_2 are isomorphic groups than we write $G_1 \cong G_2$.

If $f : G \rightarrow H$ is a group homomorphism. Then the **kernel** of f denoted $\ker(f)$ is the set of elements of G that map to the identity in H;

$$\ker(F) = \{g \in G \mid f(g) = 1\}.$$

We denote the image of f, which is a subset of H, as $\text{im}(f)$. The following result is called the **group isomorphism theorem** and we sketch a proof in the exercises.

Theorem 9.2.4 (Group Isomorphism Theorem).

(a) *Let G_1 and G_2 be groups and $f : G_1 \rightarrow G_2$ a group homomorphism. Then $\ker(f)$ is a normal subgroup of G_1, $\text{im}(f)$ is a subgroup of G_2 and*

$$G/\ker(f) \cong \text{im}(f).$$

(b) *Conversely, suppose that N is a normal subgroup of a group G. Then there exists a group H and a homomorphism $f : G \rightarrow H$ such that $\ker(f) = N$ and $\text{im}(f) = H$. In particular $H = G/N$ and the map $f : G \rightarrow G/N$ is called the **canonical homomorphism**. This is given by $f(g) = gN$.*

9.3 Examples of Groups

As already mentioned, groups arise in many diverse areas of mathematics. In this section and the next we present specific examples of groups.

First of all any ring or field under addition forms an abelian group. Hence, for example $(\mathbb{Z}, +), (\mathbb{Q}, +), (\mathbb{R}, +), (\mathbb{C}, +)$ where these are respectively the integers, the rationals, the reals and the complex numbers, all are infinite abelian groups. If $\mathbb{Z}_n$ is the modular ring. then for any natural number n, $\mathbb{Z}_n$ forms a finite abelian group under addition.

In a field F, the non-zero elements are all invertible and form a group under multiplication. This is called the **multiplicative group** of the field F and is usually denoted by F^*. Since multiplication in a field is commutative, the multiplicative group of a field is an abelian group.

The multiplicative group of a field is a special case of the **unit group** of a ring. If R is a ring with identity, recall that a **unit** is an element of R with a multiplicative inverse. Hence in $\mathbb{Z}$ the only units are ± 1 while in any field every non-zero element is a unit.

Lemma 9.3.1. *If R is a ring with identity then the set of units in R forms a group under multiplication called the **unit group** of R and is denoted by $U(R)$. If R is a field then $U(R) = R^*$ the multiplicative group of non-zero elements of R.*

Important for cryptography is the following fact. For a proof we refer to [CFR1].

Theorem 9.3.2. *The multiplicative group of a finite field is a cyclic group.*

If q is a prime then the modular ring $\mathbb{Z}_q$ is a field. Hence we have the corollary.

Corollary 9.3.3. *If q is a prime then the multiplicative group $\mathbb{Z}_q^*$ is a cyclic group. A generator of this group is called a **primitive element** modulo q .*

This corollary was used in constructing both the Diffie-Hellman and ElGamal protocols.

To present examples of non-abelian groups we turn to matrices. If F is a field we let

$$GL(n, F) = \{\, n \times n \text{ matrices over } F \text{ with non-zero determinant}\,\}$$

and

$$SL(n, F) = \{\, n \times n \text{ matrices over } F \text{ with determinant one}\,\}.$$

Lemma 9.3.4. *If F is a field then for $n \geq 2$, $GL(n, F)$ forms a non-abelian group under matrix multiplication and $SL(n, F)$ forms a normal subgroup.*

$GL(n, F)$ is called the n-dimensional **general linear group** over F, while $SL(n, F)$ is called the n-dimensional **special linear group** over F.

Groups play an important role in geometry. In any metric geometry an **isometry** is a mapping that preserves distance. To understand a geometry one must understand the group of isometries. We look briefly at the Euclidean geometry of the plane $\mathcal{E}^2$.

An **isometry** or **congruence motion** of $\mathcal{E}^2$ is a transformation or bijection T of $\mathcal{E}^2$ that preserves distance, that is $d(a, b) = d(T(a), T(b))$ for all points $a, b \in \mathcal{E}^2$.

Theorem 9.3.5. *The set of congruence motions of $\mathcal{E}^2$ forms a group called the **Euclidean group**. We denote the Euclidean group by $\mathcal{E}$.*

Lemma 9.3.6. *If D is a geometric figure in $\mathcal{E}^2$ then the set of symmetries of D forms a subgroup of $\mathcal{E}$ called the **symmetry group** of D denoted by $Sym(D)$.*

Example 9.3.7. Let T be an equilateral triangle. Then there are exactly six symmetries of T (see exercises). These are:

I = the identity

r = a rotation of 120° around the center of T

r^2 = a rotation of 270° around the center of T

f = a reflection over the perpendicular bisector of one of the sides

fr = the composition of f and r

fr^2 = the composition of f and r^2

$Sym(T)$ is called the **dihedral group** D_3. In the next section we will see that it is isomorphic to S_3, the symmetric group on 3 symbols.

This can be generalized to any regular n-gon. If D is a regular n-gon, then the symmetry group D_n has $2n$ elements and is called the **dihedral group** of order $2n$. It is generated by elements r and f which satisfy the relations $r^n = f^2 = 1, f^{-1}rf = r^{n-1}$, where r is a rotation of $\frac{2\pi}{n}$ about the center of the n-gon and f is a reflection across a line of symmetry of the regular n-gon.

Hence, D_4, the symmetries of a square, has order 8 and D_5, the symmetries of a regular pentagon, has order 10.

Groups most often appear as groups of transformations or permutations on a set. In this section we will take a short look at permutation groups.

Definition 9.3.8. If A is a set, a **permutation** on A is a one-to-one mapping of A onto itself. We denote by S_A the set of all permutations on A.

Theorem 9.3.9. *For any set A, S_A forms a group under composition called the **symmetric group** on A. If $|A| > 2$ then S_A is non-abelian. Further if A, B have the same cardinality, then $S_A \cong S_B$.*

If $A_1 \subset A$ then those permutations on A that map A_1 to A_1 form a subgroup of S_A called the **stabilizer** of A_1 denoted stab(A_1).

Lemma 9.3.10. *If $A_1 \subset A$ then stab$(A_1) = \{f \in S_A | f : A_1 \rightarrow A_1\}$ forms a subgroup of S_A.*

A **permutation group** is any subgroup of S_A for some set A.

We now look at finite permutation groups. Let A be a finite set say $A = \{a_1, a_2, \ldots, a_n\}$. Then each $f \in S_A$ can be pictured as

$$f = \begin{pmatrix} a_1 & \cdots & a_n \\ f(a_1) & \cdots & f(a_n) \end{pmatrix}.$$

For a_1 there are n choices for $f(a_1)$. For a_2 there are only $n-1$ choices since f is one-to-one. This continues down to only one choice for a_n. Using the multiplication principle, the number of choices for f and therefore the size of S_A is

$$n(n-1)\cdots 1 = n!.$$

We have thus proved the following theorem.

Theorem 9.3.11. *If $|A| = n$ then $|S_A| = n!.$*

For a set with n elements we denote S_A by S_n, called the **symmetric group on n symbols**.

Example 9.3.12. Write down the six elements of S_3 and give the multiplication table for the group.

Name the three elements 1, 2, 3. The six elements of S_3 are then:

$$1 = \begin{pmatrix} 1 & 2 & 3 \\ 1 & 2 & 3 \end{pmatrix}, \quad a = \begin{pmatrix} 1 & 2 & 3 \\ 2 & 3 & 1 \end{pmatrix}, \quad b = \begin{pmatrix} 1 & 2 & 3 \\ 3 & 1 & 2 \end{pmatrix},$$

$$c = \begin{pmatrix} 1 & 2 & 3 \\ 2 & 1 & 3 \end{pmatrix}, \quad d = \begin{pmatrix} 1 & 2 & 3 \\ 3 & 2 & 1 \end{pmatrix}, \quad e = \begin{pmatrix} 1 & 2 & 3 \\ 1 & 3 & 2 \end{pmatrix}.$$

The multiplication table for S_3 can be written down directly by doing the required composition. For example,

$$ac = \begin{pmatrix} 1 & 2 & 3 \\ 2 & 3 & 1 \end{pmatrix}\begin{pmatrix} 1 & 2 & 3 \\ 2 & 1 & 3 \end{pmatrix} = \begin{pmatrix} 1 & 2 & 3 \\ 3 & 2 & 1 \end{pmatrix} = d.$$

To see this, note that $a : 1 \to 2, 2 \to 3, 3 \to 1; c : 1 \to 2, 2 \to 1, 3 \to 3$ and so $ac : 1 \to 3, 2 \to 2, 3 \to 1$.

It is somewhat easier to construct the multiplication table if we make some observations. First, $a^2 = b$ and $a^3 = 1$. Next, $c^2 = 1, d = ac, e = a^2c$ and finally $ac = ca^2$.

From these relations the following multiplication table can be constructed

	1	a	a^2	c	ac	a^2c
1	1	a	a^2	c	ac	a^2c
a	a	a^2	1	ac	a^2c	c
a^2	a^2	1	a	a^2c	c	ac
c	c	a^2c	ac	1	a^2	a
ac	ac	c	a^2c	a	1	a^2
a^2c	a^2c	ac	c	a^2	a	1

To see this, consider, for example, $(ac)a^2 = a(ca^2) = a(ac) = a^2c$.

More generally, we can say that S_3 has a **presentation** given by

$$S_3 = \langle a, c; a^3 = c^2 = 1, ac = ca^2 \rangle.$$

By this we mean that S_3 is **generated by** a, c, or that S_3 has **generators** a, c and the whole group and its multiplication table can be generated by using the **relations** $a^3 = c^2 = 1, ac = ca^2$.

We now consider the case where a group G has a single generator that is a cyclic group.

Definition 9.3.13. A group G is **cyclic** if there exists a $g \in G$ such that $G = \langle g \rangle$.

In this case $G = \{g^n \mid n \in \mathbb{Z}\}$, that is G consists of all the powers of the element g. If there exists an integer m such that $g^m = 1$, then there exists a smallest such positive integer say n. It follows that $g^k = g^l$ if and only if $k \equiv l \pmod{n}$. In this situation the distinct powers of g are precisely

$$\{1 = g^0, g, g^2, \ldots, g^{n-1}\}.$$

It follows that $|G| = n$. We then call G a **finite cyclic group**. If no such power exists then all the powers of G are distinct and G is an infinite cyclic group.

We show next that any two cyclic groups of the same order are isomorphic.

Theorem 9.3.14.
(a) *If $G = \langle g \rangle$ is an infinite cyclic group then $G \cong (\mathbb{Z}, +)$ that is the integers under addition.*
(b) *If $G = \langle g \rangle$ is a finite cyclic group of order n then $G \cong (\mathbb{Z}_n, +)$ that is the integers modulo n under addition.*
It follows that for a given order there is only one cyclic group up to isomorphism.

Theorem 9.3.15. *Let $G = \langle g \rangle$ be a finite cyclic group of order n. Then every subgroup of G is also cyclic. Further if $d \mid n$ there exists a unique subgroup of G of order d.*

Theorem 9.3.16. *Let $G = \langle g \rangle$ be an infinite cyclic group. Then a subgroup H is of the form $H = \langle g^t \rangle$ for a positive integer t. Further if t_1, t_2 are positive integers with $t_1 \neq t_2$ then $\langle g^{t_1} \rangle$ and $\langle g^{t_2} \rangle$ are distinct.*

Theorem 9.3.17. *Let $G = \langle g \rangle$ be a cyclic group. Then:*
(a) *If $G = \langle g \rangle$ is finite of order n then g^k is also a generator if and only if $(k, n) = 1$. That is the generators of G are precisely those powers g^k where k is relatively prime to n.*
(b) *If $G = \langle g \rangle$ is infinite then the only generators are g, g^{-1}.*

9.4 Generators and Group Presentations

Crucial to using non-abelian group theory in cryptography is the concept of a **group presentation**. Roughly for a group G a presentation consists of a set of **generators** X for G, so that $G = \langle X \rangle$, and a set of **relations** between the elements of X from which, in principle, the whole group table can be constructed. In this chapter we make this concept precise. As we will see, every group G has a presentation, but it is mainly in the case where the group is finite, or countably infinite, that presentations are most useful. Historically, the idea of group presentations arose out of the attempt to describe the countably infinite fundamental groups that came out of low dimensional topology. The study of groups using group presentations is called **combinatorial group theory**.

Before looking at group presentations in general, we revisit two examples of finite groups, and then a class of infinite groups.

Consider the symmetric group on 3 symbols, S_3. We saw that it has the following 6 elements:

$$1 = \begin{pmatrix} 1 & 2 & 3 \\ 1 & 2 & 3 \end{pmatrix}, \quad a = \begin{pmatrix} 1 & 2 & 3 \\ 2 & 3 & 1 \end{pmatrix}, \quad b = \begin{pmatrix} 1 & 2 & 3 \\ 3 & 1 & 2 \end{pmatrix},$$

$$c = \begin{pmatrix} 1 & 2 & 3 \\ 2 & 1 & 3 \end{pmatrix}, \quad d = \begin{pmatrix} 1 & 2 & 3 \\ 3 & 2 & 1 \end{pmatrix}, \quad e = \begin{pmatrix} 1 & 2 & 3 \\ 1 & 3 & 2 \end{pmatrix}.$$

Notice that $a^3 = 1, c^2 = 1$ and that $ac = ca^2$. We claim that

$$\langle a, c; a^3 = c^2 = (ac)^2 = 1 \rangle$$

is a presentation for S_3. First it is easy to show that $S_3 = \langle a, c \rangle$. Indeed

$$1 = 1, \quad a = a, \quad b = a^2, \quad c = c, \quad d = ac, \quad e = a^2c$$

and so clearly a, c generate S_3.

Now from $(ac)^2 = acac = 1$ we get that $ca = a^2c$. This implies that if we write any sequence (or word in our later language) in a and c we can also rearrange it so that the only powers of a are a and a^2, the only powers of c are c and all a terms precede c terms. For example

$$aca^2cac = aca(acac) = a(ca) = a(a^2c) = (a^3)c = c.$$

Therefore using the three relations form the presentation above each element of S_3 can be written as $a^\alpha c^\beta$ with $\alpha = 0, 1, 2$ and $\beta = 0, 1$. From this the multiplication of any two elements can be determined.

Now let us look at D_3, the symmetry group on an equilateral triangle. In the last section we saw that $|D_3| = 6$ and can be generated by a rotation r of $120°$ around the center of the triangle and a flip f across any of the medians. We then had the relations $r^3 = f^2 = 1, fr = r^2f$. From these relations the group table can be constructed. Further, the relation $fr = r^2f$ can be derived from the relation $(rf)^2 = 1$ combined with $r^3 = f^2 = 1$. Hence a presentation for D_3 is given by

$$D_3 = \langle r, f; r^3 = f^2 = (rf)^2 = 1 \rangle.$$

Notice that except for the change of letters this is the same as the presentation for S_3. The map $a \rightarrow r, c \rightarrow f$ gives an isomorphism between S_3 and D_3.

We can also see that D_3 is isomorphic to S_3 via the following argument. Each symmetry on an equilateral triangle permutes the vertices. Hence each symmetry on an equilateral triangle corresponds to a permutation on 3 symbols. It follows that D_3 can be considered as a subgroup of S_3. However, $|D_3| = 6 = |S_3|$ and hence they must be the same.

In general, we let D_n stand for the symmetry group of a regular n-gon. It is called the **dihedral group** of order $2n$. Exactly this same type of argument used for an equilateral triangle applies to all the dihedral groups D_n. Therefore in general $|D_n| = 2n$ and the group is generated generated by a rotation r of angle $\frac{2\pi}{n}$ about the center of the n-gon and a reflection f about any line of symmetry. The rotation r would have order n so that $r^n = 1$. The reflection f would have order 2 so that $f^2 = 1$. The element rf is then a reflection about the rotated line which is also a line of symmetry. Therefore $(rf)^2 = 1$. Exactly as for D_3 the relation $(rf)^2 = 1$ implies that $fr = r^{-1}f = r^{n-1}f$. This allows us to always place r terms in front of f terms in any word on r and f. Therefore the elements of D_n are always of the form

$$r^\alpha f^\beta, \quad \alpha = 0, 1, 2, \ldots, n - 1, \quad \beta = 0, 1$$

and further the relations $r^n = f^2 = (rf)^2 = 1$ allow us to rearrange any word in r and f into this form. It follows that $|\langle r, f \rangle| = 2n$ and hence $D_n = \langle r, f \rangle$ together with the relations above. Putting these comments all together we obtain:

Theorem 9.4.1. *If D_n is the symmetry group of a regular n-gon then $|D_n| = 2n$ and a presentation for D_n is given by*

$$D_n = \langle r, f; r^n = f^2 = (rf)^2 = 1 \rangle.$$

We now give one class of infinite examples. If G is an infinite cyclic group, so that $G \cong \mathbb{Z}$, then $G = \langle g; \rangle$ is a presentation for G. That is G has a single generator with no relations.

A direct product of n copies of $\mathbb{Z}$ is called a **free abelian group** of rank n. We will denote this by $\mathbb{Z}^n$. A presentation for $\mathbb{Z}^n$ is then given by

$$\mathbb{Z}^n = \langle x_1, x_2, \ldots, x_n; x_i x_j = x_j x_i \text{ for all } i, j = 1, \ldots, n \rangle.$$

9.5 Free Groups and Group Presentations

Crucial to the concept of a group presentation is the idea of a **free group**.

Definition 9.5.1. A group F is **free on a subset** X if every map $f : X \to G$ with G a group can be extended to a unique homomorphism $f : F \to G$. X is called a **free basis** for F. In general, a group F is a free group, if it is free on some subset X. If X is a free basis for a free group F we write $F = F(X)$.

We first show that given any set X there does exist a free group with free basis X. Let $X = \{x_i\}_{i \in I}$ be a set (possibly empty). We will construct a group $F(X)$ which is free with free basis X. First let X^{-1} be a set disjoint from X but bijective to X. If $x_i \in X$ then the corresponding element of X^{-1} under the bijection we denote x_i^{-1} and say that x_i and x_i^{-1} are **associated**. The set X^{-1} is called the **set of formal inverses** from X and we call $X \cup X^{-1}$ the **alphabet**. Elements of the alphabet are called **letters**, hence a letter has the form $x_i^{\epsilon_i}$ where $\epsilon_i = \pm 1$. A **word** in X is a finite sequence of letters from the alphabet. That is a word has the form

$$w = x_{i_1}^{\epsilon_{i_1}} x_{i_2}^{\epsilon_{i_2}} \cdots x_{i_n}^{\epsilon_{i_n}},$$

where $x_{i_j} \in X$ and $\epsilon_{i_j} = \pm 1$. If $n = 0$ we call it the **empty word** which we will denote e. Words of the form $x_i x_i^{-1}$ or $x_i^{-1} x_i$ are called **trivial words**. We let $W(X)$ be the set of all words on X.

If $w_1, w_2 \in W(X)$ we say that w_1 is **equivalent** to w_2, denoted $w_1 \sim w_2$, if w_1 can be converted to w_2 by a finite string of insertions and deletions of trivial words. For example if $w_1 = x_3 x_4 x_4^{-1} x_2 x_2$ and $w_2 = x_3 x_2 x_2$ then $w_1 \sim w_2$. It is straightforward to verify that this is an equivalence relation on $W(X)$ (see exercises). Let $F(X)$ denote the set of equivalence classes in $W(X)$ under this relation, hence $F(X)$ is a set of equivalence classes of words from X.

A word $w \in W(X)$ is said to be **freely reduced** or **reduced** if it has no trivial subwords (a subword is a connected sequence within a word). Hence in the example above $w_2 = x_3 x_2 x_2$ is reduced but $w_1 = x_3 x_4 x_4^{-1} x_2 x_2$ is not reduced. In each equivalence class in $F(X)$ there is a unique element of minimal length. Further this element must be reduced or else it would be equivalent to something of smaller length. Two reduced words in $W(X)$ are either equal or not in the same equivalence class in $F(X)$. Hence $F(X)$ can also be considered as the set of all reduced words from $W(X)$.

Given a word $w = x_{i_1}^{\epsilon_{i_1}} x_{i_2}^{\epsilon_{i_2}} \cdots x_{i_n}^{\epsilon_{i_n}}$ we can find the unique reduced word $\overline{w}$ equivalent to w via the following **free reduction process**. Beginning from the left side of w we cancel each occurrence of a trivial subword. After all these possible cancellations we have a word w'. Now we repeat the process again starting from the left side. Since w has finite length eventually the resulting word will either be empty or reduced. The final reduced $\overline{w}$ is the **free reduction** of w.

Hence any freely reduced word w has a unique representation

$$w = x_{i_1}^{\epsilon_{i_1}} x_{i_2}^{\epsilon_{i_2}} \cdots x_{i_n}^{\epsilon_{i_n}}$$

where $x_{i_j} \in X, \epsilon_{i_j} = \pm 1$ and there is no further possible free reduction. The number n of letters in the freely reduced form is called the **free length** of w denoted by $|w|$.

Now we build a multiplication on $F(X)$. If

$$w_1 = x_{i_1}^{\epsilon_{i_1}} x_{i_2}^{\epsilon_{i_2}} \cdots x_{i_n}^{\epsilon_{i_n}}, \quad w_2 = x_{j_1}^{\epsilon_{j_1}} x_{j_2}^{\epsilon_{j_2}} \cdots x_{j_m}^{\epsilon_{j_m}}$$

are two words in $W(X)$ then their **concatenation** $w_1 \star w_2$ is simply pacing w_2 after w_1,

$$w_1 \star w_2 = x_{i_1}^{\epsilon_{i_1}} x_{i_2}^{\epsilon_{i_2}} \cdots x_{i_n}^{\epsilon_{i_n}} x_{j_1}^{\epsilon_{j_1}} x_{j_2}^{\epsilon_{j_2}} \cdots x_{j_m}^{\epsilon_{j_m}}.$$

If $w_1, w_2 \in F(x)$ then we define their product as

$$w_1 w_2 = \text{equivalence class of } w_1 \star w_2.$$

That is we concatenate w_1 and w_2, and the product is the equivalence class of the resulting word. It is easy to show that if $w_1 \sim w_1'$ and $w_2 \sim w_2'$ then $w_1 \star w_2 \sim w_1' \star w_2'$ so that the above multiplication is well-defined. Equivalently we can think of this product in the following way. If w_1, w_2 are reduced words then to find $w_1 w_2$ first concatenate and then freely reduce. Notice that if $x_{i_n}^{\epsilon_{i_n}} x_{j_1}^{\epsilon_{j_1}}$ is a trivial word then it is cancelled when the concatenation is formed. We say then that there is **cancellation** in forming the product $w_1 w_2$. Otherwise the product is formed without cancellation.

Theorem 9.5.2. *Let X be a non-empty set and let $F(X)$ be as above. Then $F(X)$ is a free group with free basis X. Further if $X = \emptyset$ then $F(X) = \{1\}$, if $|X| = 1$ then $F(X) \cong \mathbb{Z}$ and if $|X| \geq 2$ then $F(X)$ is non-abelian.*

Proof. We first show that $F(X)$ is a group and then show that it satisfies the universal mapping property on X. We consider $F(X)$ as the set of reduced words in $W(X)$ with the multiplication defined above. Clearly the empty word acts as the identity element 1. If $w = x_{i_1}^{\epsilon_{i_1}} x_{i_2}^{\epsilon_{i_2}} \cdots x_{i_n}^{\epsilon_{i_n}}$ and $w_1 = x_{i_n}^{-\epsilon_{i_n}} x_{i_{n-1}}^{-\epsilon_{i_{n-1}}} \cdots x_{i_1}^{-\epsilon_{i_1}}$ then both $w \star w_1$ and $w_1 \star w$ freely reduce to the empty word and so w_1 is the inverse of w. Therefore each element of $F(X)$ has an inverse. Therefore to show that $F(X)$ forms a group we must show that the multiplication is associative. Let

$$w_1 = x_{i_1}^{\epsilon_{i_1}} x_{i_2}^{\epsilon_{i_2}} \cdots x_{i_n}^{\epsilon_{i_n}}, \quad w_2 = x_{j_1}^{\epsilon_{j_1}} x_{j_2}^{\epsilon_{j_2}} \cdots x_{j_m}^{\epsilon_{j_m}}, \quad w_3 = x_{k_1}^{\epsilon_{k_1}} x_{k_2}^{\epsilon_{k_2}} \cdots x_{k_p}^{\epsilon_{k_p}}$$

be three freely reduced words in $F(X)$. We must show that

$$(w_1 w_2)w_3 = w_1(w_2 w_3).$$

To prove this we use induction on m, the length of w_2. If $m = 0$ then w_2 is the empty word and hence the identity and it is certainly true. Now suppose that $m = 1$ so that $w_2 = x_{j_1}^{\epsilon_{j_1}}$. We must consider exactly four cases.

Case (1): There is no cancellation in forming either $w_1 w_2$ or $w_2 w_3$. That is $x_{j_1}^{\epsilon_{j_1}} \neq x_{i_n}^{-\epsilon_{i_n}}$ and $x_{j_1}^{\epsilon_{j_1}} \neq x_{k_1}^{-\epsilon_{k_1}}$. Then the product $w_1 w_2$ is just the concatenation of the words and so is $(w_1 w_2)w_3$. The same is true for $w_1(w_2 w_3)$. Therefore in this case $w_1(w_2 w_3) = (w_1 w_2)w_3$.

Case (2): There is cancellation in forming $w_1 w_2$ but not in forming $w_2 w_3$. Then if we concatenate all three words the only cancellation occurs between w_1 and w_2 in either $w_1(w_2 w_3)$ or in $(w_1 w_2)w_3$ and hence they are equal. Therefore in this case $w_1(w_2 w_3) = (w_1 w_2)w_3$.

Case (3): There is cancellation in forming $w_2 w_3$ but not in forming $w_1 w_2$. This is entirely analogous to Case (2) so therefore in this case $w_1(w_2 w_3) = (w_1 w_2)w_3$.

Case (4): There is cancellation in forming $w_1 w_2$ and also in forming $w_2 w_3$. Then $x_{j_1}^{\epsilon_{j_1}} = x_{i_n}^{-\epsilon_{i_n}}$ and $x_{j_1}^{\epsilon_{j_1}} = x_{k_1}^{-\epsilon_{k_1}}$. Here

$$(w_1 w_2)w_3 = x_{i_1}^{\epsilon_{i_1}} \cdots x_{i_{n-1}}^{\epsilon_{i_{n-1}}} x_{k_1}^{\epsilon_{k_1}} x_{k_2}^{\epsilon_{k_2}} \cdots x_{k_p}^{\epsilon_{k_p}}.$$

On the other hand,

$$w_1(w_2 w_3) = x_{i_1}^{\epsilon_{i_1}} \cdots x_{i_n}^{\epsilon_{i_n}} x_{k_2}^{\epsilon_{k_2}} \cdots x_{k_p}^{\epsilon_{k_p}}.$$

However, these are equal since $x_{i_n}^{\epsilon_{i_n}} = x_{k_1}^{\epsilon_{k_1}}$. Therefore, in this final case, we have $w_1(w_2 w_3) = (w_1 w_2)w_3$. It follows inductively from these four cases that the associative law holds in $F(X)$, and therefore $F(X)$ forms a group.

Now suppose that $f : X \to G$ is a map from X into a group G. By the construction of $F(X)$ as a set of reduced words, this can be extended to a unique homomorphism. If $w \in F$ with $w = x_{i_1}^{\epsilon_{i_1}} \cdots x_{i_n}^{\epsilon_{i_n}}$ then define $f(w) = f(x_{i_1})^{\epsilon_{i_1}} \cdots f(x_{i_n})^{\epsilon_{i_n}}$. Since multiplication in $F(X)$ is concatenation this defines a homomorphism and again form the construction of $F(X)$ its the only one extending f. This is analogous to constructing a linear transformation from one vector space to another by specifying the images of a basis. Therefore $F(X)$ satisfies the universal mapping property of Definition 9.5.1 and hence $F(X)$ is a free group with free basis X.

The final parts are straightforward. If X is empty the only reduced word is the empty word and hence the group is just the identity. If X has a single letter then $F(X)$ has a single generator and is therefore cyclic. It is easy to see that it must be torsion-free and therefore $F(X)$ is infinite cyclic, that is $F(x) \cong \mathbb{Z}$. Finally if $|X| \geq 2$ let $x_1, x_2 \in X$. Then $x_1 x_2 \neq x_2 x_1$ and both are reduced. Therefore $F(X)$ is non-abelian. $\square$

The proof of the above theorem provides another way to look at free groups.

Theorem 9.5.3. *F is a free group if and only if there is a generating set X such that every element of F has a unique representation as a freely reduced word on X.*

The structure of a free group is entirely dependent on the cardinality of a free basis. In particular the cardinality of a free basis $|X|$ for a free group F is unique and is called the **rank of F**. If $|X| < \infty$, F is of **finite rank**. If F has rank n and $X = \{x_1, \ldots, x_n\}$ we say that F is free on $\{x_1, \ldots, x_n\}$. We denote this by $F(x_1, x_2, \ldots, x_n)$.

Theorem 9.5.4. *If X and Y are sets with the same cardinality, that is, $|X| = |Y|$, then $F(X) \cong F(Y)$, the resulting free groups are isomorphic. Further if $F(X) \cong F(Y)$ then $|X| = |Y|$.*

Proof. Suppose that $f : X \rightarrow Y$ is a bijection from X onto Y. Now $Y \subset F(Y)$ so there is a unique homomorphism $\phi : F(X) \rightarrow F(Y)$ extending f. Since f is a bijection it has an inverse $f^{-1} : Y \rightarrow X$ and since $F(Y)$ is free there is a unique homomorphism ϕ_1 from $F(Y)$ to $F(X)$ extending f^{-1}. Then $\phi\phi_1$ is the identity map on $F(Y)$ and $\phi_1\phi$ is the identity map on $F(X)$. Therefore ϕ, ϕ_1 are isomorphisms with $\phi = \phi_1^{-1}$.

Conversely suppose that $F(X) \cong F(Y)$. In $F(x)$ let $N(X)$ be the subgroup generated by all squares in $F(X)$ that is

$$N(X) = \langle g^2; g \in F(X) \rangle.$$

Then $N(X)$ is a normal subgroup and the factor group $F(X)/N(X)$ is abelian where every non-trivial element has order 2 (see exercises). Hence $F(X)/N(X)$ can be considered as a vector space over $\mathbb{Z}_2$, the finite field of order 2, with X as a vector space basis. Hence $|X|$ is the dimension of this vector space. Let $N(Y)$ be the corresponding subgroup of $F(Y)$. Since $F(X) \cong F(Y)$ we would have $F(X)/N(X) \cong F(Y)/N(Y)$ and therefore $|Y|$ is the dimension of the vector space $F(Y)/N(Y)$. Therefore $|X| = |Y|$ from the uniqueness of dimension of vector spaces. $\square$

Expressing an element of $F(X)$ as a reduced word gives a **normal form** for elements in a free group F. This solves what is termed the **word problem** for free groups. Another important concept is the following: a freely reduced word $W = x_{v_1}^{e_1} x_{v_2}^{e_2} \cdots x_{v_n}^{e_n}$ is **cyclically reduced** if $v_1 \neq v_n$ or if $v_1 = v_n$ then $e_1 \neq -e_n$. Clearly then every element of a free group is conjugate to an element given by a cyclically reduced word. This provides a method to determine conjugacy in free groups.

Theorem 9.5.5. *In a free group F two elements g_1, g_2 are conjugate if and only if a cyclically reduced word for g_1 is a cyclic permutation of a cyclically reduced word for g_2.*

The theory of free groups has a large and extensive literature. We close this section by stating several important properties. Proofs for these results can be found in [MKS], [LS], or [CRR].

Theorem 9.5.6. *A free group is torsion-free.*

From Theorem 9.4.2 we can deduce:

Theorem 9.5.7. *An abelian subgroup of a free group must be cyclic.*

Finally a celebrated theorem of Nielsen and Schreier states that a subgroup of a free group must be free.

Theorem 9.5.8 (Nielsen-Schreier). *A subgroup of a free group is itself a free group.*

Combinatorially F is free on X if X is a set of generators for F and there are no non-trivial relations. In particular:

There are several different proofs of this result (see [MKS], [LS], or [CRR]) with the most straightforward being topological in nature. Later we will give an outline of a simple topological proof.

Nielsen, using a technique now called Nielsen transformations in his honor, first proved this theorem about 1920 for finitely generated subgroups. Schreier shortly after found a combinatorial method to extend this to arbitrary subgroups. Complete versions of the original combinatorial proof appear in [MKS], [LS], [CRR], and in the notes by Johnson [Joh].

Schreier's combinatorial proof also allows for a description of the free basis for the subgroup. In particular, let F be free on X, and $H \subset F$ a subgroup. Let $T = \{t_\alpha\}$ be a complete set of right coset representatives for F modulo H with the property that if $t_\alpha = x_{v_1}^{e_1} x_{v_2}^{e_2} \cdots x_{v_n}^{e_n} \in T$, with $\epsilon_i = \pm 1$, then all the initial segments $1, x_{v_1}^{e_1}, x_{v_1}^{e_1} x_{v_2}^{e_2}$, etc. are also in T. Such a system of coset representatives can always be found and is called a **Schreier system** or **Schreier transversal** for H. If $g \in F$ let $\overline{g}$ represent its coset representative in T and further define for $g \in F$ and $t \in T$, $S_{tg} = tg(\overline{tg})^{-1}$. Notice that $S_{tg} \in H$ for all t, g. We then have:

Theorem 9.5.9 (Explicit Form of Nielsen-Schreier). *Let F be free on X and H a subgroup of F. If T is a Schreier transversal for F modulo H then H is free on the set $\{S_{tx} | t \in T, x \in X, S_{tx} \neq 1\}$.*

Example 9.5.10. Let F be free on $\{a, b\}$ and $H = F(X^2)$ the normal subgroup of F generated by all squares in F.

Then $F/F(X^2) = \langle a, b; a^2 = b^2 = (ab)^2 = 1 \rangle = \mathbb{Z}_2 \times \mathbb{Z}_2$. It follows that a Schreier system for F modulo H is $\{1, a, b, ab\}$ with $\bar{a} = a, \bar{b} = b$ and $\overline{ba} = ab$. From this it can be shown that H is free on the generating set

$$x_1 = a^2, \quad x_2 = bab^{-1}a^{-1}, \quad x_3 = b^2, \quad x_4 = abab^{-1}, \quad x_5 = ab^2a^{-1}.$$

The theorem also allows for a computation of the rank of H given the rank of F and the index. Specifically:

Corollary 9.5.11. *Suppose F is free of rank n and $|F : H| = k$. Then H is free of rank $nk - k + 1$.*

From the example we see that F is free of rank 2, H has index 4 so H is free of rank $2 \cdot 4 - 4 + 1 = 5$.

9.6 Group Presentations

The significance of free groups stems from the following result which is easily deduced from the definition, and will lead us directly to a formal definition of a group presentation. Let G be any group and F the free group on the elements of G considered as a set. The identity map $f : G \to G$ can be extended to a homomorphism of F onto G, therefore:

Theorem 9.6.1. *Every group G is a homomorphic image of a free group. That is let G be any group. Then $G = F/N$ where F is a free group.*

In the above theorem, instead of taking all the elements of G we can consider just a set X of generators for G. Then G is a factor group of $F(X)$, $G \cong F(X)/N$. The normal subgroup N is the kernel of the homomorphism from $F(X)$ onto G. We use Theorem 9.4.1 to formally define a group presentation.

If H is a subgroup of a group G then the **normal closure** of H denoted by $N(H)$ is the smallest normal subgroup of G containing H. This can be described alternatively in the following manner. The normal closure of H is the subgroup of G generated by all conjugates of elements of H.

Now suppose that G is a group with X a set of generators for G. We also call X a **generating system** for G. Now let $G = F(X)/N$ as in Theorem 9.3.1 and the comments after it. N is the kernel of the homomorphism $f : F(X) \to G$. It follows that if r is a free group word with $r \in N$ then $r = 1$ in G (under the homomorphism). We then call r a **relator** in G and the equation $r = 1$ a **relation** in G. Suppose that R is a subset of N such that $N = N(R)$, then R is called a set of **defining relators** for G. The equations $r = 1, r \in R$, are a set of **defining relations** for G. It follows that any relator in G is a product of conjugates of elements of R. Equivalently $r \in F(X)$ is a relator in G if and only if r can be reduced to the empty word by insertions and delations of elements of R and trivial words.

Definition 9.6.2. Let G be a group. Then a **group presentation** for G consists of a set of generators X for G and a set R of defining relators. In this case we write $G = \langle X; R \rangle$. We often write the presentation in terms of defining relations as $G = \langle X; r = 1, r \in R \rangle$.

From Theorem 9.3.1 it follows immediately that every group has a presentation. However, in general there are many presentations for the same group. If $R \subset R_1$ then R_1 is also a set of defining relators.

Lemma 9.6.3. *Let G be a group. Then G has a presentation.*

If $G = \langle X; R \rangle$ and X is finite then G is said to be **finitely generated**. If R is finite, G is **finitely related**. If both X and R are finite, G is **finitely presented**.

Using group presentations we get another characterization of free groups.

Theorem 9.6.4. *F is a free group if and only if F has a presentation of the form $F = \langle X; \rangle$.*

Mimicking the construction of a free group from a set X, we can show that to each presentation corresponds a group. Suppose that we are given a supposed presentation $\langle X; R \rangle$ where R is given as a set of words in X. Consider the free group $F(X)$ on X. Define two words w_1, w_2 on X to be equivalent if w_1 can be transformed into w_2 using insertions and deletions of elements of R and trivial words. As in the free group case this is an equivalence relation. Let G be the set of equivalence classes. If we define multiplication, as before, as concatenation followed by the appropriate equivalence class, then G is a group. Further, each $r \in R$ must equal the identity in G, so that $G = \langle X; R \rangle$. Notice that here there may be no unique reduced word for an element of G.

Theorem 9.6.5. *Given $\langle X; R \rangle$ where X is a set and R is a set of words on X. Then there exists a group G with presentation $\langle X; R \rangle$.*

We now give some examples of group presentations.

Example 9.6.6. A free group of rank n has a presentation

$$F_n = \langle x_1, \ldots, x_n; \rangle.$$

Example 9.6.7. A free abelian group of rank n has a presentation

$$\mathbb{Z}^n = \langle x_1, \ldots, x_n; x_i x_j x_i^{-1} x_j^{-1}, \ i = 1, \ldots, n, \ j = 1, \ldots, n \rangle.$$

Example 9.6.8. A cyclic group of order n has a presentation

$$\mathbb{Z}_n = \langle x; \ x^n = 1 \rangle.$$

Example 9.6.9. The dihedral groups of order $2n$ representing the symmetry group of a regular n-gon has a presentation

$$\langle r, f; \ r^n = f^2 = (rf)^2 = 1 \rangle.$$

We looked at this example in Section 9.3.

9.6.1 The Modular Group

In this section we give a more complicated example that will be used in group based cryptography.

If R is any ring with an identity then the set of invertible $n \times n$ matrices with entries from R forms a group under matrix multiplication called the **n-dimensional general linear group over R** (see [Rot]). This group is denoted by $GL_n(R)$. Since $\det(A)\det(B) = \det(AB)$ for square matrices A, B, it follows that the subset of $GL_n(R)$ consisting of those matrices of determinant 1 forms a subgroup. This subgroup is called the **special linear group over R** and is denoted by $SL_n(R)$. In this section we concentrate on $SL_2(\mathbb{Z})$, or more specifically a quotient of it, $PSL_2(\mathbb{Z})$ and find presentations for this group.

The group $SL_2(\mathbb{Z})$ consists of 2×2 integral matrices of determinant one:

$$SL_2(\mathbb{Z}) = \left\{ \begin{pmatrix} a & b \\ c & d \end{pmatrix}; \ a, b, c, d \in \mathbb{Z}, ad - bc = 1 \right\}.$$

$SL_2(\mathbb{Z})$ is called the **homogeneous modular group** and an element of $SL_2(\mathbb{Z})$ is called a **unimodular matrix**.

If G is any group, recall that its **center,** $Z(G)$ consists of those elements of G which commute with all elements of G, that is,

$$Z(G) = \{g \in G \mid gh = hg, \forall h \in G\}.$$

$Z(G)$ is a normal subgroup of G, and hence we can form the factor group $G/Z(G)$. For $G = SL_2(\mathbb{Z})$, the only unimodular matrices that commute with all others are $\pm I = \pm \begin{pmatrix} 1 & 0 \\ 0 & 1 \end{pmatrix}$. Therefore

$$Z(SL_2(\mathbb{Z})) = \{I, -I\}.$$

The quotient

$$SL_2(\mathbb{Z})/Z(SL_2(\mathbb{Z})) = SL_2(\mathbb{Z})/\{I, -I\}$$

is denoted $PSL_2(\mathbb{Z})$ and is called the **projective special linear group** or **inhomogeneous modular group**. More commonly $PSL_2(\mathbb{Z})$ is just called the **Modular Group** and denoted by M.

M arises in many different areas of mathematics including number theory, complex analysis, Riemann surface theory and the theory of automorphic forms and functions. M is perhaps the most widely studied single finitely presented group. Complete discussions of M and its structure can be found in the books [Fin], [New], and [CRR].

Since $M = PSL_2(\mathbb{Z}) = SL_2(\mathbb{Z})/\{I, -I\}$ it follows that each element of M can be considered as $\pm A$ where A is a unimodular matrix. A **projective unimodular matrix** is then

$$\pm \begin{pmatrix} a & b \\ c & d \end{pmatrix}, \quad a, b, c, d \in \mathbb{Z}, \quad ad - bc = 1.$$

The elements of M can also be considered as linear fractional transformations over the complex numbers

$$z' = \frac{az + b}{cz + d}, \quad a, b, c, d \in \mathbb{Z}, \quad ad - bc = 1, \quad \text{where } z \in \mathbb{C} \cup \{\infty\}.$$

Thought of in this way, M forms a **Fuchsian group**, which is a discrete group of isometries of the non-Euclidean hyperbolic plane. The book by Katok [Kat] gives a solid and clear introduction to such groups. This material can also be found in condensed form in [FR].

We now determine presentations for both $SL_2(\mathbb{Z})$ and $M = PSL_2(\mathbb{Z})$.

Theorem 9.6.10. *The group* $SL_2(\mathbb{Z})$ *is generated by the elements*

$$X = \begin{pmatrix} 0 & -1 \\ 1 & 0 \end{pmatrix} \quad and \quad Y = \begin{pmatrix} 0 & 1 \\ -1 & -1 \end{pmatrix}.$$

Further, a complete set of defining relations for the group in terms of these generators is given by

$$X^4 = Y^3 = YX^2Y^{-1}X^{-2} = I.$$

It follows that $SL_2(\mathbb{Z})$ *has the* **presentation**

$$\langle X, Y; X^4 = Y^3 = YX^2Y^{-1}X^{-2} = I \rangle.$$

Proof. We first show that $SL_2(\mathbb{Z})$ is generated by X and Y, that is, every matrix A in the group can be written as a product of powers of X and Y. Let

$$U = \begin{pmatrix} 1 & 1 \\ 0 & 1 \end{pmatrix}.$$

Then a direct multiplication shows that $U = XY$, and we show that $SL_2(\mathbb{Z})$ is generated by X and U which implies that it is also generated by X and Y. Further,

$$U^n = \begin{pmatrix} 1 & n \\ 0 & 1 \end{pmatrix}$$

so that U has infinite order.

Let $A = \begin{pmatrix} a & b \\ c & d \end{pmatrix} \in SL_2(\mathbb{Z})$. Then we have

$$XA = \begin{pmatrix} -c & -d \\ a & b \end{pmatrix} \quad \text{and} \quad U^k A = \begin{pmatrix} a + kc & b + kd \\ c & d \end{pmatrix}$$

for any $k \in \mathbb{Z}$. We may assume that $|c| \leq |a|$. Otherwise start with XA rather than A. If $c = 0$ then $A = \pm U^q$ for some q. If $A = U^q$ then certainly A is in the group generated by X and U. If $A = -U^q$ then $A = X^2 U^q$ since $X^2 = -I$. It follows that here also A is in the group generated by X and U.

Now suppose $c \neq 0$. Apply the Euclidean algorithm to a and c in the following modified way:

$$a = q_0 c + r_1$$
$$-c = q_1 r_1 + r_2$$
$$r_1 = q_2 r_2 + r_3$$
$$\cdots$$
$$(-1)^n r_{n-1} = q_n r_n + 0$$

where $r_n = \pm 1$ since $(a, c) = 1$. Then

$$XU^{-q_n} \cdots XU^{-q_0} A = \pm U^{q_{n+1}} \quad \text{with} \quad q_{n+1} \in \mathbb{Z}.$$

Therefore

$$A = X^m U^{q_0} X U^{q_1} \cdots X U^{q_n} X U^{q_{n+1}}$$

with $m = 0, 1, 2, 3$; $q_0, q_1, \ldots, q_{n+1} \in \mathbb{Z}$ and $q_0, \ldots, q_n \neq 0$. Therefore X and U and hence X and Y generate $SL_2(\mathbb{Z})$.

We must now show that

$$X^4 = Y^3 = YX^2 Y^{-1} X^{-2} = I$$

are a complete set of defining relations for $SL_2(\mathbb{Z})$, or equivalently, that every relation on these generators is derivable from these. It is straightforward to see that X and Y do satisfy these relations. Assume then that we have a relation

$$S = X^{\epsilon_1} Y^{\alpha_1} X^{\epsilon_2} Y^{\alpha_2} \cdots Y^{\alpha_n} X^{\epsilon_{n+1}} = I$$

with all $\epsilon_i, \alpha_j \in \mathbb{Z}$. Using the set of relations

$$X^4 = Y^3 = YX^2 Y^{-1} X^{-2} = I$$

we may transform S so that

$$S = X^{\epsilon_1} Y^{\alpha_1} \cdots Y^{\alpha_m} X^{\epsilon_{m+1}}$$

with $\epsilon_1, \epsilon_{m+1} = 0, 1, 2$ or 3 and $\alpha_i = 1$ or 2 for $i = 1, \ldots, m$ and $m \geq 0$. Multiplying by a suitable power of X we obtain

$$Y^{\alpha_1} X \cdots Y^{\alpha_m} X = X^\alpha = S_1$$

with $m \geq 0$ and $\alpha = 0, 1, 2$ or 3. Assume that $m \geq 1$ and let

$$S_1 = \begin{pmatrix} a & -b \\ -c & d \end{pmatrix}.$$

We show by induction that

$$a, b, c, d \geq 0, \quad b + c > 0$$

or

$$a, b, c, d \leq 0, \quad b + c < 0.$$

This claim for the entries of S_1 is true for

$$YX = \begin{pmatrix} 1 & 0 \\ -1 & 1 \end{pmatrix} \quad \text{and} \quad Y^2 X = \begin{pmatrix} -1 & 1 \\ 0 & -1 \end{pmatrix}.$$

Suppose it is correct for $S_2 = \begin{pmatrix} a_1 & -b_1 \\ -c_1 & d_1 \end{pmatrix}$. Then

$$YXS_2 = \begin{pmatrix} a_1 & -b_1 \\ -(a_1 + c_1) & b_1 + d_1 \end{pmatrix} \quad \text{and}$$

$$Y^2 X S_2 = \begin{pmatrix} -a_1 - c_1 & b_1 + d_1 \\ c_1 & d_1 \end{pmatrix}.$$

Therefore the claim is correct for all S_1 with $m \geq 1$. This gives a contradiction, for the entries of X^α with $\alpha = 0, 1, 2$ or 3 do not satisfy the claim. Hence $m = 0$ and S can be reduced to a trivial relation by the given set of relations. Therefore they are a complete set of defining relations and the theorem is proved. □

Corollary 9.6.11. *The Modular Group* $M = \mathrm{PSL}_2(\mathbb{Z})$ *has the presentation*

$$M = \langle x, y; x^2 = y^3 = 1 \rangle.$$

Further x, y can be taken as the linear fractional transformations

$$x : z' = -\frac{1}{z} \quad \text{and} \quad y : z' = -\frac{1}{z + 1}.$$

Proof. The center of $\mathrm{SL}_2(\mathbb{Z})$ is $\pm I$. Since $X^2 = -I$ setting $X^2 = I$ in the presentation for $\mathrm{SL}_2(\mathbb{Z})$ gives the presentation for M. Writing the projective matrices as linear fractional transformations gives the second statement. □

This corollary says that M is the **free product** (see Section 9.9) of a cyclic group of order 2 and a cyclic group of order 3.

We note that there is an elementary alternative proof to the above Corollary as far as showing that $X^2 = Y^3 = 1$ are a complete set of defining relations. As linear fractional transformations we have

$$X(z) = -\frac{1}{z}, \quad Y(z) = -\frac{1}{z + 1}, \quad Y^2(z) = -\frac{z + 1}{z}.$$

Now let

$$\mathbb{R}^+ = \{x \in \mathbb{R} \mid x > 0\} \quad \text{and} \quad \mathbb{R}^- = \{x \in \mathbb{R} \mid x < 0\}.$$

Then

$$X(\mathbb{R}^-) \subset \mathbb{R}^+ \quad \text{and} \quad y^\alpha(\mathbb{R}^+) \subset \mathbb{R}^-, \ \alpha = 1, 2.$$

Let $S \in \Gamma$. Using the relations $X^2 = Y^3 = 1$ and a suitable conjugation we may assume that either $S = 1$ is a consequence of these relations or that

$$S = Y^{\alpha_1} X Y^{\alpha_2} \cdots X Y^{\alpha_n}$$

with $1 \leq \alpha_i \leq 2$ and $\alpha_1 = \alpha_n$.

In this second case if $x \in \mathbb{R}^+$ then $S(x) \in \mathbb{R}^-$ and hence $S \neq 1$.

This type of *ping-pong argument* can be used in many examples (see [LS], [CRR], and [Jit]). As another example consider the unimodular matrices

$$A = \begin{pmatrix} 0 & 1 \\ -1 & 2 \end{pmatrix}, \quad B = \begin{pmatrix} 0 & -1 \\ 1 & 2 \end{pmatrix}.$$

Let $\overline{A}, \overline{B}$ denote the corresponding linear fractional transformations in the modular group M. We have

$$A^n = \begin{pmatrix} -n+1 & n \\ -n & n+1 \end{pmatrix}, \quad B^n = \begin{pmatrix} -n+1 & -n \\ n & n+1 \end{pmatrix} \quad \text{for } n \in \mathbb{Z}.$$

In particular $\overline{A}$ and $\overline{B}$ have infinite order. Now

$$\overline{A}^n(\mathbb{R}^-) \subset \mathbb{R}^+ \quad \text{and} \quad \overline{B}^n(\mathbb{R}^+) \subset \mathbb{R}^-$$

for all $n \neq 0$. The ping-pong argument used for any element of the type

$$S = \overline{A}^{n_1} \overline{B}^{m_1} \cdots \overline{B}^{m_k} \overline{A}^{n_{k+1}}$$

with all $n_i, m_i \neq 0$ and $n_1 + n_{k+1} \neq 0$ shows that $S(x) \in \mathbb{R}^+$ if $x \in \mathbb{R}^-$. It follows that there are no non-trivial relations on $\overline{A}$ and $\overline{B}$ and therefore the subgroup of M generated by $\overline{A}, \overline{B}$ must be a free group of rank 2.

9.7 Presentations of Subgroups

Given a group presentation $G = \langle X; R \rangle$ it is possible to find a presentation for a subgroup H of G. The procedure to do this is called the **Reidemeister-Schreier process** and is a consequence of the explicit version of the Nielsen-Schreier theorem (Theorem 9.4.8). We give a brief description. A complete description and a verification of its correctness is found in [MKS], [LS], [Bau], or [CRR].

Let G be a group with the presentation $\langle a_1, \ldots, a_n; R_1 = \cdots = R_k = 1 \rangle$. Let H be a subgroup of G and T a Schreier system for G modulo H defined analogously as above.

Reidemeister-Schreier Process

Let G, H and T be as above. Then H is generated by the set

$$\{S_{ta_v} \mid t \in T, a_v \in \{a_1, \dots, a_n\}, S_{ta_v} \neq 1\}$$

with a complete set of defining relations given by conjugates of the original relators rewritten in terms of the subgroup generating set.

In order to actual rewrite the relators in terms of the new generators we use a mapping τ on words on the generators of G called the **Reidemeister rewriting process**. This map is defined as follows: If

$$W = a_{v_1}^{e_1} a_{v_2}^{e_2} \cdots a_{v_j}^{e_j} \quad \text{with } e_i = \pm 1 \text{ defines an element of } H$$

$$\text{then} \quad \tau(W) = S_{t_1, a_{v_1}}^{e_1} S_{t_2, a_{v_2}}^{e_2} \cdots S_{t_j, a_{v_j}}^{e_j}$$

where t_i is the coset representative of the initial segment of W preceding a_{v_i} if $e_i = 1$ and t_i is the representative of the initial segment of W up to and including $a_{v_i}^{-1}$ if $e_i = -1$. The complete set of relators rewritten in terms of the subgroup generators is then given by

$$\{\tau(tR_i t^{-1})\} \quad \text{with } t \in T \text{ and } R_i \text{ runs over all relators in } G.$$

We present two examples; one with a finite group and then an important example with a free group which shows that a countable free group contains free subgroups of arbitrary ranks.

Example 9.7.1. Let $G = A_4$ be the alternating group on 4 symbols. Then a presentation for G is

$$G = A_4 = \langle a, b; a^2 = b^3 = (ab)^3 = 1 \rangle.$$

Let $H = A_4'$ the commutator subgroup. We use the above method to find a presentation for H. Now

$$G/H = A_4/A_4' = \langle a, b; a^2 = b^3 = (ab)^3 = [a, b] = 1 \rangle = \langle b; b^3 = 1 \rangle.$$

Therefore $|A_4 : A_4'| = 3$. A Schreier system is then $\{1, b, b^2\}$. The generators for A_4' are then

$$X_1 = S_{1a} = a, \quad X_2 = S_{ba} = bab^{-1}, \quad X_3 = S_{b^2a} = b^2ab$$

while the relations are
(1) $\tau(aa) = S_{1a}S_{1a} = X_1^2$
(2) $\tau(baab^{-1}) = X_2^2$
(3) $\tau(b^2aab^{-2}) = X_3^2$
(4) $\tau(bbb) = 1$
(5) $\tau(bbbbb^{-1}) = 1$
(6) $\tau(b^2bbbb^{-2}) = 1$
(7) $\tau(ababab) = S_{1a}S_{ba}S_{b^2a} = X_1X_2X_3$

(8) $\tau(bababab b^{-1}) = S_{ba} S_{b^2 a} S_{1a} = X_2 X_3 X_1$

(9) $\tau(b^2 ababab b^{-2}) = S_{b^2 a} S_{1a} S_{ba} = X_3 X_1 X_2$

Therefore after eliminating redundant relations and using that $X_3 = X_1 X_2$ we get as a presentation for A_4',

$$\langle X_1, X_2; X_1^2 = X_2^2 = (X_1 X_2)^2 = 1 \rangle.$$

Example 9.7.2. Let $F = \langle x, y; \rangle$ be the free group of rank 2. Let H be the commutator subgroup. Then

$$F/H = \langle x, y; [x, y] = 1 \rangle = \mathbb{Z} \times \mathbb{Z}$$

a free abelian group of rank 2. It follows that H has infinite index in F. As Schreier coset representatives we can take

$$t_{m,n} = x^m y^n, \quad m = 0, \pm 1, \pm 2, \ldots, \quad n = 0, \pm 1, \pm 2, \ldots.$$

The corresponding Schreier generators for H are

$$x_{m,n} = x^m y^n x^{-m} y^{-n}, \quad m = 0, \pm 1, \pm 2, \ldots, \quad n = 0, \pm 1, \pm 2, \ldots.$$

The relations are only trivial and therefore H is free on the countable infinitely many generators above. It follows that a free group of rank 2 contains as a subgroup a free group of countably infinite rank. Since a free group of countable infinite rank contains as subgroups free groups of all finite ranks it follows that a free group of rank 2 contains as a subgroup a free subgroup of any arbitrary finite rank.

Theorem 9.7.3. *Let F be free of rank 2. Then the commutator subgroup F' is free of countable infinite rank. In particular a free group of rank 2 contains as a subgroup a free group of any finite rank n.*

Corollary 9.7.4. *Let n, m be any pair of positive integers $n, m \geq 2$ and F_n, F_m free groups of ranks n, m respectively. Then F_n can be embedded into F_m and F_m can be embedded into F_n.*

9.8 Group Decision Problems

We have seen that given any group G there exists a presentation for it, $G = \langle X; R \rangle$. In the other direction given any presentation $\langle X; R \rangle$ we have seen that there is a group with that presentation. In principle every question about a group can be answered via a presentation. However, things are not that simple. Max Dehn, in his pioneering work on combinatorial group theory about 1910, introduced the following three fundamental **group decision problems** (see [FR]):

(1) **Word Problem:** Suppose G is a group given by a finite presentation. Does there exist an algorithm to determine if an arbitrary word w in the generators of G defines the identity element of G?

(2) **Conjugacy Problem:** Suppose G is a group given by a finite presentation. Does there exist an algorithm to determine if an arbitrary pair of words u, v in the generators of G define conjugate elements of G?

(3) **Isomorphism Problem:** Does there exist an algorithm to determine given two arbitrary finite presentations whether the groups they present are isomorphic or not?

All three of these problems have negative answers in general. That is for each of these problems, one can find a finite presentation for which these questions cannot be answered algorithmically (see [LS]). Attempts for solutions, and for solutions in restricted cases, have been of central importance in combinatorial group theory. For this reason, combinatorial group theory has always searched for, and studied, classes of groups in which these decision problems are solvable.

For finitely generated free groups there are simple and elegant solutions to all three problems. If F is a free group on $x_1, \ldots, x_n$ and W is a freely reduced word in $x_1, \ldots, x_n$ then $W \neq 1$ if and only if $|W| \geq 1$. Since freely reducing any word to a freely reduced word is algorithmic this provides a solution to the word problem.

Further a freely reduced word $W = x_{v_1}^{e_1} x_{v_2}^{e_2} \cdots x_{v_n}^{e_n}$ is **cyclically reduced** if $v_1 \neq v_n$ or if $v_1 = v_n$ then $e_1 \neq -e_n$. Clearly then every element of a free group is conjugate to an element given by a cyclically reduced word called a cyclic reduction. This leads to a solution to the conjugacy problem. Suppose V and W are two words in the generators of F and $\overline{V}, \overline{W}$ are respective cyclic reductions. Then V is conjugate to W if and only if $\overline{V}$ is a cyclic permutation of $\overline{W}$.

Finally two finitely generated free groups are isomorphic if and only if they have the same rank.

The three problems listed above are only the basic decision problems and other algorithmic problems concerning presentations can be considered. The conjugacy problem asks to algorithmically determine if two elements given in terms of the generators are conjugate. The **conjugator search problem** asks: given a group presentation for G and two elements g_1, g_2 in G that are known to be conjugate to determine algorithmically a conjugator, that is an element h such that $h^{-1} g_1 h = g_2$. It is known, as with the conjugacy problem itself, that the conjugator search problem is undecidable in general.

The computational difficulty of solving various group decision problems will play the role of a hard problem used to construct a one-way function in several non-abelian group based cryptosystems. We will return to this idea in Chapters 10 and 11.

The book by Myasnikov, Shpilrain and Ushakov [MSU1] has discussions of the complexity of many of these group decision problems.

9.9 Group Amalgams

In analyzing the structure of an infinite discrete group via combinatorial group theory, **group products** or **group amalgams** are the key constructions. These constructions will not play a major role in the application of combinatorial group theory to cryptography but will occasionally arise. In this section we briefly introduce the basic definitions.

The general idea of group amalgams is to decompose (if possible) an infinite group G into an amalgam (in a way which we will describe shortly) of some of its subgroups. These subgroups are then called the **factors** of G. Information about G can then be deduced from the corresponding information on the factors. Thus amalgam decompositions play a role in infinite group theory similar to a prime factorization theorem, although the amalgam decomposition of a group G need not be unique.

There are essentially two different types of group amalgams: **free products with amalgamation** and **HNN groups**. An infinite group, however, may decompose as both a free product with amalgamation or in a different manner as an HNN group. **Free products** are a special case of free products with amalgamation, so we discuss these first.

Before beginning, we note that there are two main approaches to the theory of group amalgams. The first is a classical combinatorial approach which deals primarily with presentations for the group and its factors. The second approach is a geometric-topological technique which depends on how the group acts (as a group of isometries) on a graph. The second method is due to Bass and Serre. We refer to [LS], [Ser], or [CRR] for a more complete discussion of all of these.

Free products of groups are closely related to free groups in both form and properties. Let $A = \langle a_1, \ldots; R_1 = 1, \ldots \rangle$ and $B = \langle b_1, \ldots; S_1 = 1, \ldots \rangle$ be two groups. We consider the groups A and B to be disjoint. Then:

Definition 9.9.1. The **free product** of A and B denoted $A * B$ is the group G with the presentation $\langle a_1, \ldots, b_1, \ldots; R_1 = 1, \ldots, S_1 = 1, \ldots \rangle$, that is, the generators of G consist of the disjoint union of the generators of A and B with relators taken as the disjoint union of the relators R_i of A and S_j of B. A and B are called the **factors** of G.

In an analogous manner the concept of a free product can be extended to an arbitrary collection of groups.

Definition 9.9.2. If $A_\alpha = \langle$ gens A_α; rels $A_\alpha \rangle, \alpha \in \mathfrak{I}$, is a collection of groups, then their free product $G = *A_\alpha$ is the group whose generators consist of the disjoint union of the generators of the A_α and whose relators are the disjoint union of the relators of the A_α.

We saw as an example in Section 9.6.1 that the Modular group $M = \mathrm{PSL}(2, \mathbb{Z})$ is a free product of a cyclic group of order 2 and a cyclic group of order 3, that is $M \cong \mathbb{Z}_2 \star \mathbb{Z}_3$.

Free products exist and are non-trivial. We have:

Theorem 9.9.3. *Let $G = A * B$. Then the maps $A \to G$ and $B \to G$ are injections. The subgroup of G generated by the generators of A has the presentation $\langle$ gens A; rels A$\rangle$, that is, is isomorphic to A. Similarly for B. Thus A and B can be considered as subgroups of G. In particular $A * B$ is non-trivial if A and B are.*

Free products share many properties with free groups. First of all there is a categorical formulation of free products. Specifically we have:

Theorem 9.9.4. *A group G is the free product of its subgroups A and B if A and B generate G and given homomorphisms $f_1 : A \to H, f_2 : B \to H$ into a group H there exists a unique homomorphism $f : G \to H$ extending f_1 and f_2.*

Secondly each element of a free product has a **normal form** related to the reduced words of free groups. If $G = A * B$ then a **reduced sequence** or **reduced word** in G is a sequence $g_1 g_2 \cdots g_n$ with $g_i \neq 1$, each g_i in either A or B and g_i, g_{i+1} not both in the same factor. Then:

Theorem 9.9.5. *Each element $g \in G = A * B$ has a unique representation as a reduced sequence. The length n is unique and is called the **syllable length**. The case $n = 0$ is reserved for the identity.*

A reduced word $g_1 \cdots g_n \in G = A * B$ is called **cyclically reduced** if either $n \leq 1$ or $n \geq 1$ and g_1 and g_n are from different factors. Certainly every element of G is conjugate to a cyclically reduced word.

From this we obtain several important properties of free products which carry over to more general amalgams.

Theorem 9.9.6. *An element of finite order in a free product is conjugate to an element of finite order in a factor. In particular a finite subgroup of a free product is entirely contained in a conjugate of a factor.*

Theorem 9.9.7. *If two elements of a free product commute then they are both powers of a single element or are contained in a conjugate of an abelian subgroup of a factor.*

Finally, the following theorem of Kurosh extends the Nielsen-Schreier theorem to free products.

Theorem 9.9.8 (Kurosh). *A subgroup of a free product is also a free product. Explicitly if $G = A * B$ and $H \subset G$ then*

$$H = F * (*A_\alpha) * (*B_\beta)$$

where F is a free group and $(*A_\alpha)$ is a free product of conjugates of subgroups of A and $(*B_\beta)$ is a free product of conjugates of subgroups of B.

We note that the rank of F as well as the number of the other factors can be computed. A complete discussion of these is in [MKS] and [LS].

We now extend this construction to free products with amalgamation. Let $A = \langle a_1, \ldots; R_1 = 1, \ldots \rangle$ and $B = \langle b_1, \ldots; S_1 = 1, \ldots \rangle$ be two groups with $H \subset A, K \subset B$ proper subgroups and $f : H \to K$ an isomorphism. Again we assume that A and B are disjoint. Then:

Definition 9.9.9. The **free product of A and B amalgamating H to K** is the group G with the presentation

$$G = \langle a_1, \ldots, b_1, \ldots; R_1 = 1, \ldots, S_1 = 1, \ldots, H = f(H) \rangle$$

that is the group G has as generators the disjoint union of the generators of A and B and has as relations the disjoint union of the relations of A and B together with an additional set of relations giving the subgroup isomorphism. Identifying H with its isomorphic image we say that G is the **free product of A and B with H amalgamated** denoted

$$G = A *_H B.$$

The groups A and B are called the **factors** of G.

A group G is a (non-trivial) **free product with amalgamation** or **amalgamated free product** if $G = G_1 *_H G_2$ for some groups G_1 and G_2 both non-trivial and some proper non-trivial subgroup H in G_1 and also in G_2.

Taking $H = \{1\}$ we obtain a free product. Therefore free products are just special cases of free products with amalgamation. As with free products, the factors inject into G and their intersection, as subgroups of G, is H, the amalgamated subgroup.

Theorem 9.9.10. *Let $G = A *_H B$. Then $A \to G$ and $B \to G$ are injections. The subgroup of G generated by the generators of A has the presentation $\langle gens\ A; rels\ A \rangle$. Similarly for B. Thus A and B can be considered as subgroups of G and $A \cap B = H$.*

The proof of this theorem depends upon a **normal form** for elements of free products with amalgamation. Let $G = A *_H B$ and let L_1 be a set of left coset representatives for A modulo H and let L_2 be a set of left coset representatives for B modulo H, normalized in both cases by taking 1 to represent H. Then a **reduced sequence** or **reduced word** or **normal form** in $G = A *_H B$ is a sequence of the form

$$g_1 g_2 \cdots g_n h$$

where $h \in H$, $1 \neq g_i \in L_1 \cup L_2$ and $g_1 \cdots g_n$ is a reduced word in the free product $A * B$, that is $g_{j+1} \notin L_i$ if $g_j \in L_i$.

Theorem 9.9.11 (Normal Form Theorem for Free Products with Amalgamation). *If $G = A *_H B$ then every $g \in G$ has a unique representation as a reduced sequence.*

Extending the concept from free products, a reduced word $g_1 \cdots g_n h$ in $G = A *_H B$ is called **cyclically reduced** if either $n \leq 1$ or $n \geq 2$ and g_1 and g_n are from different factors. Certainly every element of G is conjugate to a cyclically reduced word. From this we obtain properties analogous to those in free groups and free products. Specifically:

Theorem 9.9.12.

(1) *An element of $G = A *_H B$ of finite order must be conjugate to an element of finite order in one of the factors. Thus a finite subgroup or more generally a bounded subgroup must be entirely contained in a conjugate of a factor.*

(2) *An abelian subgroup of $G = A *_H B$ is*

(a) *a conjugate of an abelian subgroup of A or B or*

(b) *a countable ascending union of conjugates of subgroups of H or*

(c) *a direct product of an infinite cyclic group and a conjugate of a subgroup of H.*

The concept of a free product with amalgamation can be extended in a straightforward manner to more than two factors.

Definition 9.9.13. Let $\{G_i\}, i \in I$, be a family of groups. Let A be a group and for each $i, f_i : A \to G_i$ a monomorphism. Then the **free product of the G_i amalgamating A** is the quotient group of the free product $G = *_i G_i$ modulo the normal subgroup of $*_i G_i$ generated by all relations $f_i(a) = f_j(a)$ with $a \in A$ and $i, j \in I$.

As in the case of two factors, each G_i injects into G, and each element can be expressed as a normal form.

Before moving on to HNN groups we mention that there is also a categorical formulation of free products with amalgamation. Specifically;

Theorem 9.9.14. *Suppose G is a group, G_1, G_2 subgroups and A a group together with injections $\theta_1 : A \to G_1, \theta_2 : A \to G_2$. Then $G = G_1 *_A G_2$ if for every group H and every pair of homomorphisms $f_1 : G_1 \to H, f_2 : G_2 \to H$ making the following diagram commute*

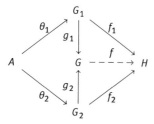

there exists a unique homomorphism $f : G \to H$ extending f_1, f_2.

Our second basic amalgam construction is that of an HNN group. This construction has properties very nearly parallel to those of free products with amalgamation. As pointed out in [Ser], they are really two different aspects of the same idea.

Definition 9.9.15. Let G be a group, $\{A_i\}, i \in I$, a family of subgroups of G, and for each $i \in I, f_i : A_i \to G$ a monomorphism. Then an **HNN extension of G** is a group G^* of the form

$$G^* = \langle t_i, i \in I, \text{ gens } G; \text{ rels } G, t_i^{-1} A_i t_i = f_i(A_i), i \in I \rangle.$$

G is called the **base**, $\{t_i\}_{i \in I}$ the **free part** or **stable letters**, and $\{A_i, f_i(A_i)\}$ the **associated subgroups**.

G^* is an **HNN group** if it can be expressed as an HNN extension of some base.

The base group G embeds in the HNN extension in the obvious manner.

Theorem 9.9.16. *Let G^* be an HNN extension of base G. Then G is embedded in G^* by $g \to g$, that is the subgroup of G^* generated by the generators of G, has the presentation $\langle$ gens G; rels $G \rangle$. Further the free part $\{t_i\}$ is a basis for a free subgroup of G^*.*

As in all previous cases, this depends on a normal form for elements of HNN groups. However, this is somewhat more intricate than that for free products with amalgamation.

Suppose that G^* is an HNN extension of G with associated subgroups $\{A_i, f_i(A_i)\}$ and suppose we choose a fixed set of left coset representatives for A_i and $f_i(A_i)$ in G where all A_i and $f_i(A_i)$ are represented by 1. Then a **normal form** in G^* is a sequence

$$g_1 t_{i_1}^{e_1} g_2 t_{i_2}^{e_2} \cdots t_{i_k}^{e_k} g_{k+1}, \quad e_j = \pm 1,$$

where $g_1, \ldots, g_{k+1}$ are elements of G such that for $j \le k$ if $e_j = 1$ then g_j is a left coset representative for A_i in G while if $e_j = -1$ then g_j is a left coset representative for $f_i(A_i)$ in G and $e_j = e_{j+1}$ whenever $g_{j+1} = 1$ and $i_j = i_{j+1}$.

Theorem 9.9.17. *Every element w in G^* has a unique representation as a normal form.*

From this as before we obtain a classification of torsion elements as well as a classification of abelian subgroups.

Theorem 9.9.18. *Let G^* be an HNN extension of G.*
(1) *Elements of finite order in G^* are conjugate to elements of finite order in the base G. Further finite subgroups must be contained in conjugates of the base.*
(2) *An abelian subgroup H of G^* is one of the following:*
  (a) *A subgroup of a conjugate of the base.*
  (b) *A countable ascending union of subgroups of conjugates of the associated subgroups.*
  (c) *An HNN group with presentation $\langle t', H'; \text{ rels } H', t'^{-1} H' t' = H'' \rangle$ with $H'' \subset H'$ and H' is the intersection of the abelian subgroup H with finitely many conjugates of the associated subgroups.*

We note that it is possible for a group to be both an HNN group and a free product with amalgamation. Consider the group

$$G = \langle a, t, u; a^2 = (at)^3 = [t, u] = 1 \rangle.$$

Let $G_1 = \langle a, t; a^2 = (at)^3 = 1 \rangle$. This is a free product of a cyclic group of order 2 generated by a and a cyclic group of order 3 generated by at. Therefore t has infinite order in G. Further let $G_2 = \langle t, u; [t, u] = 1 \rangle$ a free abelian group of rank 2. The identification

$t \to t$ is then an isomorphism and G is a free product of G_1 and G_2 with the infinite cyclic subgroup generated by t amalgamated.

Now write G as $\langle u, a, t; a^2 = (at)^3 = 1, u^{-1}tu = t\rangle$. Again let $G_1 = \langle a, t; a^2 = (at)^3 = 1\rangle = \mathbb{Z}_2 * \mathbb{Z}_3$. Then G is an HNN extension of G_1 with the single pair of associated subgroups $\langle t\rangle$ and $\langle f(t)\rangle$ where $f(t) = t$.

We note that the above presentation is a presentation for the groups $PE_2(O_d)$, the two dimensional projective elementary matrix group with entries in O_d, where O_d is the ring of integers in the quadratic imaginary number field $Q(\sqrt{-d})$ and $d \neq 1, 2, 3, 7, 11$ (see [Fin]).

HNN groups were originally developed by G. Higman, H. Neumann and B. Neumann (whence the name) in order to prove several important embedding theorems. In particular:

Theorem 9.9.19 (see [Rot]). *Every countable group can be embedded in a two generator group.*

The concept and theory of of SQ-universality developed from this theorem.

Definition 9.9.20. A group G is **SQ-universal** if every countable group can be embedded isomorphically as a subgroup of a quotient of G.

Thus the Higman, Neumann, Neumann theorem above says that a free group of rank 2 is SQ-universal. Many linear groups as well as groups arising from low-dimensional topology are SQ-universal. SQ-universality might be thought of as a measure of "largeness" of an infinite group.

As mentioned at the beginning of this section, amalgam structures have a geometric-topological interpretation. We refer to [LS], [Ser], and [CRR] for an explanation of these.

9.10 Exercises

9.1 (a) Let G be a group and $H \subset G$. Prove that H is a subgroup of G if and only if
   (1) $H \neq \emptyset$,
   (2) H is closed with respect to the operation in G and inverses, that is, if $a, b \in H$ then $ab \in H$ and $a^{-1}, b^{-1} \in H$.
   (b) Let G be a group and $H \subset G$ and assume H is non-empty. Prove that H is a subgroup of G if and only if $a, b \in H$ then $ab^{-1} \in H$. (This is sometimes called the one-step subgroup test.)
   (c) Show that the set of even integers forms a subgroup under addition of the additive group of $\mathbb{Z}$.

9.2 (a) Let R be a ring and let $U(R)$ denote the set of elements with multiplicative inverses, that is, the set of units in R. Prove that $U(R)$ forms a group. This is called the unit group of R.

(b) If R is a field what is $U(R)$?

9.3 If F is a field show that the $n \times n$ matrices of non-zero determinant, $GL(n, F)$, form a group and that $SL(n, F)$, those matrices of determinant 1, is a normal subgroup.

9.4 Let G be a group and let $a \in G$. Let

$$C_G(a) = \{x \in G \mid ax = xa\}.$$

Sometimes we omit the subscript G and just write $C(a)$. Prove that $C(a) \subset G$ is a subgroup of G. This subgroup is called the **centralizer of A in G**.

9.5 Let G be a group and define

$$Z(G) = \{g \in G \mid gg_1 = g_1g \text{ for all } g_1 \in G\}.$$

That is $Z(G)$ is the set of elements in G that commute with every element of G. Prove that $Z(G)$ forms a normal subgroup of G. This subgroup is called the **center** of G.

9.6 (a) Show that the intersection of any non-empty collection of subgroups of a group G is a subgroup of G.

(b) Show that if H, K are subgroups of G and the union $H \cup K$ is also a subgroup of G then either $H \subset K$ or $K \subset H$.

9.7 What are the generators of the additive group of the finite ring $\mathbb{Z}_{12}$?

9.8 Show that in S_3, the symmetric group on 3 symbols, we have $(ab)^2 = 1$ implies $ab = ba^2$.

9.9 Let G be a group and G' be the subgroup of G generated by all commutators $[a, b] = aba^{-1}b^{-1}$. Show that this is a normal subgroup and the quotient group G/G' is abelian.

The subgroup G' is called the **commutator subgroup** of G, and the quotient G/G' is called the **abelianization** of G.

9.10 (a) Let $F = \langle a, b; \rangle$ be a free group of rank 2. Show, using the Reidemeister-Schreier procedure, that the commutator subgroup F' is free of infinite rank.

(b) Let F_n denote a free group of rank n and F_ω denote a free group of countably infinite rank. Show that part (a) implies the following "snake eating its tail" situation where $2 < n < m < \infty$;

$$F_\omega \subset F_2 \subset F_n \subset F_m \subset F_\omega \subset F_2 \cdots.$$

9.11 Prove the following things about a free group (you may assume finite rank):

(a) A free group is **torsion-free** that is every non-trivial element has infinite order.

(b) Two non-trivial elements commute only if they both are powers of a single element. (Hint: use induction on the word lengths.)

(c) Suppose W_1, W_2 are two words in the generators of F and their inverses. Then, if W_1 is a cyclic permutation of W_2, they represent conjugate elements.

9.12 Let $M = \text{PSL}(2, \mathbb{Z})$ be the modular group. We saw that

$$M = \langle x, y; x^2 = y^3 = 1 \rangle.$$

(a) Let M' be the commutator subgroup of M. Show that M' is free of rank 2. Hint: Use the Reidemeister-Schreier procedure. The abelianization of M is cyclic of order 6, and you can use as coset representatives $(xy)^i$ where $i = 0, 1, 2, 3, 4, 5$.

(b) Explain why part (a) shows that within M there are free groups of every possible countable rank. This becomes important for certain cryptographic applications.

9.13 Let $G = A \star B$ be the free product of two groups A and B. Show that if two elements $g, h \in G$ commute then either they are both powers of a single element or both are in the same conjugate subgroup of one of the factors. Hint: Use induction on the syllable length.

10 Non-Commutative Group Based Cryptography

10.1 Group Based Methods

The public key cryptosystems and public key exchange protocols that we have discussed, such as the RSA algorithm, Diffie-Hellman, ElGamal and elliptic curve methods, are number theory based, and hence depend on the structure of abelian groups. As computing machinery has gotten stronger, and computational techniques have become more sophisticated and improved, there have been successful attacks on both RSA and Diffie-Hellman for smaller and specialized parameters (RSA and Diffie-Hellman moduli). Further there are quantum algorithms that specifically break both RSA and Diffie-Hellman. As a consequence, when and if a workable quantum computer is developed, these cryptographic methods will have to be altered. We will not discuss quantum algorithms or quantum computing in this book.

Because of these attacks and the growing strength of computers, there is a feeling that these number theoretic techniques are theoretically susceptible to attack. Somehow the relatively simple structure of abelian groups opens up the possibility of weaknesses in cryptographic protocols. As a result there has been an active line of research to develop cryptosystems and key exchange protocols using non-commutative cryptographic platforms. This line of investigation has been given the broad title of **non-commutative algebraic cryptography**. Since most of the cryptographic platforms are groups this is also known as **group based cryptography**. The book by Myasnikov, Shpilrain and Ushakov [MSU1] provides an overview of group based cryptographic methods tied to complexity theory.

Up to this point the main sources for non-commutative cryptographic platforms have been non-abelian groups. In cryptosystems based on these objects, algebraic properties of the platforms are used prominently in both devising cryptosystems and in cryptanalysis. In particular, the non-solvability of certain algorithmic problems in finitely presented groups, such as the conjugator search problem, has been crucial in encryption and decryption (see Chapter 11).

The main sources for non-abelian groups are combinatorial group theory and linear group theory, that is matrix groups. Braid group cryptography (see Chapter 11), where encryption is done within the classical braid groups, is one prominent example. The one-way functions in braid group systems are based on the difficulty of solving group theoretic decision problems such as the conjugacy problem and conjugator search problem. We discuss these later in the chapter. Although braid group cryptography had initial spectacular success, various potential attacks have been identified. Borovik, Myasnikov, Shpilrain [BMS] and others have studied the statistical aspects of these attacks and have identified what are termed black holes in the platform groups, the outsides of which present cryptographic problems. Baumslag. Fine and Xu (in [BFX1], [BFX2], and [Xu]) suggested potential cryptosystems using a combination of combinatorial group theory and linear groups and a general scheme for the these types of cryptosystems was given. In [BFX2] a version of this scheme using the classical modular group as a platform was presented. A cryptosystem using the homogeneous modular group SL(2, $\mathbb{Z}$) was developed by Yamamura (see [Yam]) but was subsequently shown to have loopholes (see [GP], [Ste], and [HGS]). In [BFX2] several attacks based on these loopholes were closed.

The extension of the cryptographic ideas to non-commutative platforms involves the following ideas:

(1) general algebraic techniques for developing cryptosystems,
(2) potential algebraic platforms (specific groups, rings, etc.) for implementing the techniques, and
(3) cryptanalysis and security analysis of the resulting systems.

The main source for non-commutative platforms are non-abelian groups, and the main method for handling non-abelian groups in cryptography is combinatorial group theory. The basic idea in using combinatorial group theory for cryptography is that elements of groups can be expressed as words in some alphabet. If there is an easy method to rewrite group elements in terms of these words, and further the technique used in this rewriting process can be supplied by a secret key, then a cryptosystem can be created.

10.2 Initial Group Theoretic Cryptosystems – the Magnus Method

One of the earliest descriptions of using a non-abelian group in cryptography appeared in a paper by W. Magnus in the early 1970's (see [Mag]). This was what is now called a **free group cryptosystem**. The seminal idea of using the difficulty of group theory decision problems in infinite non-abelian groups as one-way functions in cryptography was first developed by Magyarik and Wagner in 1985. Neither of these two methods proved successful as workable encryption methods yet their introduction ushered in a subsequent complete theory and other ideas. In this section we describe Magnus' idea and in the next subsection the Wagner-Magyarik method.

In the paper [Mag], W. Magnus studied rational representations of Fuchsian groups, as well as non-parabolic and Neumann subgroups of the classical modular group M. Recall that $M = \mathrm{PSL}_2(\mathbb{Z})$ (see Chapter 9). That is, M consists of the 2×2 projective integral matrices:

$$M = \left\{ \pm \begin{pmatrix} a & b \\ c & d \end{pmatrix} \mid ad - bc = 1, a, b, c, d \in \mathbb{Z} \right\}.$$

Equivalently, M can be considered as the set of integral linear fractional transformations with determinant 1:

$$z' = \frac{az + b}{cz + d} \quad \text{with } ad - bc = 1 \quad \text{and } a, b, c, d \in \mathbb{Z}.$$

Magnus proved the following theorem.

Theorem 10.2.1 (see [Mag] and [BFX1]). *The matrices*

$$\pm \begin{pmatrix} 1 & 1 \\ 1 & 2 \end{pmatrix}, \quad \pm \begin{pmatrix} 1 + 4t^2 & 2t \\ 2t & 1 \end{pmatrix}, \quad t = 1, 2, 3, \ldots$$

freely generate a free subgroup F of infinite index in M. Further, distinct elements of F have distinct first columns.

Since the entries in the generating matrices are positive we can do the following. Choose a set

$$T_1, \ldots, T_n$$

of projective matrices from the set above with n large enough to encode a desired plaintext alphabet $\mathcal{A}$. Any message would be encoded by a word

$$W(T_1, \ldots, T_n)$$

with non-negative exponents. This represents an element g of F. The two elements in the first column determine W and therefore g. Receiving W then determines the message uniquely.

Pure free cryptography as Magnus proposed is subject to many attacks. We will discuss this further in Section 10.3.

10.2.1 The Wagner-Magyarik Method

The idea of using the difficulty of group theory decision problems in devising hard one-way functions for cryptographic purposes was first developed by Magyarik and Wagner in 1985. They devised a public key protocol based on the difficulty of the solution to the word problem. Although this was a seminal idea, their basic cryptosystem was really unworkable and not secure in the form they presented. Wagner and Magyarik outlined a conceptual public key cryptosystem based on the hardness of the word problem for finitely presented groups. At the same time, they gave a specific example of such a system. Gonzalez-Vasco and Steinwandt proved that their approach is vulnerable to so-called reaction attacks. In particular, for the proposed instance it is possible to retrieve the private key just by watching the performance of a legitimate recipient.

The general scheme of the Wagner and Magyarik public key cryptosystem is as follows. Let X be a finite set of generators, and let R and S be finite sets of relators on X. Consider the two groups G, G_0 with presentations

$$G = \langle X; R \rangle \quad \text{and} \quad G_0 = \langle X; R \cup S \rangle.$$

The group G_0 is then a homomorphic image of G. We assume first that G has a hard word problem so that the word problem in G is not solvable in polynomial time. We next assume that the homomorphic image G_0 has a word problem solvable in polynomial time, that is an easy word problem.

Choose two words W_0 and W_1 which are not equivalent in G_0 (and hence not equivalent in G since G_0 is a homomorphic image of G). The public key is the presentation $\langle X; R \rangle$ and the chosen words W_0 and W_1. To encrypt a single bit $\in \{0, 1\}$, pick W_i and transform it into a ciphertext word W by repeatedly and randomly applying Tietze transformations to the presentation $\langle X; R \rangle$. To decrypt a word W, run the algorithm for the word problem of G_0 in order to decide which of $W_i W^{-1}$ is equivalent to the empty word for the presentation $\langle X; R \cup S \rangle$ The private key is the set S. As pointed out by Gonzales-Vasco and Steinwandt, this is not sufficient and Wagner and Magyarik are not clear on this point. The public key should be a deterministic polynomial-time algorithm for the word problem of $G_0 = \langle X; R \cup S \rangle$. Just knowing S does not automatically and explicitly give us an efficient algorithm (even if such an algorithm exists).

Although the Wagner-Magyarik protocol was not workable as a public key system, the idea opened the door for using similar types of encryption involving group theoretic decision problems.

10.3 Free Group Cryptosystems

The simplest example of a non-abelian group based cryptosystem is perhaps a **free group cryptosystem**. This can be described in the following manner.

Consider a free group F on free generators $x_1, \ldots, x_r$. Then each element g in F has a unique expression as a word $W(x_1, \ldots, x_r)$. Let $W_1, \ldots, W_k$ with $W_i = W_i(x_1, \ldots, x_r)$ be a set of words in the generators $x_1, \ldots, x_r$ of the free group F. At the most basic level, to construct a cryptosystem, suppose that we have a plaintext alphabet $\mathcal{A}$. For example suppose $\mathcal{A} = \{a, b, \ldots\}$ are the symbols needed to construct meaningful messages in English. To encrypt, use a substitution ciphertext

$$\mathcal{A} \mapsto \{W_1, \ldots, W_k\}$$

given by $a \mapsto W_1, b \mapsto W_2, \ldots$. Then, for a word $W(a, b, \ldots)$ in the plaintext alphabet, form the free group word $W(W_1, W_2, \ldots)$. This represents an element g in F. Send out g as the secret message.

In order to implement this scheme we need a concrete representation of g and then for decryption a way to rewrite g back in terms of $W_1, \ldots, W_k$. This concrete representation is the idea behind **homomorphic cryptosystems**.

The decryption algorithm in a free group cryptosystem then depends on the **Reidemeister-Schreier rewriting process**. As described in Chapter 9, this is a method to rewrite elements of a subgroup of a free group in terms of the generators of that subgroup. Recall that roughly it works as follows. Assume that $W_1, \ldots, W_k$ are free generators for some subgroup H of a free group F on $\{x_1, \ldots, x_n\}$. Each W_i is then a reduced word in the generators $\{x_1, \ldots, x_n\}$. A **Schreier transversal** for H is a set $\{h_1, \ldots, h_t, \ldots\}$ of (left) coset representatives for H in F of a special form (see [MKS]). Any subgroup of a free group has a Schreier transversal. The Reidemeister-Schreier process allows one to construct a set of generators $W_1, \ldots, W_k$ for H by using a Schreier transversal. Further given the Schreier transversal from which the set of generators for H was constructed, the **Reidemeister-Schreier Rewriting Process** allows us to algorithmically rewrite an element of H. Given such an element expressed as a word $W = W(x_1, \ldots, x_r)$ in the generators of F this algorithm rewrites W as a word $W^*(W_1, \ldots, W_k)$ in the generators of H.

The knowledge of a Schreier transversal and the use of Reidemeister-Schreier rewriting facilitates the decoding process in the free group case but is not essential. Given a known set of generators for a subgroup the Stallings Folding Method to develop a subgroup graph can also be utilized to rewrite in terms of the given generators. The paper by Kapovich and Myasnikov [KMy] is now a standard reference for this method in free groups.

Pure free group cryptosystems are subject to various attacks and can be broken easily. However, a public key free group cryptosystem using a free group representation in the Modular group was developed by Baumslag, Fine and Xu (see [BFX1] and [BFX2]). The most successful attacks on free group cryptosystems are called **length based attacks**. The general idea in a length based attack is that an attacker multiplies a word in ciphertext by a generator to get a shorter word which then could possibly be decoded. We refer to [Gar] for more on length based attacks.

Baumslag, Fine and Xu in [BFX1] described the following general encryption scheme using free group cryptography. A further enhancement was discussed in the paper [BFX2].

We start with a finitely presented group

$$G = \langle X; R \rangle$$

where $X = \{x_1, \ldots, x_n\}$ and a faithful representation

$$\rho : G \mapsto \overline{G}.$$

$\overline{G}$ can be any one of several different kinds of objects; linear group, permutation group, power series ring etc.

We assume that there is an algorithm to re-express an element of $\rho(G)$ in $\overline{G}$ in terms of the generators of G. That is if $g = W(x_1, \ldots, x_n) \in G$ where W is a word in the these generators and we are given $\rho(g) \in \overline{G}$ we can algorithmically find g and its expression as the word $W(x_1, \ldots, x_n)$.

Once we have G we assume that we have two free subgroups K, H with

$$H \subset K \subset G.$$

We assume that we have fixed Schreier transversals for K in G and for H in K both of which are held in secret by the communicating parties Bob and Alice. Now based on the fixed Schreier transversals we have sets of Schreier generators constructed from the Reidemeister-Schreier process for K and for H.

$$k_1, \ldots, k_m, \ldots \quad \text{for } K$$

and

$$h_1, \ldots, h_t, \ldots \quad \text{for } H.$$

Notice that the generators for K will be given as words in $x_1, \ldots, x_n$ the generators of G while the generators for H will be given as words in the generators $k_1, k_2, \ldots$ for K. We note further that H and K may coincide and that H and K need not in general be free but only have a unique set of normal forms so that the representation of an element in terms of the given Schreier generators is unique.

We will encode within H, or more precisely within $\rho(H)$. We assume that the number of generators for H is larger than the set of characters within our plaintext alphabet. Let $\mathcal{A} = \{a, b, c, \ldots\}$ be our plaintext alphabet. At the simplest level we choose a starting point i, within the generators of H, and encode

$$a \mapsto h_i, \quad b \mapsto h_{i+1}, \ldots, \quad \text{etc.}$$

Suppose that Bob wants to communicate the message $W(a, b, c, \ldots)$ to Alice where W is a word in the plaintext alphabet. Recall that both Bob and Alice know the various Schreier transversals which are kept secret between them. Bob then encodes $W(h_i, h_{i+1}, \ldots)$ and computes in $\overline{G}$ the element $W(\rho(h_i), \rho(h_{i+1}), \ldots)$ which he sends to Alice. This is sent as a matrix if $\overline{G}$ is a linear group or as a permutation if $\overline{G}$ is a permutation group and so on.

Alice uses the algorithm for $\overline{G}$ relative to G to rewrite $W(\rho(h_i), \rho(h_{i+1}), \ldots)$ as a word $W^*(x_1, \ldots, x_n)$ in the generators of G. She then uses the Schreier transversal for K in G to rewrite using the Reidemeister-Schreier process W^* as a word $W^{**}(k_1, \ldots, k_s)$ in the generators of K. Since K is free or has unique normal forms this expression for the element of K is unique. Once she has the word written in the generators of K she uses the transversal for H in K to rewrite again, using the Reidemeister-Schreier process, in terms of the generators for H. She then has a word $W^{***}(h_i, h_{i+1}, \ldots)$ and using $h_i \mapsto a, h_{i+1} \mapsto b, \ldots$ decodes the message.

In an actual implementation an additional *random noise factor* is added. This is explained in more detail below.

10.3.1 An Implementation Within the Classical Modular Group

We now describe an implementation of this process using for the base group G the classical modular group $M = \mathrm{PSL}(2, \mathbb{Z})$. Further, this implementation uses a poly-alphabetic cipher which is secure. This was introduced originally in [BFX1] and [BFX2].

The system in the modular group M works as follows. A list of finitely generated free subgroups $H_1, \ldots, H_m$ of M is public and presented by their systems of generators (presented as matrices). In a full practical implementation it is assumed that m is large. For each H_i we have a Schreier transversal

$$h_{1,i}, \ldots, h_{t(i),i}$$

and a corresponding ordered set of generators

$$W_{1,i}, \ldots, W_{m(i),i}$$

constructed from the Schreier transversal by the Reidemeister-Schreier process. It is assumed that each $m(i) \gg l$ where l is the size of the plaintext alphabet, that is, each subgroup has many more generators than the size of the plaintext alphabet. Although Bob and Alice know these subgroups in terms of free group generators what is made public are generating systems given in terms of matrices.

The subgroups on this list and their corresponding Schreier transversals can be chosen in a variety of ways. For example the commutator subgroup of the Modular Group is free of rank 2 and some of the subgroups H_i can be determined from homomorphisms of this subgroup onto a set of finite groups.

Suppose that Bob wants to send a message to Alice. Bob first chooses three integers (m, q, t) where

$$m = \text{choice of the subgroup } H_m$$
$$q = \text{starting point among the generators of } H_m$$
$$\text{for the substitution of the plaintext alphabet}$$
$$t = \text{size of the message unit}.$$

We clarify the meanings of q and t. Once Bob chooses m, to further clarify the meaning of q, he makes the substitution

$$a \mapsto W_{m,q}, \quad b \mapsto W_{m,q+1}, \ldots.$$

Again the assumption is that $m(i) \gg l$ so that starting almost anywhere in the sequence of generators of H_m will allow this substitution. The message unit size t is the number of coded letters that Bob will place into each coded integral matrix.

Once Bob has made the choices (m, q, t) he takes his plaintext message $W(a, b, \ldots)$ and groups blocks of t letters. He then makes the given substitution above to form the corresponding matrices in the Modular Group;

$$T_1, \ldots, T_s.$$

We now introduce a *random noise factor*. After forming $T_1, \ldots, T_s$ Bob then multiplies on the right each T_i by a random matrix in M say R_{T_i} (different for each T_i). The only restriction on this random matrix R_{T_i} is that there is no free cancellation in forming the product $T_i R_{T_i}$. This can be easily checked and ensures that the freely reduced form for $T_i R_{T_i}$ is just the concatenation of the expressions for T_i and R_{T_i}. Next he sends Alice the integral key (m, q, t) by some public key method (RSA, Anshel-Goldfeld etc.). He then sends the message as s random matrices

$$T_1 R_{T_1}, \quad T_2 R_{T_2}, \ldots, \quad T_s R_{T_s}.$$

Hence what is actually being sent out are not elements of the chosen subgroup H_m but rather elements of random right cosets of H_m in M. The purpose of sending coset elements is two-fold. The first is to hinder any geometric attack by masking the subgroup. The second is that it makes the resulting words in the Modular Group generators longer – effectively hindering a brute force attack.

To decode the message Alice first uses public key decryption to obtain the integral keys (m, q, t). She then knows the subgroup H_m, the ciphertext substitution from the generators of H_m and how many letters t each matrix encodes. She next uses the algorithms described in Section 10.2 to express each $T_i R_{T_i}$ in terms of the free group generators of M say $W_{T_i}(y_1, \ldots, y_n)$. She has knowledge of the Schreier transversal, which is held secretly by Bob and Alice, so now uses the Reidemeister-Schreier rewriting process to start expressing this freely reduced word in terms of the generators of H_m. The Reidemeister-Schreier rewriting is done letter by letter from left to right (see [MKS]). Hence when she reaches t of the free generators she stops. Notice that the string that she is rewriting is longer than what she needs to rewrite in order to decode as a result of the random polynomial R_{T_i}. This is due to the fact that she is actually rewriting not an element of the subgroup but an element in a right coset. This presents a further difficulty to an attacker. Since these are random right cosets it makes it difficult to pick up statistical patterns in the generators even if more than one message is intercepted. In practice the subgroups should be changed with each message.

The initial key (m, q, t) is changed frequently. Hence as mentioned above this method becomes a type of polyalphabetic cipher. Polyalphabetic ciphers have historically been very difficult to decode.

10.3.2 A Variation Using the Magnus Representation

We introduce a variation of this method using the Magnus representation in a formal power series ring in non-commuting variables over a field. This was introduced in [BBFR].

This encryption method will use the ring of formal power series

$$R\langle\langle x_1, \ldots, x_n \rangle\rangle$$

over a ring R in non-commuting variables $x_1, \ldots, x_n$. Although this can be done in an even more general context, for this study we concentrate on rational formal power series, that is, we consider the ring R to be the field of rational numbers $\mathbb{Q}$.

For the remainder of this section we let

$$H = \mathbb{Q}\langle\langle x_1, \ldots, x_n \rangle\rangle$$

be the formal power series ring in non-commuting variables $x_1, \ldots, x_n$ over $\mathbb{Q}$. One of our primary tools for developing encryption methods will be based upon a faithful representation of a finitely generated free group within a quotient of H. This representation was introduced by W. Magnus [Mag] and is now known as the **Magnus representation**. If $n \geq 2$ this then provides free subgroups of all countable ranks within this quotient of H. Further by imposing additional relations we can obtain representations of free nilpotent groups.

We first describe the Magnus representation and give a proof. The proof will lead us to two algorithms for describing when certain polynomials lie in the image of this representation. Further we describe the unit group of this quotient.

First let $d > 1$ be an integer and impose the relations

$$x_1^d = x_2^d = \cdots = x_n^d = 0$$

on H. We call the resulting quotient $\overline{H}$.

Notice that the elements of $\overline{H}$ are power series where each of the non-commuting variables $x_1, \ldots, x_n$ appears with exponent $< d$. The faithful representation of a free group is given in terms of the polynomials

$$\alpha_1 = 1 + x_1, \quad \alpha_2 = 1 + x_2, \ldots, \quad \alpha_n = 1 + x_n.$$

Notice that in the formal power series ring, H, we have the well-known expansion

$$\frac{1}{1 + x_i} = 1 - x_i + x_i^2 - x_i^3 + \cdots.$$

Therefore each α_i is invertible in H and hence invertible in $\overline{H}$. However, within $\overline{H}$ the inverse is a polynomial of degree $< d$ and so within $\overline{H}$

$$\frac{1}{1 + x_i} = 1 - x_i + x_i^2 - x_i^3 + \cdots + (-1)^{d-1} x_i^{d-1}.$$

Therefore each α_i is in the unit group $U(\overline{H})$ of $\overline{H}$ and therefore the set $\{\alpha_1, \ldots, \alpha_n\}$ generates a multiplicative subgroup of $U(\overline{H})$. Note also that if d, the defining power, is kept secret, then inverses are unknown.

Magnus's result is the following.

Theorem 10.3.1. *The elements*

$$\alpha_1 = 1 + x_1, \ldots, \quad \alpha_n = 1 + x_n$$

freely generate a subgroup of $U(\overline{H})$. Therefore the map given by

$$y_1 \rightarrow \alpha_1, \ldots, \quad y_n \rightarrow \alpha_n$$

provides a faithful representation of the free group on $y_1, \ldots, y_n$ into $\overline{H}$.

We present a proof, since as mentioned the proof will lead us to an algorithm necessary for our encryption methods.

Proof. Notice from the comment above that each α_i is invertible in $\overline{H}$. Therefore each α_i is in the unit group $U(\overline{H})$ of $\overline{H}$, and thus the set $\{\alpha_1, \ldots, \alpha_n\}$ generates a multiplicative subgroup of $U(\overline{H})$. We show that no non-trivial freely reduced word in the α_i can be the identity, and hence the group they generate must be a free group.

From the binomial expansion we have for any non-zero integer n, positive or negative,

$$(1 + x_i)^n = 1 + nx\alpha_i + \text{terms in higher powers.}$$

Now let

$$W(\alpha_1, \ldots, \alpha_n) = \alpha_{i_1}^{n_1} \alpha_{i_2}^{n_2} \cdots \alpha_{i_k}^{n_k}$$

be a freely reduced word in the α_i with each $|n_i| \geq 1$ and $\alpha_{i_j} \neq \alpha_{i_{j+1}}$ for $j = 1, \ldots, k - 1$. For later reference we call k the **block length**. In the ring $\overline{H}$ we then have

$$W(\alpha_1, \ldots, \alpha_n) = (1 + x_{i_1})^{n_1} \cdots (1 + x_{i_k})^{n_k}$$

and hence $W(\alpha_1, \ldots, \alpha_n)$ is of the form

$$(1 + n_1 x_{i_1} + \text{higher power terms in } x_{i_1}) \cdots (1 + n_k x_{i_k} + \text{higher power terms in } x_{i_k}).$$

The variables are non-commuting, so that in analyzing this product we see that there is a unique squarefree monomial of maximal block length k. That is, there is a unique monomial

$$n_1 n_2 \cdots n_k (x_{i_1} x_{i_2} \cdots x_{i_k}).$$

We stress here that this monomial is of maximal block length, since this will be important in the subsequent algorithm.

Since each $n_i \neq 0$ this term must appear and therefore $W(\alpha_1, \ldots, \alpha_n) \neq 1$. It follows that the group generated by $\alpha_1, \ldots, \alpha_n$ is freely generated by them. $\square$

The proof of the faithfulness of the Magnus representation leads us to several algorithms for dealing with the image in the power series ring. We will employ these algorithms in our cryptosystems. For the remainder of this section we will let $\overline{F}$ denote the free subgroup of $\overline{H}$ generated by the α_i.

The first algorithm provides a method, given a polynomial in $\overline{H}$, which is written in reduced form, that we know to be in $\overline{F}$, to write its unique free group decomposition. That is, given a polynomial

$$f = f(x_1, \ldots, x_n)$$

in the non-commuting variables $x_1, .., x_n$ that we know to be in $\bar{F}$ to rewrite f as

$$f = W(\alpha_1, \ldots, \alpha_n).$$

In general there is no factoring algorithm in $\bar{H}$.

For any monomial $x_{i_1} \cdots x_{i_k}$ in $\bar{H}$ we call k the block length of the monomial in analogy with that of a free group word.

Theorem 10.3.2 (Algorithm to Recover the Free Group Decomposition of Elements in $\bar{F}$). *Suppose $f = f(x_1, \ldots, x_n) \in \bar{H}$ and it is known that $f \in \bar{F}$. There is an algorithm that rewrites f in terms of the free generators $\alpha_1, \ldots, \alpha_n$, that is, the algorithm uniquely expresses f as a free group word*

$$f = W(\alpha_1, \ldots, \alpha_n).$$

The algorithm works as follows:

Step 1: *In f locate the monomial $nx_{i_1} \cdots x_{i_k}$ of maximal block length, where $n \in \mathbb{Z} \setminus \{0\}$, where each variable that appears in f appears linearly in this monomial. This number k gives the block length for the corresponding free group word. Further, the free group word must have the form*

$$\alpha_{i_1}^{n_1} \cdots \alpha_{i_k}^{n_k}$$

with each n_i a divisor of n.

Step 2: *Starting with $j = 1$ and $f_0 = f$, for each divisor n_i of n (both positive and negative), sequentially form $(1 + x_{i_j})^{-n_j} f_{j-1}$. In exactly one monomial of this product, the maximal block length will be $k - 1$ and there will be a unique monomial of block length $k - 1$ containing each variable in f linearly, except perhaps x_{i_1}. We then write*

$$f_{j-1} = (1 + x_{i_j})^{n_j} f_j$$

where f_j is also in $\bar{F}$.

Step 3: *Continue in this manner until we reach the identity. The free group decomposition of f is then*

$$f = (1 + x_{i_1})^{n_1} \cdots (1 + x_{i_k})^{n_k} = \alpha_{i_1}^{n_1} \cdots \alpha_{i_k}^{n_k}.$$

Proof. Since we know that $f \in \bar{F}$, we know that there is a unique free group decomposition

$$f = \alpha_{i_1}^{n_1} \cdots \alpha_{i_k}^{n_k} = (1 + x_{i_1})^{n_1} \cdots (1 + x_{i_k})^{n_k}.$$

Hence, as in the proof that the representation is faithful, there is a unique monomial

$$nx_{i_1} \cdots x_{i_k}$$

of maximal block length, where $n \in \mathbb{Z} \setminus \{0\}$, and where each variable that appears in f appears linearly in this monomial. Again, as in the proof of Theorem 10.3.1, the number k gives the block length for the corresponding free group word.

Now, since the free group representation is unique, we have for each divisor n_i of n

$$(1 + x_{i_1})^{-n_i} f = (1 + x_{i_1})^{n_1 - n_i} (1 + x_{i_2})^{n_2} \cdots (1 + x_{i_k})^{n_k}.$$

Hence only for $n_i = n_1$ will this term cancel. Consequently, there is exactly one such divisor such that $(1 + x_i)^{-n_i} f$ will now have maximal block length $k - 1$ and have a unqiue monomial of the prescribed type. It follows then that one and only one such product will reduce f to a word of shorter block length. □

A modification of the above algorithm can be used to determine if a general element of $\overline{H}$ is actually in $\overline{F}$.

Theorem 10.3.3 (Algorithm to Determine if $f \in \overline{H}$ is in $\overline{F}$). *Suppose*

$$f = f(x_1, \ldots, x_n) \in \overline{H}.$$

There is an algorithm that determines whether or not $f \in \overline{F}$ and if it is, rewrites f in terms of the free generators $\alpha_1, \ldots, \alpha_n$. The algorithm works as follows:
Step 1: *If the constant term of f is not 1 then $f \notin \overline{F}$. Further, if f has any non-integral coefficients then $f \notin \overline{F}$.*
Step 2: *Assume f passes Step 1. If f does not contain a unique monomial $nx_{i_1} \cdots x_{i_k}$ of maximal block length in f, where $n \in \mathbb{Z} \setminus \{0\}$, such that each variable that appears in f appears linearly in this monomial, then $f \notin \overline{F}$.*
Step 3: *Suppose f passes Steps 1 and 2. Then locate the monomial in f with the properties described in Step 2. If $f \in \overline{F}$ then k gives the block length for the corresponding free group word. Further, the free group word must have the form*

$$\alpha_{i_1}^{n_1} \cdots \alpha_{i_k}^{n_k}$$

with each n_i a divisor of n.
Step 4: *Starting with $j = 1$ and $f_0 = f$, for each divisor n_i of n (both positive and negative) sequentially form the product $(1 + x_{i_j})^{-n_i} f_{j-1}$. If in some monomial of this product the maximal block length is $k - 1$ and there is no new monomial having the properties described above then return $f \notin \overline{F}$. Otherwise continue.*
Step 5: *Finally, if we arrive at the identity then $f \in \overline{F}$ and the algorithm yields the free product decomposition of f.*

Proof. The proof follows in exactly the same manner as the proof of Theorem 10.3.2. □

For certain cryptographic applications we need the full unit group $U(\overline{H})$ of $\overline{H}$. Over $\mathbb{Q}$ it can be described as follows.

Theorem 10.3.4. *The unit group $U(\overline{H})$ over $\mathbb{Q}$ consists precisely of those polynomials with non-zero constant term.*

Proof. There are two ways to look at the proof of this. Algebraically, suppose that the defining power is $d > 1$ and $P(x) \in \overline{H}$ with non-zero constant term. Then $P(x)$ is relatively prime to the polynomials x_i^d and so is invertible in the factor ring in the standard way.

Analytically if $P(x) \in \overline{H}$ with non-zero constant term let $P^*(x)$ be the corresponding polynomial in H. Then $P^*(x)$ can be made into part of a convergent power series $P^{**}(x)$ in

$$\mathbb{C}\langle\langle x_1, \ldots, x_n \rangle\rangle.$$

Since $P^{**}(0) \neq 0$ this power series is analytic at 0 and so its inverse is analytic at 0 and so has a convergent power series around 0, say $Q^{**}(x)$. The image of $Q^{**}(x)$ in $\overline{H}$ would then be the inverse of $P(x)$.

Conversely, if $P(x) \in \overline{H}$ is invertible, it must have non-zero constant term. □

Before we continue we mention one final item concerning multiplication within $\overline{H}$. In general there is no factoring algorithm. However, if $f \in \overline{H}$ is known and $g = fe$ with $e \in \overline{F}$ is known then we can find e. We say that e can be peeled off fe. The algorithm to do this is essentially the same as the above two algorithms. We briefly explain. Suppose we are given f and fe. Then in fe there is a unique monomial extending the monomials in f exactly as in the proof of Theorem 10.3.2. By identifying this monomial we can find the free group decomposition of e and hence find e.

10.4 Cryptographic Protocols Using Non-Abelian Groups

As we discussed in Chapter 4, there are many different cryptographic tasks beyond secure confidential message transmission. Recall that a **cryptographic protocol** is the set of all methods to perform such a cryptographic task. Wagner and Magyarik introduced the idea of using group theoretic decision problems in public key encryption protocols. Their method was not workable but many possibly secure group based public key cryptosystems were subsequently developed. The two most prominent, developed at approximately the same time, were the **Ko-Lee protocol** and the **Anshel-Anshel-Goldfeld protocol**. We will describe these in detail in the next chapter. In this section we look at alternative cryptographic protocols that use group based methods. The following cryptographic tasks were introduced in Chapter 4.

(1) **Authentication:** An authentication procedure is the process of determining that a message, supposedly from a given person, both does come from that person and has not been tampered with. Included in authentication are the concepts of **hash functions** and **digital signatures**;

(2) **Key Exchange and Key Transport:** In a key exchange, two people, usually called Bob and Alice, exchange a secret shared key to be used in some symmetric or asymmetric encryption. In a key transport, one party transports to another a secret key that is to be used.

(3) **Secret Sharing:** This is a method where some secret is to be shared by k people but not available to any proper subset of them. There is a beautiful simple solution to the general problem given by Shamir. We will describe this later in the book.

(4) **Zero Knowledge Proof:** A zero-knowledge proof is an argument that convinces someone that you have solved a problem, for example a combinatorial problem, without giving away the solution. This is tied to authentication.

In the next three sections we present examples of these protocols using non-abelian groups. We first look at a digital signature protocol based on different aspects of the conjugacy problem. We next look at a password security protocol based on free group cryptography that is provably secure. Finally we present an non-abelian group based example of a secret sharing scheme.

10.5 A Non-Abelian Digital Signature Procedure

We present a digital signature procedure based on non-abelian groups developed by Ko, Lee *et al.* (see [KCCL]). In describing this protocol we must first introduce an additional group theoretic decision problems. In Section 9.5 we discussed the three basic group decision problems for a finitely presented group G: the **word problem**, the **conjugacy problem**, and the **isomorphism problem**. Recall that in a finitely presented group G the **conjugacy problem** asks if there exists an algorithm to decide whether or not an arbitrary pair of words u and v in the generators of G are conjugate? That is, is there an $x \in G$ such that $x^{-1}ux = v$? To distinguish this from certain other decision problems using conjugacy we call this the **decision conjugacy problem**. For a finitely presented group G the **search conjugacy problem** is the following. Given $u, v \in G$ that we know to be conjugate is there an algorithm to find $z \in G$ satisfying $z^{-1}uz = v$?

In the following we use the notation u^z for $z^{-1}uz$.

With these ideas here is the Ko-Lee digital signature scheme.

Let G be a non-abelian group in which the search conjugacy problem is infeasible and the decision conjugacy problem is solvable. Let $\{0, 1\}^*$ be the set of all $0, 1$ sequences and let $h : \{0, 1\}^* \rightarrow G$ be a hash function.

Key Generation: Alice wants to sign and send a message, m, to Bob. Alice begins by choosing two conjugate elements $u, v \in G$ with conjugator a. The conjugate pair (u, v) is public information while the conjugator a is Alice's secret key.

Signature Generation: Alice chooses arbitrary $b \in G$, and computes $\alpha = u^b$ and $y = h(m\alpha)$. Then a signature σ on the message m is the triple (α, β, γ) where $\beta = y^b$ and $\gamma = y^{a^{-1}b}$. She sends this to Bob for verification and acceptance.

Verification: Upon receiving the signature, Bob checks whether or not the following hold:

(1) $\exists c_1 \in G$ such that $u = \alpha^{c_1}$.

(2) $\exists c_2, c_3 \in G$ such that $y = \beta^{c_2}$ and $y = \gamma^{c_3}$.

(3) $\exists c_4 \in G$ such that $uy = (\alpha\beta)^{c_4}$.
(4) $\exists c_5 \in G$ such that $vy = (\alpha\gamma)^{c_5}$.
Bob accepts the signature if and only if 1-4 hold.

The security of this scheme lies in the assumption that, given a pair of conjugate elements $u, v \in G$, finding elements α, β, γ such that (1)–(4) above hold is infeasible. If the conjugator a can be found, then $(\alpha, \beta, \gamma) = (u^b, y^b, y^{a^{-1}b})$ satisfy properties (1)–(4) for any $b \in G$. Hence the conjugacy search problem has to be infeasible.

We mention that there is a digital signature scheme proposed by Anjaneyulu *et al.* [ARR] that uses a non-commutative platform but outside of group theory. In this proposal the basic cryptographic platform is a non-commutative division semiring and uses what is termed the polynomial symmetrical decomposition problem for the one way function. The reader may refer to the paper [ARR] for details.

10.6 Password Security Using Combinatorial Group Theory

Closely related to digital signatures is the problem of **secure password verification**. With the increased use of bank cards and internet credit card transactions there is at present more than ever a need for secure password identification. For many online purchases, this is being carried out by a **challenge response system** (see [Wik]) accompanying the password. In the simplest systems this takes the form of secondary password questions such as the user's mother's maiden name or place of birth. There are inherent difficulties with these types of challenge response systems. First of all there is the trivial problem of the users remembering their responses. More critical is the problem that this type of information for many people is readily available and easily found or guessed by would-be attackers or eavesdroppers.

Challenge response systems are also subject to man-in-the-middle attacks and replay attacks (see [CR]). There have been several attempts to alleviate these problems, including zero-knowledge password proofs and challenged responses somewhat based on RSA as well as timed out responses (see CRAM-MD5, Password Authenticated Key Agreement, [Har] and [Wik]).

In this section we present an alternative method for challenge response password verification using combinatorial group theory. In particular this method depends upon the difficulty of solving the word problem within a given finitely presented group without knowing the presentation and the difficulty of solving systems of equations within free groups. This latter problem has been proved to be NP-hard. The method uses the **group randomizer system** which is a computer program that is a subset of MAGNUS, a much larger computer algebra system designed to handle algorithmic problems in combinatorial group theory, MAGNUS was developed at CAISS, the Center for Algorithms and Interactive Scientific Software, a research laboratory housed at City College of the City University of New York and under the direction of the first author.

The group randomizer system can be placed on a simple hand held computer device presently under development at CAISS. The system can also be used from computer to computer. More information can be obtained from the CAISS website.

These group theoretic techniques have several major advantages over other challenge response systems. We will call the password presenter the **prover** or **Peggy the Prover** and the presentee the **verifier** or **Vic the Verifier**. The methods we present can be used for **two-way authentication**, that is to both verify the prover to the verifier and to verify the verifier to the prover. To each user in conjunction with a standard password there will be assigned a finitely presented group with a solvable word problem. We call this the **challenge group**. This will be done randomly by the group randomizer system and will be held in secret by the prover and the verifier. Cryptographically we assume the adversary can steal the encrypted form of the group theoretic responses. Probabilistically this does not present a problem. Each challenge response set of questions forms a virtual one time key pad as we will explain. Therefore the adversary must steal three things – the original password, the challenge group and the group randomizer. Hence there is almost total security in the challenge response system. Further there is an infinite supply of finitely presented groups to use as challenge groups and an infinite supply of challenge response questions that never have to be duplicated. We will explain these in the section on this protocol's security. Finally the method is symmetric between the verifier and the prover, so while the verifier verifies the prover's password simultaneously the prover verifies that he or she is dealing with the verifier.

The theoretical security of the system is provided by several results in asymptotic group theory. In particular, a result of Lysenok [LM] implies that stealing the challenge group is NP-hard while a result of Jitsukawa [Jit] says that the asymptotic density of using homomorphisms (see Section 10.5) to attack the group randomizer protocol is zero.

The whole password protocol depends upon the **group randomizer system**. This is a computer program that can handle several elementary tasks involving finitely presented groups. It is a subset of MAGNUS, a large computer algebra system developed at CAISS, the Center for Algorithms and Interactive Scientific Software, in order to do computations in infinite group theory. At present there are various versions of the group randomizer, including a portable hand held version now under development.

The scope of the a particular group randomizer system will depend on the type of login protocol or cryptographic protocol desired. At the most basic level the group randomizer system has the ability to do the following things:

(1) Recognize a finite presentation of a finitely presented group with a solvable word problem and manipulate arbitrary words in the alphabet of generators according to the rewriting rules of the presentation. In particular if the group is automatic the group randomizer can rewrite an arbitrary word in the generators in terms of its group normal form.

(2) Given a finite presentation of a group with a solvable word problem recognize whether two free group words have the same value in the given group when considered in terms of the given generators of the group.

(3) Randomly generate free group words on an alphabet of any finite size

(4) Recognize and store sets of free group words $W_1, \ldots, W_k$ on an alphabet $x_1, \ldots, x_n$ and rewrite words $W(W_1, \ldots, W_k)$ as the corresponding word in $x_1, \ldots, x_n$.

(5) Given a free group of finite rank on $x_1, \ldots, x_n$ and a set of words $W_1, \ldots, W_k$ on an alphabet $x_1, \ldots, x_n$ solve the membership problem in F relative to $H = \langle W_1, \ldots, W_k \rangle$, the subgroup of F generated by $W_1, \ldots, W_k$.

(6) Given a stored finitely presented group or a stored set of free group words the randomizer can accept a random free group word and rewrite it as a normal form in the finitely presented group in the former case or as a word in the ambient free group in the latter case.

We now present several variations on secure password verification using the group randomizer. First we give an overall outline of the protocol.

(1) General Outline of the Authentication Protocol

This is a symmetric key cryptographic authentication protocol. Both the prover and verifier use a single private key to both encrypt and decrypt within the authentication process. At the first step the prover and verifier must communicate directly, either face-to-face or by a public key method, to set the private shared secret. This is the model now used for most password/password back-up schemes. We assume that both the prover and verifier have a group randomizer system. For security analysis we assume that an adversary or eavesdropper has access to the encrypted form of the transmission but is passive in that the adversary will not change any transmissions.

Step (1): The prover and verifier communicate directly to setup a common shared secret (P, G) where P is a standard password and G is a challenge group. Each prover's challenge group is unique to that prover. The challenge group is a finitely presented group with a solvable word problem and satisfying the strong generic free group property (see Section 10.5). The password is chosen by the prover while the challenge group is randomly chosen by the group randomizer system.

Step (2): The prover presents the password to the verifier. The group randomizer of the verifier presents a group theoretic "question" (see parts (2) and (3)) concerning the challenge group G to the prover. The assumption is that this "question" is difficult in the sense that it is infeasible to answer it if the group G is unknown. The question is then answered by the group randomizer. This is repeated a finite number of times. If the answers are correct the prover (and the password) is verified.

Step (3): The protocol is then repeated from the viewpoint of the prover, authenticating the verifier to the prover.

(2) Free Subgroup Method

We assume that both the prover and the verifier has a group randomizer. Each prover has a standard password. Suppose that F is a free group on $\{x_1, \ldots, x_n\}$. The prover's password is linked to a finitely generated subgroup of a free group given as words in the generators, that is, the prover's password is linked to $W_1, \ldots, W_k$ where each W_i is a word in $x_1, \ldots, x_n$. The group

$$G = \langle W_1, \ldots, W_k \rangle$$

is called the **challenge group**. In general $k \neq n$. The prover does not need to know the generators. The randomizer can randomly choose words from this subgroup and then freely reduce them. The prover has the challenge group or subgroup also stored in its randomizer.

The prover submits his or her standard password to the prover. This activates the verifier's randomizer to the prover's set of words. The verifier now submits a random free group word on $y_1, \ldots, y_k$ to the prover's randomizer say $W(y_1, \ldots, y_k)$. The prover's randomizer treats this as $W(W_1, \ldots, W_k)$ and then reduces it in terms of the free group generators $x_1, \ldots, x_n$ and rewrites it as $W^*(x_1, \ldots, x_n)$. The verifier checks that this is correct, that is, $W(W_1, \ldots, W_k) = W^*(x_1, \ldots, x_n)$ on the free group on $x_1, \ldots, x_n$. If it is the verifier continues and does this three (or some other finite number) of times. There is one proviso. The verifier submits a word to the prover only once, so that a submitted word can never be reused. The prover's randomizer will recognize if it has (this is a verification to the prover of the verifier).

To verify that the verifier is legitimate the process is repeated from the prover's randomizer to the verifier.

An attacker only has access to the transmitted words. Given a series of free group words there is an essential probability of zero of reconstructing the subgroup (see Section 10.5). To prevent an attacker using an already used word to gain access, the group randomizer system allows a free group word, submitted as a challenge word, to be used only once. If an attacker gets access to the verifier and submits an already submitted word or vice versa from the prover this will red flag the attempt. We also suggest that if there is a previously used word, indicating perhaps an attack, the group randomizer should change the prover's group. The beauty of this system is that this can be done extremely easily; change several of the words for example. Essentially this presents an essential one-time keypad each time the prover presents the password. The map $y_i \rightarrow W_i$ is a homomorphism and an attacker can manipulate various equations in an attempt to solve. Presumably if there are enough equations the words $W_1, \ldots, W_k$ can be discovered. However, in Section 10.5 we present a security proof based on several results in asymptotic group theory showing that this cannot happen with asymptotic density one.

We suggest a noise/diffusion enhancement. The provers challenge group generator words $W_1, \ldots, W_k$ are indexed. With each use the randomizer applies a random permutation ϕ on $\{1, \ldots, k\}$ to scramble the indices. These permutations are coded and stored both in the prover's randomizer and the verifier's. This prevents a length based attack by an eavesdropper since discovering for example what W_{37} is, is of no use since it will be indexed differently for the next use. The coded permutation is sent as part of the challenge.

(3) General Finitely Presented Group Method

This is essentially the same method, however, rather than work with an ambient free group we work with a given finitely presented group with a solvable word problem. Let $G = \langle X; R \rangle$ be the group. As before we assume that both the prover and the verifier has a group randomizer. Each prover has a standard password. Suppose that $X = \{x_1, \ldots, x_n\}$ and F is a free group on $\{x_1, \ldots, x_n\}$. The prover's password is linked to a finitely generated subgroup of G again given as words in the generators X, that is, the prover's password is linked to $W_1, \ldots, W_k$ where each W_i is a word in $x_1, \ldots, x_n$. As before, we let $k \neq n$. The randomizer can randomly choose words from this subgroup and then reduce them via the finite presentation. The verifier has the group and subgroup also stored in its randomizer.

The remainder of the procedure is exactly the same as in the free group case. The prover submits his or her standard password to the verifier. This activates the verifier's randomizer to the prover's set of words. The verifier now submits a random free group word on $y_1, \ldots, y_k$ to the prover's randomizer say $W(y_1, \ldots, y_k)$. The prover's randomizer treats this as $W(W_1, \ldots, W_k)$ and rewrites it as $W^*(x_1, \ldots, x_n)$. The verifier checks that this is correct, that is, $W(W_1, \ldots, W_k) = W^*(x_1, \ldots, x_n)$, however, this time in the group G. If it is, the verifier continues and does this three (or some other finite number) of times. There is one proviso. The verifier submits a word to the prover only once so that a submitted word can never be reused. The prover's randomizer will recognize if it has (this is a verification to the prover of the verifier).

To verify that the verifier is legitimate the process is repeated from the prover's randomizer to the verifier.

As in the free group method, an attacker only has access to the transmitted words. Given a series of group words there is probability zero of reconstructing the group, however, as in the free group method a given challenge response word is to be used only once.

10.6.1 The Strong Generic Free Group Property

Part of the theoretical security of the group randomizer protocols depends on the **strong generic free group property** and **asymptotic density**. Asymptotic density is a general method to compute densities and/or probabilities on infinite discrete sets where each individual outcome is tacitly assumed to be equally likely. The origin of asymptotic density lie in the attempt to compute probabilities on the whole set of integers where each integer is considered equally likely. The method can also be used where some probability distribution is assumed on the elements. It has been effectively applied to determining densities within infinite discrete finitely generated groups where random elements are considered as being generated from random walks on the Cayley graph of the group. The paper by Borovik, Myasnikov and Shpilrain [BMS] provides a good general description of this method in group theory. Let $\mathcal{P}$ be a group property and let G be a finitely generated group. We want to determine the measure of the set of elements which satisfy $\mathcal{P}$. For each positive integer n let B_n denote the n-ball in G. Let $|B_n|$ denote the actual size of B_n (which is an integer since G is finitely generated) or the measure of $|B_n|$ if a distribution has been placed on the elements of G. Let S be the set of elements in G satisfying $\mathcal{P}$. The asymptotic density of S is then

$$\lim_{n \to \infty} \frac{|S \cap B_n|}{|B_n|}$$

provided this limit exists. We say that the property $\mathcal{P}$ is **generic** if the asymptotic density of the set S of elements satisfying $\mathcal{P}$ equals 1.

This concept can be easily extended to properties of finitely generated subgroups. We consider the asymptotic density of finite sets of elements that generate subgroups that have a considered property. For example, to say that a group has the generic free group property we mean that

$$\lim_{m,n \to \infty} \frac{|S_m \cap B_{m,n}|}{|B_{m,n}|} = 1$$

where S_m is the collection of finite sets of elements of size m that generate a free subgroup while $B_{m,n}$ are all the m-element subsets within the n-ball. We refer to the paper [BMS] and the book [MSU1] for terminology and further definitions.

We say that a group G has the **generic free group property** if a finitely generated subgroup is generically a free group. For example, a result of Epstein [Eps] says that the group $GL(n, \mathbb{R})$ satisfies the generic free group property. A group G has the **strong generic free group property** if given randomly chosen elements $g_1, \ldots, g_n$ in G then generically they are a free basis for the free subgroup they generate. Jitsukawa [Jit] proved that free groups have the strong generic free group property. That is given k random elements $W_1, \ldots, W_k$ in the free group on $y_1, \ldots, y_n$ then with asymptotic density one the elements $W_1, \ldots, W_k$ are a free basis for the subgroup they generate. We compare this with the Nielsen-Schreier theorem that says that $W_1, \ldots, W_k$ generate a free group. In the context of the group randomizer protocols, the strong generic free group property implies that if $V_1(y_1, \ldots, y_m), \ldots, V_k(y_1, \ldots, y_m)$ have already been presented as challenge words then the probability is approximately zero that a new challenge word $V(y_1, \ldots, y_m)$ lies in the subgroup generated by $V_1, \ldots, V_k$, and hence a homomorphism attack is nullified.

The strong generic free group property has been extended to arbitrary free products of infinite groups and many other amalgams including surface groups by Fine, Myasnikov and Rosenberger [FMyR]. Let us mention some further results. Gilman, Myasnikov and Osin [GMO] showed that torsion-free hyperbolic groups have the generic free group property. Myasnikov and Ushakov [MU] showed that pure braid groups P_n with $n \geq 3$ also have the strong generic free group property. A recent result of Carstensen, Fine and Rosenberger [CFR2] shows that all Fuchsian groups of finite co-volume and all braid groups B_n with $n \geq 3$ have the strong generic free group property (see the next chapter for a discussion of braid groups).

The result of Myasnikov and Ushakov on the pure braid groups has applications to the cryptanalysis of both the Ko-Lee cryptosystem and the Anshel-Anshel-Goldfeld cryptosystem. These public key cryptosystems will be discussed in the next chapter. Both cryptosystems were usually susceptible to length based attacks if the parameters chosen in the braid groups B_n were small. The reason for this is that random choices of subgroups within the braid groups are actually free groups. This does not disqualify the braid groups as platforms but rather says that subgroups cannot be chosen entirely randomly.

Extremely useful in proving that a group has the generic or strong generic free group property is the following.

Theorem 10.6.1. *Let G be a group and N a normal subgroup. If the quotient G/N satisfies the strong generic free group property then G also satisfies the strong generic free group property.*

In [FMyR] it was shown that many group amalgams, free products, free products with amalgamation, and HNN groups (see Section 9.8) satisfy the strong generic free group property. In particular, the most general result is the following.

Theorem 10.6.2. *Let A and B be arbitrary finitely generated infinite groups and let $G = A * B$ be their free product. Let $\{x_1, \ldots, x_n\}$ be n randomly chosen elements from G. Then generically these elements are a free basis for the subgroup they generate, that is, the group G satisfies the strong generic free group property.*

This can be extended to more general amalgams in many ways (see [FMyR])

Theorem 10.6.3. *Let A and B be arbitrary finitely generated infinite groups and let $G = A * B$ be their amalgamated free product with amalgamated subgroup H. Let H_1 and H_2 be the copy of H in A and B respectively. Suppose that $A/N(H_1)$ is infinite and $B/N(H_2)$ is infinite where $N(H_i)$ is the normal closure of H_i in the respective factors. Then G satisfies the strong generic subgroup property.*

A **cyclically pinched one-relator group** is an amalgamated free product of the form

$$G = F_1 \underset{U=V}{*} F_2$$

where F_1, F_2 are finitely generated free groups and U, V are non-trivial words in the respective free groups. If U is not a power of a primitive element in F_1 and V is not a power of a primitive element in F_2 then the quotient of F_1 and F_2 by the normal closure of U and V respectively is a non-trivial, infinite one-relator group.

Corollary 10.6.4. *Let G be a cyclically pinched one-relator group as above. Assume that U and V are not a power of a primitive element in F_1 and F_2 respectively. Then G satisfies the strong generic subgroup property.*

In particular any orientable surface group of genus $g \geq 2$ falls into the class of cyclically pinched one-relator groups.

Corollary 10.6.5. *Any orientable surface group of genus $g \geq 2$ and any non-orientable surface group of genus $g \geq 4$ satisfies the strong generic subgroup property.*

The situation with HNN groups becomes even more complicated but some things can be proved as consequences of the amalgam result above. Notice first however, that any HNN group with free part of rank ≥ 2 must have a free quotient of rank ≥ 2 and hence satisfy the strong generic subgroup property. Therefore only the case where the free part has rank 1 must be considered.

Theorem 10.6.6. *Let G be an HNN extension of the group B with a presentation*

$$G = \langle t, B; rel(B), t^{-1}Ut = V \rangle$$

with U, V non-trivial isomorphic subgroups of B. Let $N_B(\langle U, V \rangle)$ be the normal closure of the subgroup $\langle U, V \rangle$ in B. Then if $B/N_B(\langle U, V \rangle)$ is infinite, G satisfies the strong generic subgroup property.

Extensions of centralizers play a large role in the study of the elementary theory of free groups. Recall that if B is a group and $U \in B$ then a **rank one extension of centralizers** of B is a group with a presentation

$$G = \langle t, B; rel(B), t^{-1}UT = U \rangle.$$

Theorem 10.6.7. *Let G be a rank one extension of centralizers of the group B. Suppose that G has a presentation*

$$G = \langle t, B; rel(B), t^{-1}Ut = U \rangle$$

where U is a non-trivial element of B. If $B/N_B(U)$ is infinite, where $N_B(U)$ is the normal closure of U in B, then G satisfies the strong generic subgroup property .

In the situation where the factors are finite we must be careful even for free products. The infinite dihedral group $\mathbb{Z}_2 * \mathbb{Z}_2$ is solvable so cannot satisfy the strong generic free group property. However, if at least one factor has order greater than 2, an analysis based on Kurosh bases yields the weaker generic free group property.

Theorem 10.6.8. *Let $G = A * B$ be a non-trivial free product. If at least one factor has order greater than 2 then G satisfies the generic free group property.*

In general, asymptotic density is not independent of finite generating systems. Indeed it is possible for a group property to be generic with respect to one finite generating system and negligible with respect to another (see [KKS]). We call a group property $\mathcal{P}$ **suitable** for a finitely generated group G if it is preserved under isomorphisms and its asymptotic density is independent of finite generating systems and **super suitable** for G if its suitable both for G and all subgroups of finite index in G. It can be proved that the strong generic free group property is suitable in any group G which has a non-abelian free quotient.

Corollary 10.6.9. *The strong generic free group property is suitable in any finitely generated group G which has a non-abelian free quotient.*

In [CFR2] it was shown that there is an interesting connection between the strong generic free group property of a group G and its subgroups of finite index. The main result of that paper is that a finitely generated group which has a non-abelian free quotient satisfies the strong generic free group property if and only if each subgroup of finite index satisfies the strong generic free group property. As a consequence of this and Theorem 10.6.10, it follows that many important classes of groups, such as finitely generated Fuchsian groups with finite co-volume and the braid groups B_n for $n \geq 3$ satisfy the strong generic free group property.

Theorem 10.6.10 (Inheritance Theorem [CFR2]). *Let G be a finitely generated group and $H \subset G$ a subgroup of finite index $[G : H] = n < \infty$. Let $\mathcal{P}$ be the strong generic free group property. Then:*
(1) *If $\mathcal{P}$ is a suitable and generic property in H then it is also suitable and generic in G.*
(2) *If $\mathcal{P}$ is a suitable and generic property in G then it is also suitable and generic in H.*

10.6.2 Security Analysis of the Group Randomizer Protocols

In order to analyze the security of the group randomizer password protocols, we make the security assumption that an adversary has access to the coded group theoretic responses. The strength of the proposed protocol include that an attacker must steal three things: the original password, the group randomizer and the challenge group. There is no access without all three. This immediately nullifies middelman attacks. If the adversary pretends to be the verifier to obtain the group words the attack is thwarted by the facts that the prover can verify the verifier and further if the attacker just transmits from the middle, nothing can be stolen since each time through a new challenge word must be used. Further the group randomizer has an infinite supply of both subgroups and challenge responses that are done randomly. In addition, since a challenge word can be used only once the protocol nullifies replay attacks. Since challenge responses are machine to machine there is an essential probability of zero of an incorrect response. The protocol shuts down with an incorrect response and hence repeat attacks are harmless.

These are in distinction to answer-driven challenge-response systems where a prover often forgets or misspells a response. In these systems a prover is usually permitted several opportunities to answer making it susceptible to both man-in-the-middle and repeat attacks.

There are two theoretical attacks that must be dealt with. Relative to these the security of the system, and hence a security proof for the protocol, is provided by several results in asymptotic group theory.

The most straightforward attack is for the adversary to collect enough challenge words and respsonses. This provides a system of equations in a free group (or a finitely presented group)

$$y_{i_1} \cdots y_{i_t} = W_i(x_1, \ldots, x_n), \quad i = 1, \ldots, m.$$

An adversary can then break the protocol by solving the system

$$z_i = W_i(x_1, \ldots, x_n)$$

to obtain the challenge group.

However, a result of Lysenok [LM] shows that solving such systems of equations in free groups (and in most finitely presented groups) is NP-hard. Hence this method of attack is impractical in most cases.

A second method of attack is based on the following. The mapping $y_i \to W_i$ is a homomorphism. If a challenge word appears in the subgroup generated by previous challenge words then an attacker can use this to answer a challenge without ever solving for the challenge group. However, the probability of succeeding with this approach is essentially zero due to Jitsukawa's result mentioned in the previous section. Each challenge word lies in a free group which has the strong generic free group property. Hence as explained in the previous section the probability is essentially zero that a new challenge word is in the subgroup generated by previous challenge words.

10.6.3 Actual Implementation of a Group Randomizer System Protocol

The actual implementation of a workable group randomizer system protocol involves several choices of parameters and subprograms. These include the following choices.
(1) The choice of the rank of the ambient free group in the group randomizer systems A and B.
(2) An enhancement program which takes randomly chosen words $W_1, \ldots, W_k$ in a free group F and finds a new set of words $V_1, \ldots, V_k$ generating the same subgroup for which the words formed in $V_1, \ldots, V_k$ have a great deal of free cancellation. This involves what is called Nielsen transformations (see [MKS]).
(3) The choice of parameter sizes for the lengths of the randomly chosen words. In an actual implementation all words in the generators will have lengths between a

and b where a and b are to be determined. All words used as test logins will have lengths between c and d with c and d to be determined.

The determination of the optimal values of a, b, c, d are being studied.

(4) The implementation of a coded permutation system on $\{1, \ldots, k\}$ where k is the rank of the challenge group and which can be sent with each challenge word.

(5) The development of an automatic reset protocol for the challenge group. In an ideal situation this can be done without actually communicating the changes between verifier and prover. That is, each randomizer system does the same protocol automatically when reset is called for.

10.7 A Secret Sharing Scheme Using Combinatorial Group Theory

In Section 4.6 we discussed general secret sharing schemes. Recall that the secret sharing problem is the following. We have a secret K and a group of n participants. This group is called the **access control group**. A **dealer** allocates shares to each participant under given conditions. If a sufficient number of participants combine their shares then the secret can be recovered. If $t \leq n$ then an (t, n)-**threshold scheme** is one with n total participants and in which any t participants can combine their shares to recover the secret but not fewer than t. The number t is called the **threshold**. The scheme is called a **secure secret sharing scheme** if, given fewer shares than than the threshold, there is no chance to recover the secret.

D. Panagopoulos [Pan] devised a secret sharing scheme based on the word problem in finitely presented groups. It is an (t, n)-threshold scheme and its main advantage over many other secret sharing schemes is that it does not require the secret message to be determined before each individual person receives his share of the secret. For this scheme it is assumed that the secret is given in the form of a binary sequence. The scheme is as follows.

Step 1: A finitely presented group $G = \langle x_1, x_2, \ldots, x_k; r_1 = \cdots = r_m = 1 \rangle$ is chosen. It is assumed that the word problem is solvable for this presentation and that $m = \binom{n}{t-1}$.

Step 2: Let $A_1, \ldots, A_m$ be an enumeration of the subsets of $\{1, \ldots, n\}$ with $t - 1$ elements. Define n subsets $R_1, \ldots, R_n$ of $\{r_1, \ldots, r_m\}$ such that $r_j \in R_i$ if and only if $i \notin A_j$ for $i = 1, \ldots, n$ and $j = 1, \ldots, m$. Then for every $j \in \{1, \ldots, m\}$, the word r_j is not contained in exactly $t-1$ of the subsets $R_1, \ldots, R_n$. It follows that r_j is contained in any union of t of them, whereas if we take any $t - 1$ of the sets $R_1, \ldots, R_n$, there exists an index j such that r_j is not contained in their union.

Step 3: Distribute to each of the n persons one of the sets $R_1, \ldots, R_n$. The set $\{x_1, \ldots, x_k\}$ is known to all participants.

Step 4: If the binary sequence to be distributed is $a_1, \ldots, a_k$, construct and distribute a sequence of elements $w_1, \ldots, w_k$ of G such that $w_i = 1$ in G if and only if $a_i = 1$ for $i = 1, \ldots, k$. The word w_i must involve most of the relations $r_1 = 1, \ldots, r_m = 1$

if $w_i = 1$. Furthermore, all of the relations must be used at some point in the construction of some element.

Then any t of the n persons can obtain the sequence $a_1, \ldots, a_k$ by taking the union of the subsets of the relations of G that they possess. Thus they obtain the presentation $G = \langle x_1, x_2, \ldots, x_k; r_1 = \cdots = r_m = 1 \rangle$ and can solve the word problem $w_i = 1$ in G for $i = 1, \ldots, k$.

A collection of fewer than t persons cannot decode the message correctly, since the union of fewer than t of the sets $R_1, \ldots, R_n$ contains some but not all of the relations $r_1, \ldots, r_m$. Thus such a collection leads to a group presentation $\tilde{G} = \langle x_1, x_2, \ldots, x_k; r_{j_1} = \cdots = r_{j_p} = 1 \rangle$ with $p < m$ and $G \neq \tilde{G}$, where $w_i = 1$ in G is, in general, not equivalent to $w_i = 1$ in $\tilde{G}$.

Notice that the secret sequence to be shared is not needed until the final step. It is possible for someone to distribute the sets $R_1, \ldots, R_m$ and decide at a later time what the sequence $a_1, \ldots, a_k$ would be. In that way the scheme can also be used so that t of the n persons can verify the authenticity of the message. In particular, the binary sequence in Step 4 may contain a predetermined subsequence (signature) along with the actual message. Then any t persons may check whether this predetermined sequence is contained in the encoded message and thus validate it.

In the paper by D. Panagopoulos (see [Pan]), he also describes some methods for attacking this scheme and makes some suggestions for possible group presentation types to use.

10.8 Exercises

10.1 Let F_2 be the free group with free generating system a, b. Show that set $U = \{a^n b a^{-n} \mid n \in \mathbb{Z}\}$ is a free generating system for a free subgroup of F_2. This shows that a free subgroup of rank 2 has as a subgroup a free subgroup of countably infinite rank.

10.2 Let F_n be the free group of rank n with free generating set $X = \{x_1, \ldots, x_n\}$. Show that each conjugation

$$x_i \to g x_i g^{-1} \quad \text{with } g \in F_n$$

can be written as a sequence of elementary Nielsen transformations (see [CRR] or [LS]).

10.3 Bob has a backup password security system as described in Section 10.7. His basic words are $W_1 = x_1^{-1} x_2^2 x_3^{-2}$, $W_2 = x_1^5 x_2^3$ and $W_3 = x_2^5 x_1^3 x_2^{-2} x_3^4$. The bank sends him the word $W = y_1^2 y_3^3 y_1$. What must his group randomizer send back?

10.4 Let $M = \mathrm{PSL}(2, \mathbb{Z})$ be the modular group. Let $\mathcal{A} = \{a, b, c, d, e, f, g\}$ be a 7 letter plaintext alphabet. Choose a free subgroup of the modular group to encrypt these.

(a) Using your basic encryption and message units of size 3, what would be the encryption matrices for the message

abbdceffgcba.

(b) Using your basic encryption and the algorithm given in Problem 10.3, what is the plaintext message for

$$\begin{pmatrix} 8 & 5 \\ 5 & 3 \end{pmatrix}, \ \begin{pmatrix} 7 & 9 \\ 3 & 4 \end{pmatrix}?$$

10.5 This is the same problem as Problem 10.5, but encrypt and decrypt with the Magnus representation using formal power series over $\mathbb{Q}$. Apply the algorithms given in that section for decryption.

10.6 Prove Epstein's theorem: Given a random finitely generated subgroup of $GL_n(\mathbb{R})$ then with probability one it is a free group. The probability is standard measure on $\mathbb{R}^{n^2}$. (Hint: Given a finite set of matrices in $GL_n(R)$ think what a relation between them would mean algebraically on the coefficients and where this would place the matrices topologically.)

10.7 The following protocol is based on the factorization search problem which is: Given two subgroups A, B of a group G and $w \in G$ to find $a \in A, b \in B$ with $w = ab$. This protocol is described in [MSU1]. For this problem you must show and explain that the protocol works.

The requirements for the protocol are the same as for Ko-Lee; a public group G and two public subgroups A, B that commute elementwise. Alice randomly chooses two private elements $a_1 \in A, b_1 \in B$ and sends $a_1 b_1$ to Bob. Bob does the same choosing $a_2 \in A, b_2 \in B$ and sends $a_2 b_2$ to Alice. The common shared secret is $K = a_2 a_1 b_1 b_2$.

11 Platform Groups and Braid Group Cryptography

11.1 Cryptographic Platforms and Platform Groups

If a cryptographic protocol is based on an algebraic object, e.g., group, ring, lattice, or finite field, then this object is called the **cryptographic platform** or **platform**. In group-based cryptography this is then a **platform group** for the cryptographic protocol. The security of the cryptographic protocol is then dependent upon the difficulty, computational or theoretic, of solving a group theoretic problem within the platform group.

To be a reasonable platform group for a group based cryptographic protocol, a group G must possess certain properties that make the protocol both efficient to implement and secure.

We assume that the group G has a finite presentation

$$G = \langle X; R \rangle = \langle x_1, \ldots, x_n; R_1 = \cdots = R_m = 1 \rangle$$

and that the protocol security is based on a group theoretic problem that we denote by $\mathcal{P}$. The first necessity is that there is an efficient way to uniquely represent and then multiply the elements of G. In most cases this requires a **normal form** for elements $g \in G$ in terms of the generators $\{x_1, \ldots, x_n\}$. A normal form is, for each $g \in G$, a unique representation in terms of the generators. For example in Chapter 9 we saw that reduced words provide normal forms for elements of free groups. Normal forms provide an effective method of disguising group elements. Without this, one can determine a secret key simply by inspection of group elements. The existence of a normal form in a group implies that the group has solvable word problem, which is also essential for these protocols. For $g \in G$ we will denote its normal form, in terms of the set of generators X, by $NF_X(g)$. To be useful in cryptography, given $g \in G$, expressed as a word in $x_1, \ldots, x_n$ the process of moving between the word and the unique normal form must be efficiently computable. Usually we require at most polynomial time in the input length of g.

In addition to the platform group having normal forms, ideally, it would also exhibit exponential growth. That is, the growth function for G,

$$\gamma : \mathbb{N} \to \mathbb{R}$$

defined by

$$\gamma(n) = \# \{w \in G \mid l(w) \leq n\}$$

has an exponential growth rate. In the definition $l(W)$ stands for the minimal number of letters needed to express W as a word in $x_1, \ldots, x_n$. Exponential growth is a necessity, since this ensures that the group will provide a large key space, making a brute force search for the secret key an infeasible algorithm.

Further the normal form must exhibit **good diffusion** in determining the normal forms of products. This means that in finding the normal forms of products it is computationally difficult to rediscover the factors, that is if we know $NF_X(g_1g_2)$ it is computationally difficult to discover g_1, g_2 or $NF_X(g_1), NF_X(g_2)$.

Other necessities for a platform group depend on the particular protocol. If the security is based on the group problem $\mathcal{P}$, such as the word problem or conjugacy problem, we have to assume that in G, the solution to $\mathcal{P}$ is computationally hard (NP-hard) or unsolvable. However, what we really want is **generically hard** that is hard on most inputs. The solution to $\mathcal{P}$ might be unsolvable but have polynomial average case complexity. In this case, if care is not taken in choosing the inputs, the solution to $\mathcal{P}$ is easy and the cryptographic protocol is broken. This does not eliminate a group G as a possible platform group but indicates that one must take great care in choosing cryptographic inputs (see [MSU1]). We will discuss this further in Section 11.3.

Among the first attempts to use non-abelian groups as platforms for public key cryptosystems were the schemes of Anshel-Anshel-Goldfeld [AAG1] and Ko, Lee *et al.* [KLCHKP]. The first protocol was developed by I. Anshel, M. Anshel and D. Goldfeld. The original version of the Ko-Lee protocol was published by K. H. Ko, S. J. Lee, J. H. Han, J. Kang and C. Park. Throughout this book we will refer to the second protocol as Ko-Lee. Both sets of authors, at about the same time, proposed using non-abelian groups and combinatorial group theory for public key exchange.

The methods of both Anshel-Anshel-Goldfeld and Ko-Lee can be considered as group theoretic analogs of the number theory based Diffie-Hellman method. The basic underlying idea is the following. If G is a group and $g, h \in G$ we let g^h denote the conjugate of g by h, that is $g^h = h^{-1}gh$. The simple observation is that $(g^{h_1})^{h_2} = g^{h_1h_2}$. Therefore writing conjugation in this exponential manner behaves like ordinary exponentiation. From this straightforward idea one can almost exactly mimic the Diffie-Hellman protocol, now within a non-abelian group.

In this chapter we first examine the Ko-Lee and Anshel-Anshel-Goldfeld protocols. Both sets of developers originally suggested using braid groups as the basic and most appropriate group theoretic platform. We will then discuss braid group cryptography, which grew out of the analysis of these two non-abelian group theoretic protocols.

We describe both protocols in a most general context, that is with a general platform group. This platform group must have a finite presentation with efficiently computable normal forms, exponential growth and good diffusion. For the Ko-Lee protocol, the platform group must also contain an anbundant collection of subgroups that commute elementwise and that can be efficiently described.

11.2 The Ko-Lee and Anshel-Anshel-Goldfeld Protocols

All of the non-abelian group based protocols depend on the difficulty of solving certain group decision problems and group theoretical computational problems. Recall that the **conjugacy problem**, also called the **decision conjugacy problem**, for a group G, or more precisely for a group presentation for G, is the following: given $g, h \in G$, determine algorithmically if they are conjugate. The conjugacy problem is unsolvable in general, that is there exists group presentations for which there does not exist an algorithm that solves the conjugacy problem. Hence a solution to the conjugacy problem is usually associated with a particular class of group presentations. For example the conjugacy problem is solvable in free groups and in torsion-free hyperbolic groups.

Relevant to the Ko-Lee protocol is the **conjugator search problem**. This is, given a group presentation for G, and two elements g_1, g_2 in G, that are known to be conjugate, to determine algorithmically a conjugator, that is, an element $h \in G$ with $g_1 = h g_2 h^{-1}$. It is known, as with the decision conjugacy problem, that the conjugator search problem is undecidable in general.

The book by Myasnikov, Shpilrain and Ushakov [MSU1] has discussions of the complexity of many of these group decision problems. We will say a bit more about this in Section 11.6.

11.2.1 The Ko-Lee Protocol

Ko, Lee *et al.* [KLCHKP] developed a public key exchange system that is a direct translation of the Diffie-Hellman protocol to a non-abelian group theoretic setting. Its security is based on the difficulty of the **conjugacy problem**. We assume that the platform group has nice unique normal forms that are easy to compute for a given group element but hard to recover the individual group elements under group multiplication. By this we mean that if $G = \langle X; R \rangle$ is a finite presentation for the group G and $g \in G$ then there is a unique expression $NF_X(g)$ called a normal form as a word in the generators X. Further given any $g \in G$ it is computationally easy to find $NF_X(g)$. On the other hand, given $g_1, g_2 \in G$ and given the normal form $NF_X(g_1 g_2)$, it is computationally difficult to recover g_1 and g_2. We say that there is **good diffusion** in terms of normal forms in forming products.

In any group G and for $g, h \in G$ the notation g^h indicates the conjugate of g by h, that is, $g^h = h^{-1}gh$. What is important for both the Ko-Lee and Anshel-Anshel-Goldfeld protocols is that relative to this notation, group conjugation behaves exactly as ordinary exponentiation. That is for groups elements $g, h_1, h_2 \in G$ we have

$$(g^{h_1})^{h_2} = g^{h_1 h_2}.$$

That this is true is a straightforward computation

$$(g^{h_1})^{h_2} = h_2^{-1} g^{h_1} h_2 = h_2^{-1} h_1^{-1} g h_1 h_2 = (h_1 h_2)^{-1} g (h_1 h_2) = g^{h_1 h_2}.$$

With this observation, the Ko-Lee protocol exactly mimics, using group conjugation, the traditional Diffie-Hellman protocol. We first start with a platform group G

satisfying the necessary requirements on normal forms. We assume further that the platform group G has a collection of large (non-cyclic) subgroups that commute elementwise. That is if A, B are two of these subgroups and $a \in A$ and $b \in B$ then $ab = ba$. It is not necessary that the subgroups themselves be abelian.

Alice and Bob choose a pair of these commuting subgroups A and B of the platform group G. A is Alice's subgroup while Bob's subgroup is B and these are secret. By assumption each element of A commutes with each element of B. Further it is not assumed that A and/or B are themselves abelian. Now the method completely mimics the classical Diffie-Hellman technique. There is a public element $g \in G$, Alice chooses a random secret element element $a \in A$ and makes public g^a, the conjugate of g by a. Bob chooses a random secret element $b \in B$ and makes public g^b the conjugate of g by b. The secret shared key is g^{ab}. Notice that $ab = ba$ since the subgroups commute. It follows then that

$$(g^a)^b = g^{ab} = g^{ba} = (g^b)^a$$

just as if these were ordinary exponents.

It follows, as in the number theoretic based Diffie-Hellman protocol, that both Bob and Alice can determine the common secret. Alice knows her secret key a and Bob's public key g^b. Hence she knows $(g^b)^a = g^{ba}$. Bob knows his secret key b and g^a is public. Hence Bob knows $(g^a)^b = g^{ab}$. However, as explained $g^{ab} = g^{ba}$. The difficulty is in the difficulty of the decision conjugacy problem.

It is known that both the decision conjugacy problem and the conjugator search problem are undecidable in general. However, there are groups where both are solvable but hard, that is the problems are solvable but are not solvable in polynomial time. These groups then become the target platform groups for the Ko-Lee protocol. Ko and Lee in their initial work suggest the use of the braid groups. We will discuss braid group cryptography later in this chapter.

Formal Setup of the Ko-Lee Key Exchange Protocol

We now summarize the formal setup for the Ko-Lee Key Exchange Protocol. After this we will show how to use the ElGamal method to construct a public key encryption system from this.

Ko-Lee Preparation

(1) We start with a platform group G. We assume that G has a finite presentation with efficiently computable normal forms that have good diffusion. Further the group G must have a large collection of subgroups that commute elementwise.

(2) We choose an element $g \in G$.

(3) We assume that Alice wants to share a common key with Bob. Alice and Bob choose subgroups A and B that elementwise commute. A is Alice's subgroup and

B is Bob's subgroup. These subgroups are kept secret and known only to Bob and Alice, respectively.

Ko-Lee Key Exchange

(1) Alice randomly chooses an $a \in A$. This element a will be her secret key. Her public key is (g, g^a) where $g^a = aga^{-1}$ is the conjugate of g by her secret key a. All public information and communication is done in terms of the normal forms of these elements.

(2) Bob randomly chooses an element $b \in B$. This element b will be his secret key. His public key is (g, g^b) where $g^b = b^{-1}gb$ is the conjugate of g by his secret key b. As with Alice all public information and communication is done in terms of the normal forms of these elements.

(3) The secret shared key is g^{ab}.

ElGamal Encryption Using the Ko-Lee Protocol

As with the standard Diffie-Hellman key exchange protocol using number theory, the Ko-Lee protocol can be changed to an encryption system via the ElGamal method. There are several different variants of non-commutative ElGamal systems. At the simplest level we assume that we have a group G appropriate for the Ko-Lee key exchange and that Alice and Bob want to communicate secretly. The element $g \in G$ is public and Alice and Bob respectively have chosen their appropriate commuting subgroups A and B. Bob has made public g^b for $b \in B$ in normal form and Alice had made public g^a for $a \in A$ also in normal form. The secret shared key is then g^{ab}. We assume that Alice wants to send an encrypted message to Bob and further we assume the encrypted message can be encoded as $h \in G$, that is as an element of the group G. Alice then sends to Bob the normal form of hg^{ab}. Bob can determine the common shared secret g^{ab}. He then multiplies hg^{ab} by $(g^{ab})^{-1}$ to obtain the secret h.

As with the number theoretic based public key cryptosystems, the Ko-Lee method can be used to provide methods for other protocols, especially authentication and digital signature protocols. In Chapter 4 and then again in Chapters 7 and 8 we used the RSA, Diffie-Hellman and Elliptic Curve techniques to develop digital signature algorithms. Mimicking these procedures we can do the same with the Ko-Lee method to develop a non-abelian group-based digital signature algorithm. We will discuss these at the end of this chapter.

11.2.2 The Anshel-Anshel-Goldfeld Protocol

We now describe another non-abelian group-based public key exchange protocol. It is somewhat similar to the Ko-Lee protocol and was developed at approximately the same time. This is the Anshel-Anshel-Goldfeld public key exchange protocol.

As in the Ko-Lee protocol we start with a group G given by a finite presentation $G = \langle X; R \rangle$. We further assume as before that there are efficiently computable normal forms relative to the presentation $\langle X; R \rangle$. That is if $g \in G$ then there is a unique expression $NF_X(g)$ as a word in the generators X and given any $g \in G$ it is computationally easy to find $NF_X(g)$. Further there is good diffusion in forming products so that given $g_1, g_2 \in G$ and given the normal form $NF_X(g_1 g_2)$ it is computationally difficult to recover g_1 and g_2. The Ko-Lee protocol required two large commuting subgroups. For communication, the Anshel-Anshel-Goldfeld protocol requires a choice of subgroups of G, but they need not commute. While the difficulty of the decision conjugacy problem provides the security for the Ko-Lee method, it is the difficulty of the search conjugator problem that provides the hard problem, and hence the security, in Anshel-Anshel-Goldfeld.

Once we have our platform group G we assume that Alice and Bob want to obtain a common shared secret or a common shared secret key. We assume that this secret key can be expressed as a group element $g \in G$. The first step is for Alice and Bob to choose random finitely generated subgroups of G by giving a set of generators for each.

$$A = \{a_1, \ldots, a_n\}, \quad B = \{b_1, \ldots, b_m\}$$

and make them public. The subgroup A is Alice's subgroup while the subgroup B is Bob's subgroup.

Alice chooses a secret group word $a = W(a_1, \ldots, a_n)$ in her subgroup while Bob chooses a secret group word $b = V(b_1, \ldots, b_m)$ in his subgroup. As before, for an element $g \in G$ we let $NF_X(g)$ denote the normal form for g. Alice knows her secret word a and knows the generators b_i of Bob's subgroup. She can then form the conjugates of the generators of Bob's subgroup B by her secret element $a \in A$. That is she can compute $b_i^a = a^{-1} b_i a$ for each b_i. She then makes public the normal forms of these conjugates

$$NF_X(b_i^a), \quad i = 1, \ldots, m.$$

Bob does the analogous thing. He knows his secret word b and the generators a_i, $i = 1, \ldots, n$ of Alice's subgroup A and hence can compute the conjugates $a_i^b = b^{-1} a_i b$ for $i = 1, \ldots, n$. He then makes public the normal forms of the conjugates

$$NF_X(a_j^b), \quad j = 1, \ldots, n.$$

The common shared secret is the commutator

$$[a, b] = a^{-1} b^{-1} a b = a^{-1} a^b = (b^a)^{-1} b.$$

Notice that this is known for both Alice and Bob. Alice knows $a^b = b^{-1}ab$ since she knows a in terms of generators a_i of her subgroup and she knows the conjugates by b since Bob has made the conjugates of the generators of A by b public. That is Alice knows $a = W(a_1, \ldots, a_n)$ and $a^b = b^{-1}ab = W(b^{-1}a_1b, \ldots, b^{-1}a_nb) = W(a_1^b, \ldots, a_n^b)$. Since Alice knows a^b she knows

$$[a, b] = a^{-1}b^{-1}ab = a^{-1}a^b.$$

In an analogous manner Bob knows $[a, b] = (b^a)^{-1}b$ since he knows his secret element b in terms of the generators $b_j, j = 1, \ldots, m$, of his subgroup B and Alice has made public the conjugates of each of his generators by her secret element a. Hence $b = V(b_1, \ldots, b_m)$ so that $b^a = V(b_1^a, \ldots, b_m^a)$ and this is known to Bob. Since Bob knows b^a and b he knows

$$[a, b] = a^{-1}b^{-1}ab = a^b b = (b^{-1})^a b = (b^a)^{-1}b.$$

Notice that in this system there is no requirement that the chosen subgroups A and B commute.

An attacker would have to know the corresponding **conjugator**, that is the element that conjugates each of the generators. This is the **conjugator search problem**. Given elements g, h in a group G, where it is known that $g^k = k^{-1}gk = h$, determine the conjugator k. It is known that this problem is undecidable in general, that is, there are groups where the conjugator cannot be determined algorithmically. On the other hand there are groups where the conjugator search problem is solvable but "difficult", that is, the complexity of solving the conjugator search problem is hard. Such groups become the ideal platform groups for the Anshel-Anshel-Goldfeld protocol.

The security in this system is then in the computational difficulty of the conjugator search problem. Anshel, Anshel, Goldfeld suggested, as did Ko-Lee, the braid groups, B_n, as potential platforms. The braid groups are a class of infinite, finitely presented groups that arise in many different contexts. The braid group B_n has a standard presentation with $n-1$ generators. The braid groups will be discussed in detail in Section 11.6.

The necessary parameters that must be decided in using the braid groups as platforms for either the Ko-Lee protocol or the Anshel-Anshel-Goldfeld protocol are then the number of generators of the braid groups used and the number of generators for the chosen subgroups. For example B_{80}, the braid group on 80 strands with 12 or more generators in the chosen subgroups might be used. It has been shown that the larger the number of strands (see Section 11.5) the harder it is to attack the protocol. The suggested use of the braid groups by both Anshel, Anshel and Goldfeld and Ko and Lee led to the development of **braid group cryptography**. There have been various attacks on the braid group cryptosystems (see Section 11.7). However, some of these attacks can be handled by changing the parameters. In general the ideas remain valid despite the attacks.

Formal Setup of the Anshel-Anshel-Goldfeld Key Exchange Protocol

We now summarize the formal setup for the Anshel-Anshel-Goldfeld Key Exchange Protocol. After this we will show how to use the ElGamal method to construct a public key encryption system from this.

Anshel-Anshel-Goldfeld Preparation

(1) We start with a platform group G. We assume that G has a finite presentation with efficiently computable normal forms that have good diffusion. Further there is a large collection of efficiently computable subgroups

(2) We assume that Alice wants to share a common key with Bob. Alice and Bob choose random finitely generated subgroups of G by giving a set of generators for each

$$A = \{a_1, \ldots, a_n\}, \quad B = \{b_1, \ldots, b_m\}$$

and make them public. The subgroup A is Alice's subgroup while the subgroup B is Bob's subgroup.

Anshel-Anshel-Goldfeld Key Exchange

(1) Alice chooses a secret group word $a = W(a_1, \ldots, a_n)$ in her subgroup. Alice knows her secret word a and knows the generators b_i of Bob's subgroup. She can then form the conjugates of the generators of Bob's subgroup B by her secret element $a \in A$. That is she can compute $b_i^a = a^{-1} b_i a$ for each b_i. She then makes public the normal forms of these conjugates

$$NF_X(b_i^a), \quad i = 1, \ldots, m.$$

(2) Bob chooses a secret group word $b = W(b_1, \ldots, b_m)$ in his subgroup. Bob knows his secret word b and knows the generators a_i of Alice's subgroup. He can then form the conjugates of the generators of Alice's subgroup A by his secret element $b \in B$. That is he can compute $a_i^b = b^{-1} a_i b$ for each a_i. He then makes public the normal forms of these conjugates

$$NF_X(a_i^b), \quad i = 1, \ldots, m.$$

(3) The secret shared key is the commutator

$$[a, b] = a^{-1} b^{-1} a b = a^{-1} a^b = (b^a)^{-1} b.$$

ElGamal Encryption Using the Anshel-Anshel-Goldfeld Protocol

As with all public key exchange protocols, the Anshel-Anshel-Goldfeld key exchange can be developed into a cryptosystem by the ElGamal method. This works essentially in the same manner as for Ko-Lee. We assume that we have a group G appropriate for the Anshel-Anshel-Goldfeld key exchange and that Alice and Bob want to communicate secretly. Alice and Bob respectively have chosen their appropriate subgroups A and B whose generators have been made public. Bob has made public the conjugates of the generators of A by his secret element $b \in B$ in normal form and Alice had made public the conjugates of the generators of B by her secret element $a \in A$, also in normal form. The secret shared key is then the commutator $[a, b]$. We assume that Alice wants to send an encrypted message to Bob, and further we assume that the encrypted message can be encoded as $h \in G$, that is, as an element of the group G. Alice then sends to Bob the normal form of $h[a, b]$. Bob can determine the common shared key $[a, b]$. He then multiplies $h[a, b]$ by $[a, b]^{-1}$ to obtain the secret h.

11.3 Some Other Group Based Cryptosystems

There have been many other public key exchange protocols developed using non-abelian groups. A large number of these systems are described in the book of Myasnikov, Shpilrain and Ushakov [MSU1]. The authors themselves have developed many of these. They use different "hard" group theoretic decision problems and many have been broken. In this section we first describe two variants of non-commutative El-Gamal systems due to Khan and Kharobaei and then a separate protocol by Shpilrain and Zapata.

The following key exchange protocol is a non-commutative El Gamal system based on the search conjugacy problem. It was proposed by Kahrobaei and Khan [KKh]. As with the Ko-Lee and Anshel-Anshel-Goldfeld protocols we start with a finitely presented platform group G given by a group presentation. As before, the major assumptions are that the elements of G have nice unique normal forms that are easy to compute for given group elements. However, it is further assumed that given normal forms for $x, y \in G$, the normal form for the product xy does not reveal x or y. Further G contains two commuting, finitely generated proper subgroups, S and T. The cryptographic goal is for Alice and Bob to establish a session key over an unsecured network.

Bob chooses a secret element $s \in S$ and an arbitrary element $b \in G$. Bob publishes b and $c = b^s$. Suppose Alice wants to send $x \in G$ as a session key to Bob. Then,

(1) Alice chooses a random $t \in T$ and sends $E = x^{(c^t)}$ to Bob along with the header $h = b^t$.

(2) Bob then calculates $(b^t)^s = (b^s)^t = c^t$.

(3) Now, Bob may calculate $E' = (c^t)^{-1}$, allowing him to decrypt the session key since $(x^{(c^t)})^{E'} = (x^{(c^t)})^{(c^t)^{-1}} = x$.

The feasibility of this scheme relies on the assumption that products and inverses in G can be computed efficiently. Determining Bob's private key s entails solving the search conjugacy problem for G. That is given c, b, and $c = b^s$, determine s. Hence, the security of this scheme is based on the assumption that there is no practical algorithm for solving the search conjugacy problem for G.

A second variation of the non-commutative ElGamal is the following key exchange protocol based on the search power conjugacy problem. It was also proposed by Kahrobaei and Khan (see [KKh]). The same assumptions are made as in the previous protocol. The cryptographic goal is again for Alice and Bob to establish a session key over an unsecured network. Now suppose that the search conjugacy problem in G is tractable. Then we may modify the above conjugacy based scheme to a power conjugacy based scheme. Now Bob chooses secret elements $s \in S$ and $n \in \mathbb{N}$ and an arbitrary element $g \in G$. Bob publishes $v = g^n$ and $w = s^{-1}gs$. Note that $w^n = s^{-1}vs$. Suppose that Alice wishes to send $x \in G$ to Bob. Then,

(1) Alice chooses random $m \in \mathbb{N}$ and $t \in T$ and sends $E = x^{-1}t^{-1}v^m tx = x^{-1}t^{-1}g^{mn}tx$ to Bob along with header $h = t^{-1}w^m t = t^{-1}s^{-1}g^m st$.

(2) Bob then computes $E' = sh^n s^{-1} = t^{-1}g^{mn}t$.

(3) Since $E = x^{-1}E'x$ and the search conjugacy problem is tractable, Bob may recover the element x. This element x serves as the common secret which can be used as a symmetric session key for secure communication.

As before, the feasibility of this scheme relies on the assumption that products and inverses in G can be computed efficiently as well as the assumption that the conjugacy problem is solvable. In order to obtain Bob's secret elements s and n from the public information g^n and w requires solving $w^n = s^{-1}g^n s$ for n and s. Hence, the security of this scheme rests on the assumption that there is no practical algorithm for solving the power search conjugacy problem for G.

The final protocol is due to Shpilrain and Zapata and depends upon the computational difficulty of solving the word problem in groups with efficiently solvable word problems. The protocol works as follows: We assume that the secret is to be sent from Bob to Alice and that the secret can be presented as a binary sequence.

(1) Alice has a pool of finite group presentations with efficiently solvable word problems. She makes these public. Shpilrain and Zapata suggest using small cancellation presentations (see [MSU1]).

(2) Alice randomly chooses a particular group presentation $\langle A; R \rangle$ from her pool and alters the presentation using Tietze transformations so that the new presentation is $\langle X'; R' \rangle$ and defines an isomorphic group. She then discards some of the relators in R' to get a subset $S' \subset R'$ and publishes the new abridged presentation $\langle X'; S' \rangle$.

(3) Bob transmits his secret binary sequence by transmitting a sequence of elements $w_1, \ldots, w_n$ of $\langle X'; S' \rangle$ where $w_i = 1$ in $\langle X'; S' \rangle$ and hence also equal to 1 in $\langle X'; R' \rangle$ if the transmitted binary digit is 1 and $w_i \neq 1$ if the transmitted binary digit is 0.

(4) Alice then recovers the secret, that is Bob's transmitted binary sequence, by first converting elements of $\langle X'; R' \rangle$ into elements of $\langle X; R \rangle$ using the isomorphism she knows. She then solves the word problem in $\langle X; R \rangle$ to obtain the proper sequence of 0 and 1.

A complete analysis of this protocol and its strengths and deficiencies can be found in [MSU1]. For more information we refer the reader to that source. Here we just note that the Shpilrain-Zapata scheme thwarts an exhaustive search attack. It also raises the general question of generating random elements in finitely presented groups. This was used in all the group based protocols.

11.4 The Shamir Three-Pass

In both the Ko-Lee and Anshel-Anshel-Goldfeld protocols what is required of the platform group G is an abundant collection of finitely generated subgroups A, B. In the Ko-Lee protocol it is required that the elements of A and B pairwise commute. This is not necessary in the Anshel-Anshel-Goldfeld protocol.

Both of these are variants of what is called a **Shamir Three-Pass**. This was developed by Shamir in a general algebraic context, that is using any algebraic object as a platform. Thus a Shamir Three-Pass can be applied to a group, non-commutative ring or a lattice. A Shamir Three-Pass works in the following manner. We describe it with a group-based platform but the ideas work equally well for any algebraic platform.

Suppose that Bob wants to communicate with Alice via an open airway. The secret key telling them which encryption system to use is encoded within the finitely generated group G. There are two subgroups A_1, A_2 which commute elementwise. A_1 is the subgroup chosen by Bob and A_2 the subgroup chosen by Alice. These subgroups are kept secret by the communicating parties. Bob wants to send a secret key $W \in G$ to Alice. He chooses two random elements $B_1, B_2 \in A_1$ and sends Alice the message (in encrypted form) $B_1 W B_2$. Alice now chooses two random elements $C_1, C_2 \in A_2$ and sends $C_1 B_1 W B_2 C_2$ back to Bob. These messages appear in the representation of G and hence for example as matrices, or as reduced words in the generators, so they do not appear as solely concatenation of letters. Since A_1 commutes elementwise with A_2 we have

$$C_1 B_1 W B_2 C_2 = B_1 C_1 W C_2 B_2.$$

Further since Bob knows his chosen elements B_1 and B_2 he can multiply by their inverses to obtain $C_1 W C_2$ which he then sends back to Alice. Since Alice knows her chosen elements C_1, C_2 she can multiply by their inverses to obtain the key W. It is assumed that for each message Bob and Alice would choose different pairs of random elements from either A_1 or A_2.

A Shamir Three-Pass is actually a **key transport protocol** rather than a key exchange protocol. A **key transport protocol** is a method that allows the sending of a key (telling for example what encryption system to use) from one user to another over a public airway. Notice that a key transport protocol, such as a Shamir Three-Pass, is not symmetric in the communicating parties. In the Shamir Three-Pass, the secret key is completely determined by Bob, who then communicates it to Alice.

Key transport protocols are in most cases designed assuming that an underlying encryption system (and usually also a signature verification system) is in place. The security of the key transport protocol will rely on the security of these auxiliary schemes. In a group theoretic protocol the encryption scheme is suggested to be done within the same group as the key transport protocol, although this is not essential. In the group theoretic key transport protocol, an attacker has knowledge of the overall group G, and a view of encrypted messages. The security lies in the difficulty of determining the elementwise commuting subgroups A_1, A_2, which are kept secret by Bob and Alice, and in the security of the actual encryption scheme.

A group G is a candidate platform group for this type of key transport protocol if it satisfies the criteria given in Section 11.1; it has a nice finite presentation $G = \langle X; R \rangle$ with efficiently computable normal forms and has either a large abelian subgroup A or two large subgroups A_1, A_2 that commute elementwise. Although the word large here is ambiguous, we mean large enough so that random choices can be made from them. In particular, for example, cyclic subgroups are inappropriate. There also should be some tie between the group used for the key exchange and the encryption method, although this is not essential. The standard braid groups, that we will describe in Section 11.6, have several possibilities for normal forms and have large commuting subgroups. Hence they are excellent candidates for this method. In [BCFRX] several additional potential platform groups were suggested. These include the full automorphism group of a finitely generated free group, the matrix group SL(4, $\mathbb{Z}$) and the surface braid groups. Shpilrain and Ushakov [SU] used this method employing Thompson's group F as a platform. Further work on this method in the surface braid groups was done by Camps [Camp].

Currently, there are many potential platform groups that have been suggested. The following are some of the proposed platforms:
- Braid groups (Ko-Lee, Anshel-Anshel-Goldfeld)
- Thompson Groups (Shpilrain-Ushakov) [SU]
- Polycyclic Groups (Eick-Kahrobaei) [EK]
- Linear Groups (Baumslag-Fine-Xu) [BFX1], [BFX2]
- Free metabelian Groups (Shpilrain-Zapata) [SZ]
- Artin Groups (Shpilrain-Zapata) [SZ]
- Grigorchuk Groups (Petrides) [Pan]
- Groups of Matrices (Grigoriev-Ponomarenko) [GP]
- Surface Braid Groups (Camps) [Camp]

Most of these are discussed in detail in [MSU1].

11.5 Hard Group Theoretic Problems and Average Case Complexity

Several of the group based protocols, whose security depends on group decision problems, have been broken. This has been especially true for the use of braid groups as platform groups. A careful analysis of these situations pinpoints the problem. The group decision problems are hard in general but easy "on average". We explain what this means.

We have already discussed algorithmically hard and algorithmically unsolvable problems. For the algorithmically hard problems this means that there is no general algorithm to solve the particular problem in polynomial time (or at all in the unsolvable case). However, what is important in cryptographic uses is not that there are specific cases where algorithms do not work but rather the average performance of an algorithm.

In the last chapter we described asymptotic density as a method to find probabilistic properties in infinite discrete groups. A similar method can be used to analyze the computational complexity of algorithms, especially the algorithms to solve the group decision problems. A complete chapter in the book of Myasnikov, Shpilrain and Ushakov [MSU1] is devoted to this procedure and its generalizations. There is also a chapter on the related notion of **generic complexity**. Here we summarize the basic ideas.

If $\mathcal{D}$ is any decision problem (more general than a group decision problem) and $\mathcal{J}$ is the set of possible inputs then this is a **computational decision problem** if there exists a probability measure μ on $\mathcal{J}$. If the set of inputs is discrete, as in the case of infinite discrete groups, then the measure must be atomic, that is determined by its value on singletons $\{\mu(x) \mid x \in \mathcal{J}\}$. Then for any subset $S \subset \mathcal{J}$,

$$\mu(S) = \sum_{x \in S} \mu(x).$$

If G is a group presentation, and $\mathcal{D}$ is one of the group decision problems, then the set of inputs $\mathcal{J}$ are sets of group elements, singletons for example in the case of the word problem or pairs of elements in the case of the decision conjugacy problem. The analysis of the complexity of an algorithm to solve one of these problems then depends on the method to randomly generate instances of the problem and the measure μ on these inputs. Given this measure, μ, we can determine the probablity that the solution is in a particular complexity class, and thus have a probability distribution on instances of complexity classes of this algorithm. For example the probability that the algorithm is solvable in polynomial time. How this is done is explained in [MSU1].

The **average case complexity** is the average complexity class (average in the probability or expected value sense, see [MSU1]) relative to this distribution depending on μ. The **worst case complexity** is the hardest instance of the solution. Gurevich (see [MSU1]) has shown that there are computationally NP-complete problems whose average case complexity is polynomial relative to some natural measure.

If the asymptotic density (in the sense of the last chapter) of being in a particular complexity class has value one then we say that is the **generic complexity**. It is the distinction between average case complexity, worst-case complexity, and generic complexity, that is of most relevance to cryptographic uses.

For a particular group G the solution of a given decision problem might be NP-hard. This means that the worst case complexity of solving this problem within G is hard. The problem then seems to look good as a potential platform for a one-way function. However, suppose the problem, even though it has worst case complexity hard, is easy (say in polynomial time) on average. Then if random inputs are used the algorithm is on average easy and using the algorithm will break the crypto uses. Hence in designing cryptosystems from group decision problems this indicates that we must be very careful in choosing inputs (for secret keys for example). A decision problem that is generically difficult is then of course preferable.

11.6 Braid Group Cryptography

The **braid groups**, B_n, are a class of non-abelian finitely presented infinite groups that arise in diverse areas of mathematics, including topology and combinatorial group theory. Both Ko-Lee and Anshel-Anshel-Goldfeld suggested the braid groups as platform groups for their respective protocols. The groups in this class possess the desired properties for the key exchange and key transport protocols; they have nice presentations with solvable word problems and solvable conjugacy problems; the solution to the conjugacy and conjugator search problem is "hard"; there are several possibilities for normal forms for elements and they have many choices for large commuting subgroups.

The suggestion to use the braid groups as platforms for the respective protocols was made to both groups of authors by Joan Birman, a leading expert on the braid groups and their group theoretic properties. She recognized that the requirements of the two protocols were met by this class.

Initially the braid groups were considered so ideal as platforms that many other cryptographic applications were framed within the braid group setting. These included **authentication** and **digital signatures**. There was so much enthusiasm about using these groups that the whole area of study was named **braid group cryptography**.

After the initial successes with braid group cryptographic schemes, there were some surprisingly effective attacks. There were essentially three types of attacks; an attack using solutions to the conjugacy and conjugator search problems, an attack using heuristic probability within B_n, and an attack based on the fact that there are faithful linear representations of each B_n. What is most surprising is that the

Anshel-Anshel-Goldfeld method was susceptible to a length based attack. In the Anshel-Anshel-Goldfeld method the **parameters** are the specific braid group B_n and the rank of the secret subgroups for Bob and Alice. A length based attack essentially broke the cryptosystem for the initial parameters suggested by Anshel, Anshel and Goldfeld. The parameters were then made larger and attacks by this method were less successful. However, this led to research on why these attacks on the conjugator search problem within B_n were successful. What was discovered was that **generically**, (see previous section and the last chapter) a random subgroup of B_n is a free group, and hence length based attacks are essentially attacks on free group cryptography and therefore successful. What this indicated was that, although randomness is important in cryptography, in using the braid groups as platforms, subgroups cannot be chosen purely randomly.

What has actually occurred is that there has been a tremendous amount of research on braid groups as a result of the potential cryptographic uses. Perhaps the introduction of braid group cryptography has had more of an effect on the theory of braid groups than on cryptography. In the next section we look at the theory of braid groups and then the cryptanalysis of braid group cryptosystems. There are many comprehensive references on braid groups and braid group cryptography. The book by J. Birman [Bir] provides the basic algebraic material on braid groups and the well-written articles by Dehornoy [Deh] and Garber [Gar] provide detailed overviews of the subject.

Braid groups arise in several different areas of mathematics and have several equivalent formulations. In the next section we describe braid groups theoretically and highlight those properties that are important for cryptographic applications. A complete topological and algebraic description can be found in the book of Joan Birman [Bir].

11.7 The Braid Groups

The first of several equivalent formulations of the braid group B_n describes it as the group generated by manipulations of braids. Here a **braid** on n strings (or n strands) is obtained by starting with n parallel strings (or strands) and intertwining them. We number the strings at each vertical position and keep track of where each individual string begins and ends. A **braid crossing** is where one strand crosses over or under another. The diagram below pictures two such braids.

It is very important that undercrossings are different from overcrossings. Hence the two braids pictured below are different braids.

We say that two braids are equivalent if it is possible to move the strings of one of the braids in space without moving the endpoints or moving through a string and obtain the other braid. The braids in Figure 11.3 are equivalent. This is an equivalence

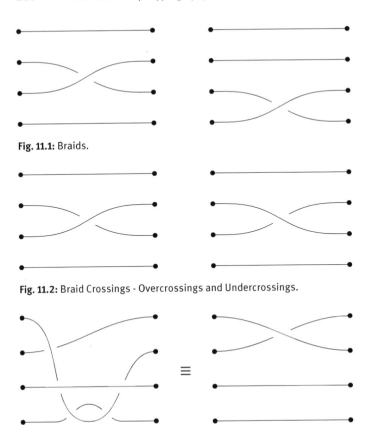

Fig. 11.1: Braids.

Fig. 11.2: Braid Crossings - Overcrossings and Undercrossings.

Fig. 11.3: Equivalence of Baids.

relation on the set of braids and it is on the equivalence classes that we will build a group structure.

On the set of equivalence classes of braids we now place an operation. We form a product of braids in the following manner. If u is the first braid and v is the second braid then uv is the braid formed by placing the starting points for the strings in v at the endpoints of the strings in u. The diagram in Figure 11.4 shows this operation. It can be proved that this operation is associative.

A braid with no crossings is called a **trivial braid** (see Figure 11.5). It clearly forms an identity for the operation on equivalence classes of braids.

The inverse of a braid is the mirror image in the horizontal plane. It is clear that if we form the product of a braid and its mirror image we get a braid equivalent to the trivial braid. With these definitions the set of all equivalence classes of braids on n strings forms a group B_n. We let σ_i denote the braid that has a single crossing from string i over string $i + 1$. Since a general braid is just a series of crossings it follows that B_n is generated by the set $\sigma_i; i = 1, \ldots, n - 1$.

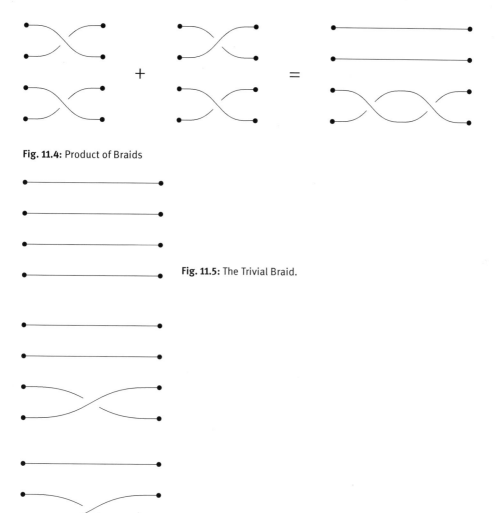

Fig. 11.4: Product of Braids

Fig. 11.5: The Trivial Braid.

Fig. 11.6: The Braid Generators.

From the description of B_n in terms of braids several things are very clear. First of all the group B_n embeds into B_{n+1} for any n and further B_n must be torison-free, that is every non-identity element must have infinite order. Not as clear, but easily proved. (see the exercises) is that for $n \geq 3$ the group B_n must contain a rank 2 free group and hence free groups of all possible ranks. In the last chapter, we introduced the concept of being **generically free**, that is with asymptotic density one any finitely generated subgroup is a free group. It can be shown that for $n \geq 3$ this is true for the braid groups B_n. This fact plays a role in the cryptanalysis of braid group based cryptosystems.

11.7.1 The Artin Presentation

There are several equivalent formulations of the braid groups (see [Bir]). An algebraic description of the braid group B_n (see [MKS] or [Bir]) can be given in the following manner, which also provides a simple solution to the word problem in B_n, for words given in terms of the standard generators described in the previous section.

Let F_n be a free group on the n generators $x_1, \ldots, x_n$ with $n > 2$. Let $\sigma_i, i = 1, \ldots, n-1$ be the automorphism of F_n given by

$$\sigma_i : x_i \mapsto x_{i+1}, \quad x_{i+1} \mapsto x_{i+1}^{-1} x_i x_{i+1}$$
$$\sigma_i; x_j \mapsto x_j, \quad j \neq i, i+1.$$

Then each σ_i corresponds precisely to the basic braid crossings in B_n. Therefore, the group B_n can be considered as the subgroup of $\mathrm{Aut}(F_n)$ generated by the automorphisms σ_i. Artin proved in [Art] (see also [MKS]) that a finite presentation for B_n is given by

$$B_n = \langle \sigma_1, \ldots, \sigma_{n-1}; \quad [\sigma_i, \sigma_j] = 1 \text{ if } |i - j| > 1,$$
$$\sigma_{i+1}\sigma_i\sigma_{i+1} = \sigma_i\sigma_{i+1}\sigma_i \quad \text{for } i = 1, \ldots, n-1 \rangle.$$

This is now called the **Artin presentation**.

The fact that B_n is contained in $\mathrm{Aut}(F_n)$ provides an elementary solution to the word problem in B_n in terms of the standard generators, since one can determine easily if an automorphism of F_n is trivial on all the generators. In the next section we will introduce normal forms for the Artin presentation which provide another solution to the word problem.

The starting point for encryption, and other cryptographic protocols using B_n, is the Artin presentation. Hence, we consider a braid group B_n as a group G with a presentation of the form

$$B_n = \langle \sigma_1, \ldots, \sigma_{n-1}; \quad [\sigma_i, \sigma_j] = 1 \text{ if } |i - j| > 1,$$
$$\sigma_{i+1}\sigma_i\sigma_{i+1} = \sigma_i\sigma_{i+1}\sigma_i \quad \text{for } i = 1, \ldots, n-1 \rangle.$$

We will say that this is the **braid group on n strands**. We will use the interpretation of a word $W(x_1, \ldots, x_n)$ in the Artin presentations as either a sequence of braid crossings, or as an automorphism of a free group, as needed. Using the Artin presentation many of the necessary properties for a good platform group are seen to be satisfied. The cubic relations $\sigma_{i+1}\sigma_i\sigma_{i+1} = \sigma_i\sigma_{i+1}\sigma_i$ are called the **braid relations**.

As a consequence of the relation $[\sigma_i, \sigma_j] = 1$, if $|i - j| > 1$, it follows that any two generators, whose indices are more than 2 apart, commute. It follows then that if $\{\sigma_{i_1}, \ldots, \sigma_{i_m}\}$ and $\{\sigma_{j_1}, \ldots, \sigma_{j_k}\}$ are two subsets of $\{\sigma_1, \ldots, \sigma_m\}$ with $\max\{\sigma_{i_1}, \ldots, \sigma_{i_m}\} + 2 < \min\{\sigma_{j_1}, \ldots, \sigma_{i_k}\}$ then the subgroups $A = \langle \sigma_{i_1}, \ldots, \sigma_{i_m} \rangle$ and $B = \langle \sigma_{j_1}, \ldots, \sigma_{j_m} \rangle$ are pairwise commutative. Hence there is an abundant collection of such subgroups within any B_n for large n.

For cryptographic applications, we will deal exclusively with either the braid interpretation of the braid groups or with the Artin presentation. The other interpretations of the braid groups are primarily topological. For example, the braid groups B_n are isomorphic to the mapping class group of a punctured disk with n punctures. Although the braid groups were introduced as such by Artin in 1925, this interpretation of a braid group as a fundamental group was present implicitly in the work of Hurwitz as far back as the late nineteenth century. More information on this topological background is available in the book of J. Birman [Bir].

Braids, and braid groups, are also tied to knots and links. If one ties the strands of a given braid together, a link, and sometimes a knot is obtained. There is a theorem of Alexander that any knot or link is obtained from tying together the strands of a braid. For more on this again see [Bir].

Finally we mention that Artin, as well as showing that B_n can be embedded in $\text{Aut}(F_n)$, established criteria for determining when an automorphism of F_n actually falls in the braid group B_n. In particular he proved.

Theorem 11.7.1. *Let* $\beta \in \text{Aut}(F_n)$ *where* F_n *is a free group of rank n on the generators* $x_1, \ldots, x_n$. *Then* $\beta \in B_n$ *if and only if* β *satisfies the following two conditions:*
(1) $\beta(x_i)$ *is conjugate to another generator. That is* $\beta(x_i) = w^{-1}x_j w$ *where* x_j *is another generator and w is any element of* F_n.
(2) $\beta(x_1 x_2 \cdots x_n) = x_1 x_2 \cdots x_n$.

11.7.2 Normal Forms Within B_n

There are several possibilities for normal forms for elements of B_n. We will consider three; the **Garside normal form, the Dehornoy handle form** and the **Birman-Ko-Lee normal form**. The Garside normal form is also called the **greedy normal form**. All are efficiently computable and have good diffusion by themselves and hence give, within the braid groups, a choice for cryptography. Unfortunately the fact that there are more than one good normal form often works against cryptographic security. In a paper by Myasnikov, Shpilrain and Ushakov [MSU1] they were able to break an implementation of the Ko-Lee protocol, with a braid group platform, by using the Dehornoy form on the encrypted Garside form (see [MSU1] for details). What this indicates is that, as in choosing braid group parameters, great care must be taken in using the various instances of normal forms and how they behave versus each other relative to diffusion.

The **Garside normal form** was discovered first and used by Garside to prove both the word problem and the conjugacy problem within B_n. Here we must think of the elements of B_n as braids and $\sigma_1, \ldots, \sigma_{n-1}$ as the basic braid generators. The set B_n^+ of **positive braids** consists of those braids that can be written as only positive powers of the σ_i. A **monoid** is a set with a single associative operation together with an identity but not necessarily inverses. One can think of a monoid as a group without inverse elements. The set B_n^+ of positive braids under braid multiplication forms a monoid.

To obtain the Garside normal form we must define the **fundamental braid** within the monoid B_n^+.

Definition 11.7.2. The **fundamental braid** $\Delta_n \in B_n^+$ is the braid given by

$$\Delta_n = (\sigma_1 \cdots \sigma_{n-1})(\sigma_1 \cdots \sigma_{n-2}) \cdots \sigma_1.$$

Geometrically, the fundamental braid is a braid on n strands where any two strands cross positively exactly one time.

The fundamental braid satisfies certain very straightforward properties.

(1) For any basic generator σ_i we have a decomposition $\Delta_n = \sigma_i A = B \sigma_i$ with $A, B \in B_n^+$. That is Δ_n is a left common multiple and a right common multiple for any generator σ_i in the monoid of positive braids.

(2) $\sigma_i \Delta_n = \Delta_n \sigma_{n-i}$

(3) Δ_n^2 generates the center of B_n^+.

In order to define the normal form and prove that it is unique we need to place a partial order on the set of braids. We say that $A \le B$ for $A, B \in B_n$ if there exists a $C \in B_n^+$ with $B = AC$. It is clear that for any $B \in B_n^+$ we have $1 \le B$ and further $A \le B$ if and only if $B^{-1} \le A^{-1}$. A braid P such that $1 \le P \le \Delta_n$ is called a **permutation braid**. Geometrically this is a braid on n strands where any two strands cross positively at most once.

For a permutation braid P we let

$$S(P) = \{i \mid P = \sigma_i P', P \in B_n^+\}$$

and

$$F(P) = \{i \mid P = P' \sigma_i, P' \in B_n^+\}.$$

It is straightforward that $S(\Delta_n) = F(\Delta_n) = \{1, 2, \ldots, n-1\}$.

If $A \in B_n^+$ is a positive braid, then a **left-weighted decomposition** of A is $A = P_1 \cdots P_k$, where each P_i is a permutation braid, and $S(P_i) \subset F(P_{i+1})$. Garside proved essentially the following theorem.

Theorem 11.7.3. *For every braid $b \in B_n$, there is a unique expression given by*

$$b = \Delta_n^k P_1 P_2 \cdots P_r$$

where $k \in \mathbb{Z}$ is maximal, P_i are permutation braids, $P_k \ne 1$ and $P_1 P_2 \cdots P_k$ is a left-weighted decomposition.

The sequence $(k; P_1, \ldots, P_r)$ is the **Garside normal form** for $b \in B_n$. The integer r is its **length** or **complexity**.

There is a simple algorithm for computing the Garside normal form which shows that it is efficiently computable. It can be described as follows:

Given a braid $w \in B_n$:

(1) Replace σ_i^{-1} by $\Delta_n^{-1} P_i$ where P_i is a permutation braid.
(2) Move any appearance of Δ_n to the left. We then get $w = \Delta_n^t A$ where A is a positive braid.
(3) Write A as a left-weighted decomposition of permutation braids by computing the various $S(P)$ and $F(P)$.

It can be shown that the complexity of this procedure is quadratic in the world length of w within B_n.

Since this is a unique expression, it does provide a normal form and hence provides another solution to the word problem within B_n. We will show in Section 11.7.5 how Garside used this form to solve the conjugacy problem.

The second normal form that we examine in B_n is the **Dehornoy normal form**, or the **handle reduction form**. As we mentioned at the beginning of this section, Myasnikov, Shplirain and Ushakov used the Dehornoy form to counteract the diffusion afforded by the Garside normal form and experimentally break a Ko-Lee encryption braid group implementation.

The Dehornoy normal form involves braid word transformations called **handle reductions**. The method turns out to be very efficiently computable in practice (see [Deh]). The underlying structure behind handle reduction is a linear ordering of braids.

Handle reduction is an extension of free reduction. The latter consists in iteratively deleting subwords of the form xx^{-1} or $x^{-1}x$. Handle reduction considers not only subwords of the form $\sigma_i \sigma_i^{-1}$ or $\sigma_i^{-1} \sigma_i$ but also more general subwords of the form $\sigma_i \cdots \sigma_i^{-1}$ or $\sigma_i^{-1} \cdots \sigma_i$ with intermediate generators between σ_i and σ_i^{-1}.

We need the concept of a **handle**.

Definition 11.7.4. A σ_i-**handle** is a braid word of the form

$$w = \sigma_i^e w_0 \sigma_{i+1}^d w_1 \sigma_i^d \cdots \sigma_{i+1}^d w_m \sigma_i^{-e}$$

with $e, d = \pm 1$, $m \geq 0$ and $w_0, \ldots, w_m$ are words containing no $\sigma_j^{\pm 1}$ with $j \leq i + 1$.

Removing a handle is called **handle reduction** and yields an equivalent braid word.

Definition 11.7.5. Let w be as in the last definition. Then a **handle reduction** of w is the word

$$w' = w_0 \sigma_{i+1}^{-e} \sigma^d \sigma_{i+1}^e w_1 \sigma_{i+1}^{-e} \cdots \sigma_{i+1}^{-e} \sigma_i^d \sigma_{i+1}^e w_m$$

that is we delete the initial and final generators $\sigma_i^{\pm 1}$ and replace each generator $\sigma_i^{\pm 1}$ with $\sigma_{i+1}^{-e} \sigma_i^{\pm 1}$.

A braid word is **handle reduced** if it contains no σ_i-handle where σ_i is the generator with minimal index occurring in w.

Braid words of the form $\sigma_i\sigma_i^{-1}$ or $\sigma_i^{-1}\sigma_i$ are handles, and reducing them means deleting these expressions. Therefore handle reduction generalizes free reduction. Reducing a handle yields an equivalent braid word. So, as in the case of free reduction, if there is a reduction sequence from a braid word w to the empty word, that is, a sequence $w = w_0, \ldots, w_N = 1$ such that, for each k, the word w_{k+1} is obtained from w_k by replacing some handle of w_k by its simple handle reduction. The process of handle reduction will always terminate in a handle reduced word and this is the Dehornoy normal form.

Theorem 11.7.6. *For every braid word w, every sequence of handle reductions from w leads in finitely many steps to a handle reduced word w_0. Moreover, a handle reduced word w_0 represents the identity on B_n if and only if it is empty.*

Handle reduction then leads to equivalent braid words and hence can be used as a normal form. Therefore the Dehornoy normal form provides a solution of the word problem.

Corollary 11.7.7. *A braid word w represents the identity element 1 in the braid group if and only if any sequence of handle reductions ultimately terminates in the empty word.*

The final normal form we define is the **Birman-Ko-Lee normal form**. This comes from a separate set of generators for B_n called **band generators** developed by Birman, Ko and Lee. For every $s > r$ the band generator a_{sr} represents a braid in which the strands s and r are swapped with strand s crossing on top and strand r crossing over every other strand apart from s. It follows that

$$a_{sr} = \sigma_{s_1} \cdots \sigma_{r+1}\sigma_r\sigma_{r+1}^{-1} \cdots \sigma_{s-1}^{-1}$$

and further $a_{i+1,i} = \sigma_i$, so it is clear that the band generators generate the whole braid group B_n.

It is straightforward that the band generators satisfy the relations
(1) $a_{ts}a_{rq} = a_{rq}a_{ts}$ if $[s,t] \cap [q,r] = \emptyset$ or $[s,t] \subset [q,r]$ and if $[p,r] \subset [s,t]$ and $[q,r] \subset [s,t]$,
(2) $a_{ts}a_{sr} = a_{tr}a_{ts} = a_{sr}a_{tr}$ for $1 \leq r < s < t \leq n$.

We then get the **Birman-Ko-Lee presentation** as

$$B_n = \langle a_{sr}; \text{ relations } (1),(2) \rangle.$$

To get a normal form with this presentation we define a fundamental word as in the Garside normal form. In particular

$$\delta_n = a_{n,n-1} \cdots a_{3,2}a_{2,1}.$$

We then get

Theorem 11.7.8. *Each braid word $w \in B_n$ has the following unique expression*

$$w = \delta_n^r A_1 A_2 \cdots A_k$$

where $A = A_1 A_2 \cdots A_k$ is positive, r is maximal and k is minimal and the A_i's are determined uniquely. They are termed **canonical factors.**

There are $C_n = \frac{(2n)!}{n!(n+1)!}$ (the n-th Catalan number) different canonical factors.

This unique form is called the **Birman-Ko-Lee normal form**.

There are further normal forms for elements in B_n. We refer to the article by Dehornoy [Deh] for a description of some of these. What we must take away from this section, is that for cryptographic purposes there is a selection of normal forms for elements of B_n. However, care must be taken as to which of these forms is chosen, and how the elements are generated as the article by Myasnikov, Shpilrain and Ushakov [MSU2] points out.

11.7.3 The Pure Braid Group for B_n

If one considers an element of the braid group B_n as a set of braid crossings and just considers where a parallel string begins and ends one obtains for each $b \in B_n$ a permutation on n letters and hence an element $\pi \in S_n$. Calling the permutation corresponding to a braid b as b_π, the mapping $b \to b_\pi$ provides a homomorphism from B_n to the symmetric group S_n. The kernel of this homomorphism is called the **pure braid group on n strands.** We denote this group by PB_N.

If σ_i is standard generator of B_n then σ_i represents a single braid crossing. It follows easily that under the map to S_n we have $\sigma_i^2 \mapsto 1$. Appending these relations to the Artin presentation we get the standard **Coxeter presentation** for the symmetric groups S_n. That is:

$$S_n = \langle \sigma_1, \ldots, \sigma_{n-1}; \quad \sigma_i^2 = 1, \quad i = 1, \ldots, n-1, \quad [\sigma_i, \sigma_j] = 1 \text{ if } |i - j| > 1,$$
$$\sigma_{i+1}\sigma_i\sigma_{i+1} = \sigma_i\sigma_{i+1}\sigma_i, \quad i = 1, \ldots, n-1 \rangle.$$

The Artin presentation has been generalized in various ways to obtain the **Coxeter Groups** and the **Right-angle Artin Groups**. We refer to [Bir] for a discussion of these generalizations. Although they lead to important mathematics they will play no role in our treatment of braid groups and cryptography.

11.7.4 Linear Representations of B_n

Soon after Artin introduced the braid groups, Burau found a linear representation of B_n into $GL(n-1, \mathbb{Z}[t, t^{-1}])$. This is known as the **Burau representation** and represents each braid as a n dimensional invertible matrix over the ring of integral polynomials in t and t^{-1}. The following matrices determine the n-dimensional Burau representation where $\sigma_i, i = 1, \ldots, n$ are the standard Artin generators. In the matrices, I_m is the appropriate sized identity matrix making the matrices $n \times n$:

$$\rho(\sigma_1) = \begin{pmatrix} -t & 1 \\ 0 & 1 \end{pmatrix} \oplus I_{n-3}$$

$$\rho(\sigma_i) = I_{i-2} \oplus \begin{pmatrix} 1 & 0 & 0 \\ t & -t & 1 \\ 0 & 1 & 0 \end{pmatrix}, \quad 1 < i < n-1$$

$$\rho(\sigma_{n-1}) = I_{n-3} \oplus \begin{pmatrix} 1 & 0 \\ t & -t \end{pmatrix}.$$

It is easy to show that these matrices satisfy the braid group relations and hence provide a representation of B_n. For a long time it was an open question whether the Burau representation was faithful or not. However, it was proved by Moody that for $n \geq 5$ it is not faithful. This was not surprising since it was known that $\mathrm{Aut}(F_n)$ was not linear.

Subsequently, Ruth Lawrence and Daan Krammer discovered a second, more complicated representation of B_n into $\mathrm{GL}(\frac{n(n-1)}{2}, \mathbb{Z}[q, t, q^{-1}, t^{-1}])$. In developing this representation they used topological methods that depended on the interpretation of the braid groups as mapping class groups of surfaces. The matrix form of this representations can be described in the following way. Let κ denote the representation

$$\kappa : B_n \rightarrow \mathrm{GL}\left(\frac{n(n-1)}{2}, \mathbb{Z}[q, t, q^{-1}, t^{-1}]\right).$$

Denote the basis elements for the module $\mathbb{Z}[q, t, q^{-1}t^{-1}]$ by x_{ij} for $1 \leq i < j \leq n$. Then we have that the action of the matrix $\kappa(\sigma_i)$, where σ_i are the Artin generators, is given by

$$\kappa(\sigma_i) : x_{ij} \mapsto \begin{cases} tq^2 x_{k,k+1} & \text{if } i = k, j = k+1, \\ (1-q)x_{i,k} + qx_{i,k+1} & \text{if } j = k, i < k, \\ x_{ik} + tq^{k-i+1}(q-1)x_{k,k+1} & \text{if } j = k+1, i < k, \\ tq(q-1)x_{k,k+1} + qx_{k+1,j} & \text{if } i = k, k+1 < j' \\ x_{k,j} + (1-q)x_{k+1,j} & \text{if } i = k+1, k+1 < j, \\ x_{i,j} & \text{if } i < j < k \text{ or } k+1 < i < j, \\ x_{ij} + tq^{k-1}(q-1)^2 x_{k,k+1} & \text{if } i < k, k+1 < j. \end{cases}$$

In 2001 Stephen Bigelow and Daan Krammer independently proved that the Lawrence-Krammer representation of dimension $n(n-1)/2$ depending on the variables q and t is faithful. Hence the braid groups are linear. This is somewhat surprising since as mentioned above $\mathrm{Aut}(F_n)$ is known to be non-linear (see [FP]). By suitably specializing the variables in the Lawrence-Krammer representation, the braid groups B_n may be realized as subgroups of general linear groups over the complex numbers.

Both of these linear representations, the Burau representation and the Lawrence-Krammer representation, can be used to attack braid group based cryptosystems. Although the Burau representation is not faithful, it has a relatively small kernel. We will say more about this in the next section.

For cryptographic purposes an extended version of the Burau respresentation is often used. This is called the **colored Burau representation** and is constructed in the following manner. This also has a small kernel.

Definition 11.7.9. The **colored Burau group** CB_n is the group
$S_n \times GL(n-1, \mathbb{Z}[t_1, t_1^{-1}, \ldots, t_n, t_n^{-1}])$ with S_n the symmetric group on n symbols and multiplication given by

$$(\pi_1, M_1) \cdot (\pi_2, M_2) = (\pi_1 \pi_2, (\pi_2^{-1} M_1) M_2)$$

where a permutation $\pi_i \in S_n$ acts on each element of a matrix
$M \in GL(n-1, \mathbb{Z}[t_1, t_1^{-1}, \ldots, t_n, t_n^{-1}])$ by taking t_i to t_{π_i}.
The **colored Burau representation** $\rho : B_n \to CB_n$ is given by

$$\rho(\sigma_i) = ((i, i+1), \rho(\sigma_i)(t_i)).$$

where σ_i are the Artin generators.

The following is straightforward and we leave the proof to the exercises.

Theorem 11.7.10.
(1) *The colored Burau group is a group under the given operations.*
(2) *The colored Burau representation preserves the braid relations and hence does give a group homomorphism.*

11.8 Cryptanalysis of Braid Group Cryptosystems

When the braid group based cryptosystems were originally introduced, there was a great deal of excitement. The difficulty of the main group theoretic problems used, the conjugacy problem and conjugator search problem, seemed hard enough in B_n to ensure security. However, with the additional research that was now done on the abstract braid groups, it became apparent that without special considerations the braid group based cryptosystems were insecure. In this section we discuss attacks on braid group based cryptosystems.

These fall into several attack categories: attacks based directly on the Garside solution to the conjugacy problem via so-called **summit sets**, **length based attacks** which use the length of braids, **representation theoretic attacks**, and finally attacks on the diffusion of group element multiplication by varying the normal forms. We briefly discuss each. More detailed discussions can be found in the excellent survey papers of Dehornoy [Deh] and Garber [Gar].

11.8.1 Attacks on the Conjugacy Search Problem

Since the security of the initial braid group cryptosystems was based on the supposed difficulty of the conjugacy problem, the conjugacy search problem and the conjugator search problem, most attacks have focused on the solutions of these problems. The conjugacy problem for B_n was originally solved by Garside and it was assumed that it was hard in the complexity sense. Therefore it was felt that using either the conjugacy problem or the conjugator search problem was secure enough for cryptographic purposes. Recently there has been significant research on the complexity of the solution to the various conjugacy search problems (see [MSU1] and [Deh]) and exposed weaknesses in the strength of the cryptosystems based on these.

The Garside solution to the conjugacy problem in B_n used what are called **summit sets**. For each $b \in B_n$ its summit set which we will denote $SS(b)$ consists of a finite set of conjugates of b. The summit set contains a subset called the **super summit set** denoted $SSS(b)$ which is algorithmically computable. El-Rifai and Morton showed how to compute for each $b \in B$ the super summit set $SSS(b)$. Further two braids b, b' are conjugate if and only if their super summit sets coincide, or equivalently, if and only if they intersect.

Theorem 11.8.1 (see [EM]). *For every $b \in B_n$ the set $SSS(b)$ is finite and is algorithmically computable. If $b, b' \in B_n$ then b and b' are conjugate if and only if $SSS(b) \cap SSS(b') \neq \emptyset$.*

Therefore to solve the conjugacy problem, and in essence break the Ko-Lee type protocols, find the super summit sets of the elements. There is recent refinement of the super summit set called the **ultra summit set** $USS(b)$ introduced by Gebhardt [Geb]. The size of $USS(b)$ is often much smaller than the size of $SSS(b)$. Long detailed discussions of the various algorithms to find the summit sets and ultra summit sets are in the paper of Garber [Gar] as well as the original papers.

11.8.2 Length Based Attacks

The Anshel-Anshel-Goldfeld protocol depends on the conjugator search problem. However, to attack the protocol it is not necessary to find the exact conjugator. If b, b' are conjugate then knowing any conjugator, that is an element w such that $wbw^{-1} = b'$ is sufficient.

Length based attacks have been used effectively in attacking the conjugator search problem. Recall that using either the Garside normal form or the Dehornoy normal form each braid has a well defined length. This gives a well defined length function on the braid groups B_n analogous to the length function on free groups.

In Anshel-Anshel-Goldfeld type protocols that depend on the conjugator search problem an attacker has access to multiple sets of pairs (a, axa^{-1}) for a secret value a taken from a set with a finite number of known generators. The common principle in length based attacks is to try to retrieve a conjugator for a pair (b, b') which is supposed to be derived form b by iteratively conjugating b' into a new element $tb't^{-1}$ so that the length is minimal. If we define the distance between two braids a, b as $d(a, b) = l(ab^{-1})$ this gives a true metric on the braid groups which satisfies the triangle inequality. The idea in a length based attack is that there is not too much cancellation is multiplying by generators σ_i and therefore if a generator σ_i is not an initial segment of a conjugator a then with non-zero probability the length of $\sigma(at_ia^{-1})\sigma^{-1}$ is larger than the length of at_ia^{-1}. Then apply each generator σ_i until the length reduces and this generator can be peeled off of the conjugator.

There has been considerable success with this technique (see [Deh] and [Mah]). Part of the success though is due to the fact that braid groups have the generic free group property (see Chapter 10). Doing length reduction is in actuality doing free group reduction. In fact many braid group based cryptosystems have been broken because the braid groups have the generic free group property.

There have been attempts to modify the conjugator search protocols to defend against length based attacks (see [Mah] for a discussion of these).

11.8.3 Representation Theoretic Attacks

Several successful attacks on braid group crytpographic protocols have used the various linear representations; Burau, colored Burau and Lawrence-Krammer. In many attacks finding secret keys is reduced, by using these representations, to linear algebra over finite fields.

Recall that the Ko-Lee protocol works as follows:

Bob and Alice each choose a subgroup within B_n that commute elementwise. A is Alice's subgroup and B is Bob's subgroup. Alice randomly chooses an $a \in A$. This element a will be her secret key. Her public key is (g, g^a) where $g^a = aga^{-1}$ is the conjugate of g by her secret key a. All public information and communication is done in terms of the normal forms of these elements.

Similarly Bob randomly chooses an element $b \in B$. This element b will be his secret key. His public key is (g, g^b) where $g^b = b^{-1}gb$ is the conjugate of g by his secret key b. As with Alice all public information and communication is done in terms of the normal forms of these elements.

The secret shared key is g^{ab}.

Cheon and Jun [CJ] show how to use the Lawrence-Krammer representation to find the secret key knowing only g^a and g^b. Suppose that $Y_a = \rho(g^a)$ and $Y_b = \rho(g^b)$ are the images of g^a and g^b under the Lawrence-Krammer representation ρ. Cheon and Kim working modulo a prime p and certain irreducible polynomials in t and q solve for the matrix M in the equations

$$Y_a M = M Y_b$$
$$\rho(\sigma_i)M = M\rho(\sigma_i) \quad \text{for } \sigma_i \in B_{2n}.$$

In this attack even though the solution to the linear equations may not be $\rho(a)$ the commutativity equations give

$$MY_bM^{-1} = \rho(b)Y_a\rho(b)^{-1} = \rho(abxb^{-1}a^{-1})$$

which can be lifted to the necessary braid.

As pointed out in [CJ] and in [Mah], this attack is effective because the matrices arising under ρ satisfy very restrictive bounds.

The effectiveness of this method shows that the Lawrence-Krammer representation allows for a solution to the Diffie-Hellman problem (on conjugation) in braid groups in polynomial time.

11.8.4 Braid Group Security Summary

These various attacks have dampened much of the initial enthusiasm over braid group based cryptosystems. There have been attempts to alter the protocols to protect against these attacks. This has raised problems due to the fact that security requirements for one protocol are often at odds with those of another. In general all atacks are less successful with larger parameters (bigger initial n in B_n and larger rank subgroups. There are more extensive discussions of these in the papers of Dehornoy [Deh] and Marburg [Mah].

There have also been other types of attacks. As mentioned earlier, Myasnikov and Ushakov [MU] broke a braid group based cryptosystem by transforming from one normal form to another.

The point of all of this is that great care must be taken in developing Ko-Lee and Anshel-Anshel-Goldfeld type protocols within the braid group setting. There have been various suggestions on how to continue.

In one direction the idea is to use variants of the basic cryptographic ideas in Ko-Lee and Anshel-Anshel-Goldfeld but use alternative groups such as Thompson's group and polycyclic groups as the platform.

In another direction there have been ideas to continue to use the braid groups as the basic platforms but to employ different hard group theoretic problems as the fundamental trapdoor. For example a cryptosystem has been developed using the shortest braid problem, that is given a word w in the Artin generators find the shortest word w' equivalent to w. Another braid cryptosystem uses the twisted conjugacy problem.

There are in depth discussions of these alternative suggestions in [MSU1] and [Gar].

Anshel, Anshel, Goldfeld and Lemieux have developed a completely different type of key exchange protocol using group actions on monoids. They call this the **Algebraic Eraser Method** and suggest an implementation using the colored Burau representation. They apply this for use in an RFID tag. The details can be found in [AAG2].

11.9 Some Other Braid Group Based Protocols

There are many other cryptographic protocols beyond pure encryption and decryption (see Chapter 4), such as authentication and digital signatures. With the initial enthusiasm over braid group cryptosytems there have been attempts to place these additional protocols within a braid group setting. To close this chapter we mention several of these.

As pointed out in Dehornoy [Deh], there have been direct attempts to copy the authentication and digital signature schemes described in Chapter 4 using braid groups as a platform. The idea is that the conjugacy problem and conjugator search problem are sufficiently difficult to allow security in these protocols. For example the following is a direct translation of a Ko-Lee authentication protocol.

Bob wants to authenticate Alice. Alice's public key is a pair (p, p') of conjugate braids in B_n with $p' = sps^{-1}$. Alice's private key is the conjugator s. For this protocol we assume that $s \in LB_n$. Suppose that h is a hash function on B_n. To authenticate Alice, Bob chooses a random braid $r \in UB_n$ and sends Alice the challenge $p'' = r^{-1}pr$. Alice sends the response $y = h(s^{-1}p''s)$ and Bob checks that $y = h(r^{-1}p'r)$.

If this is true, Bob accepts Alice's authentication. Since r and s commute we have $rp'r^{-1} = sp''s^{-1}$ so Alice can send the correct response only if she has been able to solve the Diffie-Hellman problem in B_n. Hence the security is based on the difficulty of this problem within B_n.

There have been other ways to adapt Shamir Three-Passes and various signature protocols. As with the basic encryption schemes various attacks have led to questions about the security of the braid groups as potential platforms.

11.10 Exercises

11.1 Let G be B_{20} the braid group on 19 generators $\sigma_1, \ldots, \sigma_{19}$. Let A be the subgroup generated by $\sigma_1, \sigma_2, \sigma_2, \sigma_4, \sigma_5$ and B the subgroup generated by $\sigma_{16}, \sigma_{17}, \sigma_{18}, \sigma_{19}$. Clearly A and B commute elementwise.
   (a) Let $g = \sigma_7^3 \sigma_{12} \sigma_3^{-1} \sigma_5^{-2} \sigma_{10}$ and $a = \sigma_2^4 \sigma_3^2 \sigma_1$, $b = \sigma_{17}^4 \sigma_{18}^{-1} \sigma_{17}$. Compute the Garside normal forms for g, $g^a = a^{-1}ga$, and $g^b = b^{-1}gb$.
   (b) What is the secret shared key using the Ko-Lee protocol?

11.2 Let G, A and B be as in problem 11.1. Given a, b, what is the secret sheared key using the AAG protocol?

11.3 Let PB_n stand for the pure braid group on n strands. Show that this group has a presentation with generators

$$A_{ij} = \sigma_{j-1}\sigma_{j-2} \cdots \sigma_{i+1}\sigma_i^2\sigma_{i+1}^{-1} \cdots \sigma_{j-1}^{-1}\sigma_{j-1}^{-1}$$

where $1 \leq i < j \leq n$ and relations:

$$A_{rs}A_{ij}A_{rs}^{-1} = A_{ij} \qquad\qquad\qquad\qquad \text{if } s < i \text{ or } j < r$$
$$A_{rs}A_{ij}A_{rs}^{-1} = A_{is}^{-1}A_{ij}A_{is} \qquad\qquad\qquad \text{if } i < j = r < s$$
$$A_{rs}A_{ij}A_{rs}^{-1} = A_{ij}^{-1}A_{ir}^{-1}A_{ij}A_{ir}A_{ij} \qquad\quad \text{if } i < r < j = s$$
$$A_{rs}A_{ij}A_{rs}^{-1} = A_{is}^{-1}A_{ir}^{-1}A_{is}A_{ir}A_{ij}A_{ir}^{-1}A_{is}^{-1}A_{ir}A_{is} \quad \text{if } i < r, j, s$$

11.4 Show that the pure braid group PB_3 is isomorphic to the direct product $F_2 \times \mathbb{Z}_2$.

11.5 Show that the pure braid group on n strands, $n \geq 3$, satisfies the generic free group property.

11.6 Let F_n be the free group of rank n on the free generating system $\{x_1, \ldots, x_n\}$ and $\beta \in \mathrm{Aut}(F_n)$. Show that $\beta \in B_n$ if and only if β satisfies the following two conditions.

(1) $\beta(x_i)$ is conjugate to another generator.

(2) $\beta(x_1 x_2 \cdots x_n) = x_1 x_2 \cdots x_n$.

11.7 Prove that the colored Burau group is a group under the given multiplication.

11.8 Prove that the colored Burau representation actually preserves the braid relations and hence is a group representation.

12 Further Applications Using Group Theory

12.1 Finitely Presented Groups and Cryptography

In the previous several chapters we have seen how group theory, specifically the combinatorial group theory of finitely presented groups, can be utilized in a cryptographic setting. The basic idea is that a finitely presented group can be described by a finite amount of data. This provides techniques to enormously compress and hide information. The body of knowledge, learned in more than 150 years of intense study of finitely presented groups, provides a far reaching tool when applied to cryptology.

Numbers and words are often used as a means of identification, authorization, access to information, record keeping, the control of devices and so on. They are translated into bit strings without any structure. The replacement of these numbers and words by group-theoretic objects, such as finitely presented groups, their subgroups, their elements and the maps between them, can play a role in computer science. These algebraic objects have well-defined structures, and a huge body of knowledge exists, which makes it possible to easily design all manner of objects to fit a variety of purposes, not easily done solely with bit strings.

In the case of a group defined by finitely many relations, a few symbols and a few equations can be used to encode in many very different ways an infinite amount of information (see Chapter 9 for a discussion of finitely presented groups). Indeed, finite presentations provide a concrete way of defining infinite sets with many properties, the presence or absence of which can be used to allow for the passage or otherwise of information, in a manner analogous to gates in networks. These sets inherit a group structure, and as a consequence, such gates can be controlled dynamically, and opening or closing them, can be realized by answers to an infinite number of questions which can be arranged in such a way as to be arbitrarily difficult to answer. Thus the use of finite group presentations have the potential to allow secure access and control of devices and to provide new approaches to cryptography, to authentication, to identification and authorization and to record-keeping.

This entire chapter is somewhat speculative. The objective is to suggest additional uses of group theory, some complicated and some simple, with most of the details omitted. Whether implementation of the suggestions that follow will lead to secure group-theoretic alternatives to some existing protocols is unclear. For example, we will describe a far-reaching extension of the permission mechanisms in the Unix operating system for controlling access to files and devices partly based on what we term the **entry control group**, which we will discuss next.

12.2 Group Theory for Access Control

In data base management, changing access to files can be time consuming and difficult. Here we suggest a novel approach to this by using a single finitely presented group that we call an **entry control group** to control and effect such access. This can be done in a reasonably secure manner. The discussion that follows is a very special instance of a general entry control group, and can readily be adjusted, and be used, in a number of different ways which take into account limiting access in any manner deemed useful. We note that in earlier writings on these ideas the entry control group was called an access control group. However,the terms access control group has an older, and more established, meaning in secret sharing (see Chapter 4), so the terminology entry control group was substituted.

Wreath Products

Our first example of an entry control group depends on a group theoretic construction called a **wreath product** of two groups A and B. This is a specialized product of two groups, based on a semidirect product and has been extensively studied. Theoretically, wreath products provide a method to construct many interesting examples of groups.

Given two groups A and B, there exist two variations of the wreath product: the unrestricted wreath product and the restricted wreath product. To begin we briefly describe this construction. First we define a **semidirect product**. If G is a group and A and B are subgroups of G, then G is the semidirect product of A by B if A is a normal subgroup and every element of G can be written in a unique way as a product ab, with $a \in A$ and $b \in B$.

Definition 12.2.1. Let A and B be groups and we suppose that B acts on A, that is, each element of B acts as a homomorphism on A.

Let K be the direct product of copies of A indexed by B. The elements of K are then arbitrary sequences (a_b) of elements of A indexed by B with component wise multiplication. Then the action of B on A extends in a natural way to an action of B on the group K by $b_i((a_b)) = (a_{b_i^{-1}b})$.

Then the **unrestricted wreath product**, $A \wr B$, of A by B, is the semidirect product of K by B. The subgroup K of $A \wr B$ is called the base of the wreath product.

The **restricted wreath product**, $A \wr B$, is constructed in the same way as the unrestricted wreath product except that the direct sum of copies of A indexed by B is used as the base of the wreath product. In this case the elements of K are sequences of elements in A, indexed by B, of which all but finitely many are the identity element of A.

The wreath product construction plays an important role in the structure theory of permutation groups. The structure of the wreath product $A \wr B$ depends on the action of B on A and on whether it is the restricted or unrestricted wreath product. Further the restricted wreath product is always a subgroup of the unrestricted wreath product.

There is a universal embedding theorem which states that if a group G is an extension of A by B, that is, A is a normal subgroup of G with G/A isomorphic to B then there exists a subgroup of the unrestricted wreath product of A by B which is isomorphic to G.

Access Control to Files

We now exhibit how the wreath product construction can be utilized to define an entry control group. For more material on the basic combinatorial group theory see Chapter 9.

To start, let F be the free group on a, x, y, let S be the infinite cyclic group on s and let

$$W = F \wr S$$

be the wreath product of F and the infinite cyclic group S. Furthermore, let

$$V_1 = \langle F, F^s, s^2 \rangle, \quad V_2 = \langle F, F^s, s^3 \rangle.$$

Put

$$F_i = s^{-i} F s^i \quad \text{with } i \in \mathbb{Z}.$$

Form the HNN extension H of W with stable letter t and

$$a^t = a, \quad x^t = x, \quad y^t = y, \quad a^{st} = a, \quad x^{st} = x^s, \quad y^{st} = y^s, \quad s^{2t} = s^3.$$

Then H can be presented on the generators

$$a, x, y, s, t$$

and defining relations

$$a^t = a, \quad x^t = x, \quad y^t = y, \quad a^{st} = a^s, \quad x^{st} = x^s,$$

$$y^{st} = y^s, \quad [a, a^s] = 1, \quad [a, x^s] = 1, \quad [a, y^s] = 1,$$

$$[x, x^s] = 1, \quad [x^s, y] = 1, \quad [y, y^s] = 1, \quad [x, y^s] = 1, \quad s^{2t} = s^3.$$

The reader can take this as the definition of the group H which is to be our entry control group.

Suppose now that we have data consisting of a number of files labelled by the positive integers $i = 1, \ldots$ where the i-th file is named E_i. The contents of E_i is recorded in the free group F_i, generated by the elements

$$a_i = s^{-i} a s^i, \quad x_i = s^{-i} x s^i, \quad y_i = s^{-i} y s^i.$$

Each file E_i then defines a free subgroup of the ambient free group.

The users of the database are arranged in order of the times at which they subscribe to the database, as

$$U_1, \ldots, U_i, \ldots.$$

Each user U_i is assigned an element $u(i) \in H$. U_i is to be permitted access to the files in

$$E_{i_1}, \ldots, E_{i_m},$$

if and only if $u(i)$ does *not* commute with any one of the subgroups $E_{i_1}, \ldots, E_{i_m}$. In order to effect this we set $u(i) = 1$ if U_i is not to have access to any of the files in the database. Otherwise we put

$$u_i = a_{i_1} \cdots a_{i_m}$$

which ensures that $u(i)$ has access to the files in $E_{i_1}, \ldots, E_{i_m}$. This scheme allows then for easy changes of permission to access the files in the database being set up. This is a particularly important aspect of the design of the entry control group.

The Entry Control Group Security

Later in this chapter we introduce what we term **social security groups**. These are complicated finitely presented groups. They are chosen in such a way to ensure that it is difficult to answer questions about them without additional information, which is kept private unless needed. If the free group F in the description above of the entry control group is replaced by such a social security group, say X, then access to any of files which are now residing in one of the conjugates $w^{-1}Xw$ of X is controlled by requiring that various questions about w have to have the right answers. This ensures that access to the database is secure. This follows the security provisions set out for the group theoretic password security described in Section 10.6.

Entry Permissions in Operating Systems

In many operating systems and in particular in the Unix operating system, it is usually the case to assign so-called permissions to various files to the users of the system which allow the files to be read, written to and executed as appropriate, labelled simply as *rwx*. Our Access Control Group is enormously flexible allowing for us to designate files not only with all manner of access, but also degrees of access for various users. For example, if a user wants to use a 3D printer, which is very expensive to operate, that user must be given permission to do so and the limitations of such use both partial and total can also be controlled by an appropriate entry control group.

Access Control for Websites

The idea here is to show how the entry control group can enable Alice with the ability to access any of the websites that she is interested in without having to remember passwords. We again use the entry control group H.

List the websites that Alice is interested in as $W_1, \ldots, W_m$ together with the passwords as required. Install the information about W_i, its URL and the password that Alice wants to use to gain access. W_i is in a non-trivial subgroup E_i of F_i generated by a_i. For each of these websites, say $W_{i_1}, \ldots, W_{i_m}$, assign what we might term the website locator acceptor

$$w_A = a_{i_1} \cdots a_{i_m}.$$

Notice that w_A does not commute with E_j only if $j \in \{1_1, \ldots, i_m\}$ which then provides Alice access to any one of the web sites she is interested in. Moreover if we want to ensure that Alice does not want to access any particular site say W_{i_j} for any reason she changes her website locator by omitting i_j. Of course this idea can be varied to accommodate any given set of conditions. Computations can be carried out quickly inside groups like H.

12.3 Public Key Control Groups

In this section we show how to use the complexity of a finitely presented group as the basis for a new public key protocol. In the next section we do the same using the insolvability of Hilbert's Tenth Problem.

We describe first a complicated finitely presented group H which we use as the basis for a new public key cryptosystem. Such groups are plentiful and we will give an additional example later to justify this claim.

In order to define H we repeat for ease of understanding, the usual notation for commutators and conjugation:

$$[x, y] = x^{-1} y^{-1} xy, \quad \text{and} \quad x^y = y^{-1} xy.$$

The group H is an amalgamated product of two finitely presented groups A and B given by:

$$A = \langle a, h, s; [a, h] = 1, a^{2s} = a^3, h^{2s} = h^3 \rangle, \quad B = \langle b, t; b^{2t} = b^3 \rangle.$$

The subgroups U and V to be amalgamated are free and given by:

$$U = \langle s, [a, a^s] \rangle \quad \text{and} \quad V = \langle [b, b^t], t \rangle.$$

H then is defined as follows:

$$H = \langle A * B; s = [b, b^t], [a, a^s] = t \rangle.$$

It follows that H is a 5-generator, 6 relator group.

Now let H_0 and H_1 be copies of H, with the elements in H_0 corresponding to a and h denoted by a_0, h_0 and so on. Let D be the direct product of H_0 and H_1:

$$D = H_0 \times H_1.$$

We now use von Dyck's Lemma (see [MKS]) to show that D is a quotient of H. With this in mind we define a map ϕ from the generators of H into D as follows:

$$\phi(a) = a_0 a_1, \quad \phi(h) = a_0 h_0^2 h_1^2, \quad \phi(s) = s_0 s_1, \quad \phi(b) = b_0 b_1, \quad \phi(t) = t_0 t_1.$$

It is clear that the relations not involving h are preserved. We have to check that the remaining equations are also preserved, and this is straightforward. It remains to prove that the subgroup E generated by images of the generators under ϕ is all of D. To this end, observe first that $s_0 \in E$:

$$[a_0 a_1, (a_0 h_0^2 h_1^2)^{s_0 s_1}] = [a_0, (a_0 h_0^2)^{s_0}][a_1, h_1^{2s_1}] = [a_0, a_0^{s_0}] = t_0.$$

It follows then also that $t_1 \in E$. Consequently

$$[b_0 b_1, (b_0 b_1)^{t_0}] = [b_0, b_0^{t_0}] = s_0 \in E.$$

Additionally

$$[(a_0 a_1)^2 s_0] = [a_0^2, s_0] = a_0 \in E.$$

Again it follows that $a_1 \in E$ as usual. Finally $h_0^2 h_1^2 \in E$ because $a \in E$ which implies that

$$[h_0^2 h_1^2, s_0] = [h_0^2, s_0] = h_0 \in E$$

and so also $h_1 \in E$. This ensures that $E = D$ and so D is a quotient of H, which completes the claim.

We show how the construction of H enables us to use this group as the basis of a public key cryptosystem. Suppose that F is a free group, freely generated by 2 elements. It follows that F then contains a free subgroup freely generated by 256 generators. If we put these generators in a one-to-one correspondence with the 256 ascii characters, then every message can be represented by the appropriate a product of these generators, that is, as an element of F. It follows that if a given group contains a non-abelian free subgroup, then we can encode messages as elements in that group.

In order to take advantage of the properties of the group H we make use of the vertices of a rooted binary tree T. We denote the root of the tree by 0. There are two edges emanating out of a vertex v. We denote the terminus of one of these edges by $0v$ and the terminus of the other vertex by $1v$. Now we put a copy H_v of H on each vertex v. Then there is a homomorphism ϕ_v of H_v onto the direct product $H_{0v} \times H_{1v}$ of H_{0v} and H_{1v}. Let π_{0v} be the projection of H_v onto H_{0v} and π_{1v} be the projection of H_v onto H_{1v}. The composition $\gamma_{0v} = \phi_v \pi_{0v}$ maps H_v onto H_{0v} and the composition $\gamma_{1v} = \phi_v \pi_{1v}$ maps H_v onto H_{1v}. We now attach to the two edges $0v$ and $1v$ the respective homomorphisms γ_{0v} and γ_{1v}. For each path p emanating out of 0 and ending up at the vertex v we can compose the homomorphisms attached to these edges which results in a homomorphism λ_v from H_0 to H_v.

Now if ρ_v is a path in the binary tree from the root vertex 0 to the vertex v, choose a free subgroup F_v of H_v of rank 256. Choose for each of the generators of F_v a pre-image under λ_v in H_0. So if Bob wants to send a message to Alice, he encodes it as a product P of these pre-images. Alice then applies λ_v to P and is able to decode P in H_v. The point here is that decoding a message from Bob requires knowledge of the path ρ_v which is known only to Alice. This is a public key protocol.

The discussion above has many variations. Each such variation gives rise to a new public key protocol. The interested reader can supply those variations as desired. It suffices here to roughly describe the bare bones of such a variation. To this end, in place of the group H above we can choose a non-hopfian group H, that is, a group that is isomorphic to one of its proper quotients, for example, $H = \langle a, b; a^{-1}b^2 a = b^3 \rangle$.

Let T be a rooted binary tree with initial vertex or root v_0. Assign to each vertex in T a vertex group which is a copy of H and at each edge e an epimorphism of the copy of H at the origin of e to the copy of H at terminus of e with a non-trivial kernel. Let ψ be a random walk in T starting at v_0 and ending say at v'. Let ρ_ψ be the composition of the epimorphisms on the edges that comprise the paths constituting ψ. So ψ is a homomorphism of the copy of H at v_0 to the copy of H at v'. Let N_ψ be the kernel of this homomorphism. Choose a pair of generators, say x', y' of a free subgroup of N_ψ and let x, y be respectively pre-images of x', y'. Now suppose that Alice wants Bob to send her a message. So she sends Bob the elements x and y and Bob sends her a message, that is, a word w in x and y in the given generators of the copy of H at v_0. Alice then computes the image w' of w under the homomorphism ρ_ψ and rewrites it as a word in x', y'. This then is a public key protocol - only Alice is in possession of the random walk ψ which ensures the security of this protocol.

12.4 Diophantine Control Security groups

We discuss next some matrix groups which we will refer to as **Diophantine Security Control Groups**. They can be used to give rise to some additional public key cryptosystems. They are based on the complexity of finding integral solutions of Diophantine equations.

Diophantine equations

Y. Matiyasevich proved that there is no algorithm which determines whether any Diophantine equation has a zero. The objective of this section is to take advantage of this theorem of Matiyasevich to concoct some public key cryptosystems. These cryptosystems make use of some of the properties of the group GL(2, R) of invertible 2×2 matrices over suitably chosen unitary rings. Each of the groups considered here contains a free subgroup F of rank two and therefore a free subgroup freely generated by 256 elements. If we put these elements in a one-to-one correspondence with the 256 ascii characters as usual, then this allows one to uniquely encrypt any message as a word in these generators and hence as a matrix over R. This will allows us to take advantage of two properties of Γ = PSL(2, $\mathbb{Z}$), the group of 2×2 projective integral matrices of determinant 1.

The first is that Γ is finitely presented:

$$\Gamma = \langle s, t; s^2 = (st)^3 = 1 \rangle.$$

The elements s and t can be identified with the matrices

$$s = \begin{pmatrix} 0 & 1 \\ -1 & 0 \end{pmatrix}, \quad t = \begin{pmatrix} 1 & 1 \\ 0 & 1 \end{pmatrix}.$$

Using this description of Γ it is not hard to show, as is well known, that

$$a = \begin{pmatrix} 1 & 2 \\ 0 & 1 \end{pmatrix}, \quad b = \begin{pmatrix} 1 & 0 \\ 2 & 1 \end{pmatrix}$$

freely generate a free subgroup H of Γ. Since a non-abelian free group contains subgroups of every finite rank, H contains a subgroup L of rank 256 which can be put into a one-to-one correspondence with the 256 ascii symbols. This allows us to represent any document or message uniquely as a word in the generators of L, and hence, on multiplying the matrices together, as an integral matrix of determinant 1.

The second observation needed here is that there is an algorithm to rewrite any $T \in$ PSL(2, $\mathbb{Z}$) in terms of s and t. This shows that this protocol for encrypting messages by integral matrices is not secure as it stands. We shall discuss here some variations that make use of the negative solution of Hilbert's tenth problem, as well as the existence of finitely presented groups whose word problems can be made arbitrarily complicated. These variations do appear to lead to protocols which are secure.

Augmented Rings and Specializations of Matrices

We begin this section with a definition.

Definition 12.4.1. An **augmented ring** is a unitary ring R together with a unitary homomorphism

$$\epsilon : R \longrightarrow \mathbb{Z}.$$

So a ring R with a multiplicative identity 1 becomes an augmented ring if it contains an ideal I, the augmentation ideal, and R/I is isomorphic to $\mathbb{Z}$. It follows that we can view $\mathbb{Z}$ as a subring of R and that

$$R = \mathbb{Z} \oplus I.$$

We shall need two such augmented rings, $\mathbb{Z}[x_1, \ldots, x_n]$ the ring of polynomials in any number of variables and $\mathbb{Z}[G]$ the integral group ring of a group G.

The following lemma, and its corollaries, allow one to connect such augmented rings to free groups and then to an application of Hilbert's tenth problem.

Lemma 12.4.2. *Let R and S be unitary rings and let ϕ be a homomorphism from R to S. If $\mathrm{GL}(m, R)$ is the group of all invertible $m \times m$ matrices over R, then ϕ induces a homomorphism ϕ^* of $\mathrm{GL}(m, R)$ into $\mathrm{GL}(m, S)$.*

Corollary 12.4.3. *If R is an augmented ring with augmentation ϵ, then the augmentation ϵ from R to $\mathbb{Z}$ induces a homomorphism from $\mathrm{GL}(n, R)$ to $\mathrm{GL}(n, \mathbb{Z})$.*

Corollary 12.4.4. *Suppose that ϕ is a unitary homomorphism of the unitary ring R into the unitary ring S and that X is a subset of $\mathrm{GL}(m, R)$. If $\phi^*(X)$ freely generates a free subgroup of $\mathrm{GL}(m, S)$, then X freely generates a free subgroup of $\mathrm{GL}(m, R)$.*

Finally, we have the following key consequence of the corollary.

Lemma 12.4.5. *Let R be an augmented ring and $r_1, r_2, r_3, r_4, r_5, r_6, r_7, r_8 \in R$. Furthermore, let*

$$X = \begin{pmatrix} r_1 & r_2 \\ r_3 & r_4 \end{pmatrix} \quad and \quad Y = \begin{pmatrix} r_5 & r_6 \\ r_7 & 8_8 \end{pmatrix}.$$

If $\epsilon(r_1) = 1$, $\epsilon(r_2) = 2$, $\epsilon(r_3) = 0$, $\epsilon(r_4) = 1$, $\epsilon(r_5) = 1$, $\epsilon(r_6) = 0$, $\epsilon(r_7) = 2$, $\epsilon(r_8) = 1$, and if X and Y are invertible, then they freely generate a free group.

Adapting Hilbert's Tenth Problem to Cryptography

As already noted, Yuri Matiyasevich proved that there is no algorithm which determines whether or not an integral polynomial in any number of variables has a zero. It follows that there exist polynomials whose zeros are arbitrarily large in absolute value.

In order to make use of Matiyasevich's theorem consider instead the matrices

$$A = \begin{pmatrix} f_1 & f_2 \\ f_3 & f_4 \end{pmatrix}$$

and

$$B = \begin{pmatrix} f_5 & f_6 \\ f_7 & f_8 \end{pmatrix}.$$

Here the f_i $(i = 1, \ldots, 8)$ are integral polynomials in the commuting variables $x_1, \ldots, x_n$. Now suppose that there is an augmentation ϵ from $k = \mathbb{Z}[x_1, \ldots, x_n]$ onto $\mathbb{Z}$ such that

$$\epsilon(f_1) = 1, \quad \epsilon(f_2) = 2, \quad \epsilon(f_3) = 0, \quad \epsilon(f_4) = 1,$$
$$\epsilon(f_5) = 1, \quad \epsilon(f_6) = 0, \quad \epsilon(f_7) = 2, \quad \epsilon(f_8) = 1.$$

Notice that ϵ induces a homomorphism ϵ^* from $GL(2, k)$ into $GL(2, \mathbb{Z})$ which maps A to a and B to b. So A and B freely generate a free group in $GL(2, k)$. Now let H be a free subgroup of rank 256 of $\langle a, b \rangle$ freely generated by

$$h_1, \ldots, h_{256}$$

and choose $H_i \in GL(2, k)$ to be a pre-image of h_i in $GL(2, k)$. Then

$$K = \langle H_1, \ldots, H_{256} \rangle$$

is a free group of rank 256 and so any message M can be encrypted in H and send across as a matrix of polynomials. To decrypt M, compute its image $\epsilon^*(M)$ under ϵ^*. Now $\epsilon^*(M)$ is an integral matrix and is an element of H. So it can be decomposed as a product

$$p = p(h_1, \ldots, h_{256})$$

of the matrices $h_1, \ldots, h_{256}$. Then the corresponding product

$$P = p(H_1, \ldots, H_{256})$$

translates into the original message M.

Public Key Cryptography and Polynomial Equations

The discussion above translates into a public key cryptographic protocol as follows. Bob encodes a message using X and Y. Alice then uses the secret substitution for the variables to convert the matrices X and Y into a and b respectively. Then she uses the algorithm mentioned previously to decomposes the resultant integral matrix into the unique word in a and b. Replacing a by X and b by Y allows Alice to obtain the message sent by Bob.

We make this precise in a system we call AMC1.

The Cryptosystem AMC1

The discussion above can be extended in several ways. We choose here to take an approach that makes use of the integral group ring $\mathbb{Z}[G]$ of a group G. $\mathbb{Z}[G]$ is an augmented ring with augmentation

$$\epsilon : \mathbb{Z}[G] \longrightarrow \mathbb{Z}$$

defined by

$$\Sigma_{i=1}^{n} a_i g_i \mapsto \Sigma_{i=1}^{n} a_i,$$

where here $a_i \in \mathbb{Z}$, $g_i \in G$. The kernel $I(G)$ of ϵ consists of the elements

$$\alpha = \Sigma_{i=1}^{n} a_i g_i \in \mathbb{Z}[G]$$

with coefficient sum 0, that is,

$$\Sigma_{i=1}^{n} a_i = 0.$$

The point is that, if G is any group, in particular one given by a finite presentation, then the same approach used in working with polynomials in many variables can be carried out using the elements of $\mathbb{Z}[G]$ instead, which appears to have some advantages due to the fact that most problems about groups given by a finite presentation are algorithmically undecidable. A basic observation then that comes out of this approach and which can be used to good effect to encode information in a seemingly secure manner, is the following

Lemma 12.4.6. *Let G be a group and let*

$$A = \begin{pmatrix} \alpha_1 & \alpha_2 \\ \alpha_3 & \alpha_4 \end{pmatrix}$$

and

$$B = \begin{pmatrix} \alpha_5 & \alpha_6 \\ \alpha_7 & \alpha_8 \end{pmatrix}$$

be elements in $GL(2, \mathbb{Z}[G])$. *Further suppose that*

$$\epsilon(\alpha_1) = 1, \quad \epsilon(\alpha_2) = 2, \quad \epsilon(\alpha_3) = 0, \quad \epsilon(\alpha_4) = 1,$$
$$\epsilon(\alpha_5) = 1, \quad \epsilon(\alpha_6) = 0, \quad \epsilon(\alpha_7) = 2, \quad \epsilon(\alpha_8) = 1.$$

Notice that ϵ induces a homomorphism ϵ^* from $GL(2, \mathbb{Z}[G])$ *into* $GL(2, \mathbb{Z})$ *which maps A to a and B to b, where again*

$$a = \begin{pmatrix} 1 & 2 \\ 0 & 1 \end{pmatrix}, \quad b = \begin{pmatrix} 1 & 0 \\ 2 & 1 \end{pmatrix}.$$

Then A and B freely generate a free group in $GL(2, \mathbb{Z}[G])$.

The following corollary is an immediate consequence of this lemma.

Corollary 12.4.7. *Let g and h be elements of the group G. Then the matrices*

$$A = \begin{pmatrix} 1 & g + 1 \\ 0 & 1 \end{pmatrix}$$

and

$$B = \begin{pmatrix} 1 & 0 \\ h + 1 & 1 \end{pmatrix}$$

freely generate a free subgroup of $GL(2, \mathbb{Z}[G])$.

Let G be a torsion-free group given by a finite presentation on the finite set S. Let $R = \mathbb{Z}[y]$ be the group ring of the group G over the ring of integral polynomials in a single variable y. Then we have the following simple lemma:

Lemma 12.4.8. *Let $g \in G$. Then the element $x = y(1 - g)$ is transcendental, that is, is not the root of an integral polynomial $f(x)$ if $g \neq 1$.*

The subring Λ of R, generated by y and g, is simply a polynomial ring over $\mathbb{Z}$, in the variables y and g. The assertion in the lemma follows from this observation.

   Now let w be a word in the generators S of G. We now introduce two matrices $A(w)$ and $B(w)$ in $\mathrm{GL}(2, R)$, the group of 2×2 matrices over R:

$$A(w) = \begin{pmatrix} 1 & x(w-1) \\ 0 & 1 \end{pmatrix}, \quad B(w) = \begin{pmatrix} 1 & 0 \\ x(w-1) & 1 \end{pmatrix}.$$

Observe that $A(w)$ and $B(w)$ freely generate a free group if and only if $w \neq 1$. Now form 256 words W_i in $A(w)$ and $B(w)$, for instance the conjugates of $B(w)$ by $A(w)^i$ with i ranging from 1 to 256. As usual a message M can be encoded as a word in the W_i and thence as a matrix $M(W)$ in $\mathrm{GL}(2, R)$. This matrix can be decoded in what is now the usual way since it can be viewed as a matrix over an augmented ring. Here the augmented ring is a polynomial ring in two variables over $\mathbb{Z}$. Now decoding the message requires determining whether or not $w = 1$ in S. This leads to an array of protocols the decoding of which can be made arbitrarily hard and indeed if G is a group with a complicated word problem, decoding of the messages becomes arbitrarily complicated unless $w = 1$.

12.5 The Social Security Control Groups

The Social Security Number (SSN) was introduced in 1936 to be used solely for Social Security programs. It is used for driver's licences, credit cards, banking accounts, employee files, medical records and a host of electronic transactions. Seemingly little has been done to safeguard the SSN. One of the consequences is that identity theft has increased. Computer usage also carries with it problems, with the associated login procedures protected by user-selected passwords, which are poorly chosen such as names of pets or loved ones. Many commercial electronic transactions, such as credit card purchases, banking and so on are usually protected by powerful cryptographic protocols. These include algorithms such as RSA, DES and AES. The security of such algorithms often rests upon the difficulty involved in carrying out certain computations that involve finite collections of numbers whose choice is dictated by the existence of very large primes. These finite collections can be made into finite abelian groups. It is the complexity of these finite abelian groups that make such protocols secure, although this has never been proved. Recently Agrawal, Kayal and Saxena (see [AKS]) showed that it is possible to determine whether a number is prime in a relatively uncomplicated way (technically speaking, they obtained a polynomial-time algorithm to determine primality). This raises the possibility that protocols that rest on products of two very large primes, the basis for the thinking that the use of finite abelian groups leads to secure public key cryptosystems, may indeed not be the case. Quantum computers appear also to be a threat. This raises concerns on the reliance on, on the one hand, a single number, the SSN and on the other, the complexity of a finite structure, a finite abelian group, for identification, for the safekeeping of records and for secure electronic communication. As we have already seen, finite abelian groups are very special instances of finitely presented groups, which we have already discussed.

These finitely presented groups are usually infinite. They have been intensively studied for more than 150 years and many questions about them have been proved to be very difficult, indeed at times impossible, to answer. This suggests that finitely presented groups can be made the basis of challenge-response protocols which lead naturally to a new means of identification. This suggests the possibility of replacing the SSN by a finitely presented group, here termed a **Social Security Control Group** or **SSCG**. One can extend the use of such finitely presented groups to, for example, assigning **Bank Security Control Groups** (BSCGs) to banks, giving rise to an entire family of groups, collectively referred to as **XSCGs**. The extraordinary complexity of these groups leads to new methods of encryption and by using different aspects of them making their use more secure.

We have already discussed the possible uses of finitely presented groups. The possibility of such usage touches on almost every aspect of electronic communication and identification. Not only is there a possibility of the SSCGs providing a more secure means of identification which can help to prevent identity theft, they might be useful also as universal passwords in a variety of other ways. We will discuss this further in this chapter. They can also be used to help make various hand-held such as mobile phones more secure. In addition, the algebraic, mathematical nature of an XSCG makes it possible to encode information inside the group itself, resulting in a compartmentalization of information and a means of verifying that their use is legitimate. In this way different parties, such as credit card companies, banks, HMOs and merchants, will have access only to specific, small areas of each XSG, set aside for each of them. This will enhance security and help to prevent theft. Even if one part of an XSCG is compromised, the entire group remains outside the reach of an attacker because of this compartmentalization.

It must be pointed out that buffer overflows often make it possible for successful attacks on machines servicing the electronic highway. This problem has begun to be handled and will not be discussed here.

The Complexity of Finitely Presented Groups

As we have seen throughout this book, cryptographic protocols, especially public key protocols, are based on hard to solve problems. This becomes especially relevant in discussing finitely presented groups where even what might seem like straightforward questions in group theory frequently turn out to be algorithmically unsolvable.

We give some examples, some of which we have examined earlier.

(1) There is no algorithm which decides whether or not any pair of finitely presented groups are isomorphic. We saw this as the **isomorphism problem**.

(2) There is no algorithm that decides, for example, whether or not a finitely presented group is trivial, finite or abelian.

(3) There is no algorithm that decides, for example, whether or not in a given finitely presented group two elements are conjugate. This is known as the **conjugacy problem** and was used in conjunction with the Ko-Lee encryption protocol.

(4) There are finitely presented groups such that there is no algorithm that decides whether or not any word in the generators of the group is equal to the identity. This was known as the **word problem**.

(5) There are finitely presented groups G such that there is no algorithm that can decide whether or not a given finite subset of G generates G (see Miller [Mil] for a discussion of the algorithmic unsolvability of many questions about groups and an explanation, if needed as to the terms used here).

Each finitely presented group G comes with a description which takes the form (see Chapter 9)

$$G = \langle a_1, \ldots, a_m; r_1 = \cdots = r_n = 1 \rangle.$$

This notation is shorthand for expressing G as a factor group of the free group F on $a_1, \ldots, a_m$ by the smallest normal subgroup N of F containing the elements $r_1, \ldots, r_n$. In other words G is given in terms of the isomorphism

$$G \cong F/N.$$

Notice that N consists of all products of the elements of $R = \{r_1, \ldots, r_n\}$, their inverses and their conjugates. N is often denoted by writing $N = \langle\langle R \rangle\rangle_F$ and as such is referred to as the normal closure of R. As already noted finitely presented groups are described by a finite amount of data. This data then can hide an immense amount of information. It is this facet of finitely presented groups that will be taken advantage of here. To give one further illustration of how this comes into play, if a word $w = x_1 \cdots x_n$ is a product of the a_j and their inverses in which consecutive terms are not inverses and if $w \in N$, then it can be expressed as a product of conjugates $f^{-1}rf$ of the elements $r \in R$ and their inverses. If one now considers all such words w of length at most n in N, then the minimum number of conjugates required to express them gives rise to a function $\phi(n)$, termed the **isoperimetric function** of the group $G = F/N$. It provides, in a sense, a measure of the way in which N is contained in F. It is possible to show that these functions can be arbitrarily complicated (see Birget, Olshanskii, Rips and Sapir [BORS]). Indeed they need not even be computable. This property can be used to concoct some challenge responses which give rise to functions which can be used as the basis for various cryptographic protocols. However we will not develop this further. We note that the work of Kapovich, Myasnikov, Schupp and Shpilrain (see [MSU1]) on the average-case complexity of questions in group theory indicates that if we are to take advantage of the complexity of finitely presented groups by asking questions about their elements, their subgroups and the homomorphisms between them, care must be taken to choose such questions carefully.

Security Control Groups

Group theory lends itself to the design of secure web-based transactions. The general idea is to use group-theoretical objects, their properties, and questions about them, as the basis for a challenge response authentication protocol leading to secure electronic communication and transactions. The questions that must be answered before communication is allowed, generally need some group-theoretical software packages such as GAP, MAGMA or MAGNUS to provide answers to such questions. Here we return to the idea of using a finitely presented group, which we term in the case of a replacement of the Social Security Number, a **Social Security Control Group**. This can be used not only as a means of identification, but as a flexible storage facility. It can be used as a template for other security groups. The construction of the Access Control needs to be kept in mind as some its features can be taken advantage of here. This SSCG is typical of the security control groups that we are going to use in discussing one of the avenues to follow in applying group theory to cryptography.

A Social Security Control Group

Although it is not clear how to choose a SSCG, we will discuss a possible choice here and indicate how it has the potential of satisfying many of the criteria needed if it is to be used as a replacement for the Social Security Number.

To this end, let X be a person who already has a social security number. Further suppose that the social security number for X is $v_X = n_1 \cdots n_9$, where the digits n_i lie between 0 and 9. From this we will assign to X a SSCG $\sigma(X) = \sigma(v_X)$ which we will define in the course of the discussion that follows. Such an SSCG can be adapted for use in a variety of other ways and we will discuss this later in this chapter. A key subgroup of this selection of an SSCG is a group constructed many years ago by Graham Higman and Bernhard Neumann and communicated to us by Chuck Miller.

A Finitely Presented Group Γ that Contains a Copy of Every Finite Group

The following construction gives an embedding of the group Φ of finitary permutations on a countable set into a finitely presented group Γ. For group theoretical information in this section we refer to [Rot].

Let $\Omega = \mathbb{Z} \cup \{\infty\}$ and let $t_i = (i \, \infty)$ be the transposition which interchanges i and ∞. These t_i in fact generate the group Φ of all permutations of Ω with finite support. Moreover Φ can be presented on these generators by the defining relations:

$$t_i^2 = (t_i t_j)^3 = (t_i t_j t_i t_k)^2 = 1,$$

for all i, j, k which are all pairwise different. Note that Φ is locally finite and contains a copy of every finite group.

Φ can be embedded in a finitely generated group which is the HNN-extension obtained by introducing a new generator s to Φ together with the defining relations:

$$s^{-1} t_i s = t_{i+1} .$$

Viewing the elements t_i as an infinite sequence

$$\ldots, t_{-4}, t_{-3}, t_{-2}, t_{-1}, t_0, t_1, t_2, t_3, t_4, t_5, \ldots$$

one can also consider the group generated by the subsequence of the t_i's obtained by leaving out every fourth generator:

$$\ldots, t_{-5}, t_{-3}, t_{-2}, t_{-1}, t_1, t_2, t_3, t_5, \ldots .$$

This deleted sequence in fact generates a group isomorphic to that generated by the entire sequence.

The sequence of all t_i splits into 3 orbits under the action of s^3 while the subsequence obtained by omitting every fourth t_i splits into 3 orbits under s^4. So form a further HNN-extension Γ obtained by introducing a further generator u together with the defining relations:

$$u^{-1} s^3 u = s^4,$$

$$u^{-1} t_{3i+e} u = t_{4i+e}, \quad e = 1, 2, 3 .$$

The resulting group Γ is generated by u, s, t_0, t_1, t_2, t_3 and, as will be shown below, can be presented using only finitely many of the relations involving the t_i because conjugation by the element u expands small indices to larger indices.

An explicit finite presentation of this group Γ on a convenient set of generators is as follows:

Generators:

$$u, s, t_0, t_1, t_2, t_3.$$

Defining relations:

$$s^{-1}t_i s = t_{i+1} \quad \text{for } i = 0, 1, 2$$
$$u^{-1}s^3 u = s^4$$
$$u^{-1}t_i u = t_i \quad \text{for } i = 1, 2, 3$$
$$t_0^2 = 1$$
$$(t_i t_j)^3 = 1 \quad \text{for } i, j = 0, 1, 2, 3 \text{ with } i \neq j$$
$$(t_i t_j t_i t_k)^2 = 1 \quad \text{for } i, j, k = 0, 1, 2, 3 \text{ all pairwise different}$$

The result we establish is the following:

Lemma 12.5.1. *The map* $t_i \mapsto s^{-i}t_0 s^i$ *embeds* Φ *into the finitely presented group* Γ.

In view of the above discussion it only remains to see that if we introduce the abbreviation $t_i = s^{-i}t_0 s^i$ (which is compatible with the first set of defining of relations of Γ), then all of the previous relations follow from these.

First, the following calculation shows that the relations between u and the t_i's hold: $u^{-1}t_{3i+e}u = u^{-1}s^{-3i}t_e s^{3i}u = s^{-4i}u^{-1}t_e us^{4i} = s^{-4i}t_e s^{4i} = t_{4i+e}$ for $e = 1, 2, 3$. Now $t_i^2 = 1$ for all i since $t_0^2 = 1$ so the first collection of t_i relations hold.

In the calculations that follow it will be convenient to use the notation $y \sim_x z$ to mean that y is conjugate to z by the element x, that is $x^{-1}yx = z$.

Next we prove by induction that $(t_0 t_j)^3 = 1$ for all $j \geq 1$. For $1 \leq j \leq 3$ these are among the given relations for G. So suppose inductively these relations have been deduced for indices smaller than j. If $j = 4m + d$ where $m \geq 1$ and $0 \leq d \leq 2$ then in G we have

$$(t_0 t_j)^3 = (t_0 t_{4m+d})^3 \sim_s (t_1 t_{4m+d+1})^3 \sim_{u^{-1}} (t_1 t_{3m+d+1})^3 \sim_s (t_0 t_{3m+d})^3 = 1,$$

where the last equality follows from the induction hypothesis. If $j = 4m + 3$ where $m \geq 1$ then in G

$$(t_0 t_j)^3 = (t_0 t_{4m+3})^3 \sim_{s^2} (t_2 t_{4m+5})^3 \sim_{u^{-1}} (t_2 t_{3m+4})^3 \sim_{s^{-2}} (t_0 t_{3m+2})^3 = 1$$

where the last equality follows from the induction hypothesis. By induction this establishes these relations for all j. But now by conjugating by a suitable power of s and taking a cyclic permutation if required it follows that $(t_i t_j)^3 = 1$ for all $i \neq j$.

Lastly we must show $(t_i t_j t_i t_k)^2 = 1$ for all i, j, k all pairwise different. Consider the case $i < j$ and $i < k$. For this case, after conjugating by s^{-i} and relabeling it suffices to show that $(t_0 t_j t_0 t_k)^2 = 1$ for all positive $j \neq k$. This we do by induction on the sum $j + k$. For $1 \leq j, k \leq 3$ the desired relations are among the given relations for G. Inductively assume the desired relations have been established for indices with smaller sum. Now we may suppose that either $j \geq 4$ or $k \geq 4$ and write $j = 4m + d, k = 4n + e$ where $0 \leq d, e \leq 3$. Notice that either $m > 0$ or $n > 0$.

In case $0 \le d, e < 3$ we have the following computation in G:

$$(t_0 t_j t_0 t_k)^2 = (t_0 t_{4m+d} t_0 t_{4n+e})^2 \sim_s (t_1 t_{4m+d+1} t_1 t_{4n+e+1})^2$$
$$\sim_{u^{-1}} (t_1 t_{3m+d+1} t_1 t_{3n+e+1})^2 \sim_{s^{-1}} (t_0 t_{3m+d} t_0 t_{3n+e})^2 = 1$$

by the induction hypothesis. Next consider the case $e = 3$. If $d = 0$ or $d = 1$ then

$$(t_0 t_j t_0 t_k)^2 = (t_0 t_{4m+d} t_0 t_{4n+3})^2 \sim_{s^2} (t_2 t_{4m+d+2} t_2 t_{4n+5})^2 \sim_{u^{-1}} (t_2 t_{3m+d+2} t_2 t_{3n+4})^2$$
$$\sim_{s^{-2}} (t_0 t_{3m+d} t_0 t_{3n+2})^2 = 1$$

by the induction hypothesis. If $d = 2$ or $d = 3$ then

$$(t_0 t_j t_0 t_k)^2 = (t_0 t_{4m+d} t_0 t_{4n+3})^2 \sim_{s^3} (t_3 t_{4m+d+3} t_3 t_{4n+6})^2 \sim_{u^{-1}} (t_3 t_{3m+d+2} t_3 t_{3n+5})^2$$
$$\sim_{s^{-3}} (t_0 t_{3m+d-1} t_0 t_{3n+2})^2 = 1$$

by the induction hypothesis.

The case $d = 3$ and $0 \le e \le 3$ follows by the analogous argument.

Finally, the cases in which j (respectively k) is the smallest of the three indices are dealt with by an entirely analogous argument.

A possible candidate for a Social Security Control Group

We now form the direct product

$$\sigma(X) = \gamma \times F = \langle x, y; \rangle \wr \langle u, v; v^{-1} u^2 v = u^{v(X)} \rangle.$$

So $\langle x, y; \rangle$ is the free group on x and y and $\langle u, v; v^{-1} u^2 v = u^{v(X)} \rangle$ is the so-called group $BS(2, v(X))$ groups. The Social Security Administration has a copy of $\sigma(X)$ and the software needed to answer questions about $\sigma(X)$. When an entity attempts to impersonate X, the SSAA sends questions to the impersonator I to be answered. In addition previous communications with the person I are recorded in the file located in the subgroup $F^{\sigma(I)}$ and these answers have to be answered by I. This makes it impossible for I to access any records off X. If there is a break in of the system permissions are easily altered in accordance with the means provided by the Access Control Group. We shall not go in to the questions that may be asked by an imposter, but the very nature of the SSCG makes such questions easy to generate and easy to answer given enough information.

12.6 Further Extensions of Diffie-Hellman and RSA

An XSG-version of Diffie-Hellman

The XSGs that we have in mind all contain a variety of subgroups which lie partly in the infinitary symmetric group and partly outside. In particular they all contain a finitely generated non-trivial subgroup C and a second finitely generated abelian subgroup D, given by finite presentations generators s, t, u, v and a free abelian subgroup A of infinite rank. We will think of these subgroups as public. However the structure of the groups that they generate is hidden inside the XSG's that contain them. Alice initiates contact with Bob with the aid of the challenge-response protocol discussed previously. She then selects a non-trivial word $w = w(s, t)$ in s and t. Then she randomly chooses $h_1, h_2, a \in A$ and sends Bob $h_1 w h_2$.

RSA in an appropriate ring

Let h be a large integer and let Λ be the quotient ring of the power series ring over the integers in the non-commuting variables $x_1, \ldots, x_m$ obtained on adding the relations $x_i^h = 0$ for $j = 1, \ldots, m$. The power series ring, without any of these relations, is public like the product $n = pq$ above, which defines the ring $\mathbb{Z}_n$ of integers, modulo n. The integer h is kept secret and can be viewed as the counterpart to the factorization of n in the description of RSA. The counterpart to e is an element of Λ chosen so that it has a multiplicative inverse in Λ. This inverse cannot be found without knowledge of h, which is chosen very large. Plaintext is encoded as a sum t of monomials in the x_j and sent out as te. The recipient can decode the message by computing $(te)e^{-1} = t$. The strength of RSA lies in the difficulty of finding inverses which in turn relies upon integer factorization. In these ring-theoretic settings, finding inverses is even more difficult particularly when non-commuting variables are involved.

12.7 Exercises

12.1 Let M be the modular group $PSL(2, \mathbb{Z})$. This has a group presentation

$$M = \langle x, y; x^2 = y^3 = 1 \rangle = \langle a, t; a^2 = (at)^3 = 1 \rangle$$

where

$$x = \pm \begin{pmatrix} 0 & 1 \\ -1 & 0 \end{pmatrix}, \quad y = \pm \begin{pmatrix} 0 & 1 \\ -1 & 1 \end{pmatrix}.$$

Let

$$a = x \quad \text{and} \quad t = xy = \pm \begin{pmatrix} 1 & 1 \\ 0 & 1 \end{pmatrix}.$$

We show that there is an algorithm based on the Euclidean algorithm which given a projective integral matrix T can rewrite it in terms of the standard generators x, y. Let

$$T = \pm \begin{pmatrix} \alpha & \beta \\ \gamma & \delta \end{pmatrix}$$

be a matrix in M.

(a) Show that

$$aT = \pm \begin{pmatrix} \gamma & \delta \\ -\alpha & -\beta \end{pmatrix} = \pm \begin{pmatrix} -\gamma & -\delta \\ \alpha & \beta \end{pmatrix}$$

and

$$t^k T = \pm \begin{pmatrix} \alpha + k\gamma & \beta + k\delta \\ \gamma & \delta \end{pmatrix}.$$

(b) Show that the following algorithm then expresses the matrix T as word $W(a, t)$. Then using that $t = xy$ we get T expressed as a word $W_1(x, y)$ in the standard generators x, y.

Multiplying on the left by a and then by an appropriate t^k we can make the lower left entry positive and smaller. Continuing we get eventually

$$t^{k_1} a t^{k_1} a \dots t^{k_n} a^\epsilon T = \pm \begin{pmatrix} \mu & \nu \\ 0 & \eta \end{pmatrix}$$

where $\epsilon = 1$ or 0. Since $\mu\eta = 1$ this implies that the right hand side

$$\pm \begin{pmatrix} \mu & \nu \\ 0 & \eta \end{pmatrix} = \pm \begin{pmatrix} 1 & \rho \\ 0 & 1 \end{pmatrix} = t^\rho.$$

12.2 Let $T = \pm \left(\begin{smallmatrix} 5 & 8 \\ 3 & 5 \end{smallmatrix} \right)$. Use the algorithm from the previous problem to express T as a word in the generators x, y.

12.3 Use the presentation of the modular group $M = \langle x, y; x^2 = y^3 = 1 \rangle$ to show that any element of order 2 must be conjugate within M to x and any element of order 3 must be conjugate within M to either y or y^2.

12.4 Let $A = \pm \left(\begin{smallmatrix} a & b \\ c & d \end{smallmatrix} \right)$ be an element of order 2 in the modular group PSL$(2, \mathbb{Z})$. Show that

(a) $a + d = 0$,

(b) A is conjugate in PSL$(2, \mathbb{Z})$ to the element $S = \pm \left(\begin{smallmatrix} 0 & 1 \\ -1 & 0 \end{smallmatrix} \right)$.

12.5 Let $n \in \mathbb{N}$. Show that $n = x^2 + y^2$ for some $x, y \in \mathbb{Z}$ if -1 is a quadratic residue modulo n. (Hint: Consider elements of order two and hence trace zero in the modular group.)

12.6 Let $n \in \mathbb{N}$. Show that if $n = x^2 + y^2$ for some $x, y \in \mathbb{Z}$ then -1 is a quadratic residue modulo n. (Hint: Again consider elements of order two and hence trace zero in the modular group.)

12.7 Show that the monoid SL$(2, \mathbb{N})$ is freely generated by the two matrices

$$T = \begin{pmatrix} 1 & 1 \\ 0 & 1 \end{pmatrix} \quad \text{and} \quad U = \begin{pmatrix} 1 & 0 \\ 1 & 1 \end{pmatrix}.$$

12.8 Let $W = \left(\begin{smallmatrix} a_1 & a_2 \\ a_3 & a_4 \end{smallmatrix} \right) \in$ SL$(2, \mathbb{N})$. From the previous problem, W can be expressed as a word in T, U. Let $\ell = \ell(w)$ be the length of W as a word in T and U. Show that $a_i \leq f_{\ell+1}$ where $f_{\ell+1}$ is the $\ell + 1^{\text{st}}$ Fibonacci number. As usual $f_0 = 0, f_1 = 1$ and $f_{n+1} = f_n + f_{n-1}$ for $n \geq 1$.

12.9 The **lamplighter group** L is given by the presentation

$$L = \langle a, t; (at^u at^{-u})^2 = 1 \rangle$$

where $n \in \mathbb{Z}$. Show that L is the restricted wreath product $\mathbb{Z}_2 \wr \mathbb{Z}$.

12.10 Prove Lemma 12.4.5.

12.11 Show that $\pm \left(\begin{smallmatrix} 1 & 2 \\ 0 & 1 \end{smallmatrix} \right)$ and $\pm \left(\begin{smallmatrix} 1 & 0 \\ 2 & 1 \end{smallmatrix} \right)$ freely generate a free subgroup of $\mathrm{PSL}(2, \mathbb{Z})$.

13 Commutative Gröbner Basis Methods

13.1 Commutative Gröbner Bases

Besides numbers and group elements, polynomials are natural objects to use in algebraic cryptography. Recall that a **polynomial** f is a formal linear combination $f = c_1 t_1 + \cdots + c_s t_s$ whose **coefficients** c_i are taken in a field K and where the elements t_i are **terms**, that is power products $t_i = x_1^{\alpha_{i1}} \cdots x_n^{\alpha_{in}}$ of **indeterminates** x_i. The set of all polynomials of this form is called the **polynomial ring** $P = K[x_1, \ldots, x_n]$. This set is a commutative ring with identity. In other words, polynomials can be added and multiplied in a natural way and the usual rules such as the associative and distributive laws hold.

Example 13.1.1. Using the base field $K = \mathbb{Q}$ and the indeterminates x, y, z, we can add the polynomials $f = xy + yz$ and $g = x^2 + xy$ and obtain $f + g = x^2 + 2xy + yz$. Their product equals $f \cdot g = x^3 y + x^2 y^2 + x^2 yz + xy^2 z$. Notice that we can also factor f and g in the form $f = y(x + z)$ and $g = x(x + y)$.

For cryptographic applications, we need to work effectively with polynomials. The main tool for effective calculations with polynomials and polynomial ideals are Gröbner bases. The first task we have to solve is to make equality of polynomials effective and to store them efficiently. How is it possible to detect in the preceding example that the polynomial $xy^2 z + x^2 y^2 + x^3 y + x^2 yz$ equals $f \cdot g$? Clearly, we need to order the terms in a unique way.

Definition 13.1.2. Let $x_1, \ldots, x_n$ be indeterminates, and let

$$\mathbb{T}^n = \{x_1^{\alpha_1} \cdots x_n^{\alpha_n} \mid \alpha_i \geq 0\}$$

be the set of terms in these indeterminates. A complete ordering relation σ is called a **term ordering** on $\mathbb{T}^n$ if the following conditions are satisfied.
(1) The relation σ is **multiplicative**, that is, if two terms $t_1, t_2 \in \mathbb{T}^n$ satisfy $t_1 \leq_\sigma t_2$ and if $t_3 \in \mathbb{T}^n$ then we have $t_1 t_3 \leq_\sigma t_2 t_3$.
(2) The relation σ is a **well-ordering**, that is, every decending chain $t_1 \geq_\sigma t_2 \geq_\sigma \cdots$ in $\mathbb{T}^n$ is eventually stationary.

It is easy to see that the second condition is equivalent to requiring $t \geq_\sigma 1$ for all $t \in \mathbb{T}^n$. There exist plenty of term orderings, and many of them have special useful properties. The following example introduces two well-known term orderings.

Example 13.1.3. Let $\mathbb{T}^n$ be the set of terms in the indeterminates $x_1, \ldots, x_n$.
(1) For $t = x_1^{\alpha_1} \cdots x_n^{\alpha_n}$ and $t' = x_1^{\beta_1} \cdots x_n^{\beta_n}$, we let $t \leq_{\text{lex}} t'$ if and only if one of the following conditions holds:

- $\alpha_1 < \beta_1$
- $\alpha_1 = \beta_1$ and $\alpha_2 < \beta_2$

$\vdots$

- $\alpha_1 = \beta_1, \dots, \alpha_{n-1} = \beta_{n-1}$ and $\alpha_n < \beta_n$
- $t = t'$

It is easy to check that this defines a term ordering $\leq_{\text{lex}}$ on $\mathbb{T}^n$. It is called the **lexicographic term ordering**.

(2) For practical computations, the most well-known and useful term ordering is the **degree reverse lexicographic term ordering** $\leq_{\text{drl}}$. It is defined by letting $t \leq_{\text{drl}} t'$ for $t = x_1^{\alpha_1} \cdots x_n^{\alpha_n}$ and $t' = x_1^{\beta_1} \cdots x_n^{\beta_n}$ if and only if $t = t'$ or one of the following conditions is satisfied:

- $\deg(t) = \alpha_1 + \cdots + \alpha_n < \deg(t') = \beta_1 + \cdots + \beta_n$
- $\deg(t) = \deg(t')$ and $\alpha_n > \beta_n$
- $\deg(t) = \deg(t')$, $\alpha_n = \beta_n$ and $\alpha_{n-1} > \beta_{n-1}$

$\vdots$

- $\deg(t) = \deg(t')$, $\alpha_n = \beta_n, \dots, \alpha_3 = \beta_3$ and $\alpha_2 > \beta_2$

It is straightforward to verify that these conditions indeed define a term ordering.

Given a term ordering σ, there is a natural way to represent a non-zero polynomial $f \in P$, namely to write $f = c_1 t_1 + \cdots + c_s t_s$ with $c_i \in K$ and terms $t_i \in \mathbb{T}^n$ satisfying $t_1 >_\sigma \cdots >_\sigma t_s$. Using these representations, we can now define several notions which are central to Gröbner basis theory.

Definition 13.1.4. Let σ be a term ordering on $\mathbb{T}^n$, and let $f = c_1 t_1 + \cdots + c_s t_s \in P \setminus \{0\}$ be represented as above.

(1) The term $\text{Lt}(f) = t_1$ is called the **leading term** of f with respect to σ.
(2) The field element $\text{Lc}(f) = c_1$ is called the **leading coefficient** of f with respect to σ.
(3) The polynomial $\text{Lm}(f) = c_1 t_1$ is called the **leading monomial** of f with respect to σ.

For instance, the polynomial $f = x^2 + xy + y^3$ satisfies $\text{Lt}_{\text{lex}}(f) = x^2$ and $\text{Lt}_{\text{drl}}(f) = y^3$. These definitions can be extended to polynomial ideals. A set of polynomials I is called an **ideal** in P if it is a subgroup with respect to addition and if it has the property that $f \in I$ and $g \in P$ imply $fg \in I$. Important examples of ideals are obtained as follows.

Example 13.1.5. Let I be a subset of the polynomial ring P.

(1) If there exists a polynomial $f \in P$ such that $I = \{fg \mid g \in P\}$ then I is called the **principal ideal** generated by f and is denoted by $I = \langle f \rangle$.
(2) More generally, if there exist polynomials $f_1, \dots, f_s \in P$ such that $I = \{f_1 g_1 + \cdots + f_s g_s \mid g_i \in P\}$ then I is called the ideal **generated by** $f_1, \dots, f_s$ and is denoted by $I = \langle f_1, \dots, f_s \rangle$.

As an application of Gröbner basis theory we shall see that, in fact, every ideal in P is of the type described in the preceding example. But first we generalize the notion of a leading term as follows.

Definition 13.1.6. Let σ be a term ordering on $\mathbb{T}^n$, and let I be an ideal in P. Then the ideal

$$
\begin{aligned}
\mathrm{Lt}_\sigma(I) &= \langle \mathrm{Lt}_\sigma(f) \mid f \in I \rangle \\
&= \{f_1 g_1 + \cdots + f_s g_s \mid s \geq 0, \ f_i \in I, \ g_i \in P\}
\end{aligned}
$$

is called the **leading term ideal** of I with respect to σ.

At this point we encounter a problem which is one of the main incentives to introduce Gröbner bases, namely the fact that the leading terms of a set of generating polynomials need not generate the leading term ideal of an ideal. The following example is a case in point.

Example 13.1.7. Let $P = K[x, y]$, let $\sigma = \mathtt{drl}$, and let $I = \langle f_1, f_2 \rangle$ be the polynomial ideal generated by $f_1 = x^2 - 1$ and $f_2 = xy - 1$. Then we have $x - y = yf_1 - xf_2 \in I$ and therefore $x = \mathrm{Lt}_\sigma(x - y) \in \mathrm{Lt}_\sigma(I)$. Thus the inclusion $\langle \mathrm{Lt}_\sigma(f_1), \mathrm{Lt}_\sigma(f_2) \rangle \subset \mathrm{Lt}_\sigma(I)$ is strict.

This example leads us to the following definition.

Definition 13.1.8. Let σ be a term ordering on $\mathbb{T}^n$, let $I \subseteq P$ be a polynomial ideal, and let $G \subset I$ be a subset of I. The set G is called a σ-**Gröbner basis** of I if $\mathrm{Lt}_\sigma(I) = \langle \mathrm{Lt}_\sigma(g) \mid g \in G \rangle$, that is if the leading term ideal of I is generated by the leading terms of the polynomials in G.

For instance, the polynomials in the preceding example can be extended to a Gröbner basis of I as follows.

Example 13.1.9. In the setting of the preceding example, let $f_3 = x - y \in I$ and $f_4 = f_2 - yf_3 = y^2 - 1 \in I$. Then the set $G = \{f_1, f_2, f_3, f_4\}$ satisfies $\langle \mathrm{Lt}_\sigma(f_1), \ldots, \mathrm{Lt}_\sigma(f_4) \rangle = \langle x^2, xy, x, y^2 \rangle = \langle x, y^2 \rangle$. Using f_3 and f_4, it is not difficult to show that no polynomial of the form $c_1 y + c_2$ with non-zero constants c_1 or c_2 is contained in I. Hence we have $\mathrm{Lt}_\sigma(I) = \langle x, y^2 \rangle$ and it follows that G is a σ-Gröbner basis of I.

In general, every finite set of polynomials generating an ideal I can be extended to a finite Gröbner basis of I. This is achieved by Buchberger's Algorithm. To introduce this algorithm, we first have to describe the multivariate version of polynomial division.

Proposition 13.1.10 (The Division Algorithm). *Let $P = K[x_1, \ldots, x_n]$, let σ be a term ordering on $\mathbb{T}^n$, and let $f, g_1, \ldots, g_s \in P$. Consider the following instructions.*
(1) *Let $q_1 = \cdots = q_s = r = 0$ and $h = f$.*
(2) *Repeat the following steps until $h = 0$. Then return $(q_1, \ldots, q_s)$ and r and stop.*
(3) *Repeat the following step as often as possible.*

(4) *Find the smallest $i \in \{1,\ldots,s\}$ such that $\mathrm{Lt}_\sigma(h)$ is a multiple of $\mathrm{Lt}_\sigma(g_i)$. If such an i exists, replace q_i by $q_i + \mathrm{Lm}_\sigma(h)/\mathrm{Lm}_\sigma(g_i)$ and h by $h - (\mathrm{Lm}_\sigma(h)/\mathrm{Lm}_\sigma(g_i))\,g_i$.*

(5) *Replace r by $r + \mathrm{Lm}_\sigma(h)$ and h by $h - \mathrm{Lm}_\sigma(h)$.*

This is an algorithm which computes $(q_1,\ldots,q_s) \in P^s$ and $r \in P$ such that $f = q_1 g_1 + \cdots + q_s g_s + r$ and such that the following conditions hold.

(1) *No term of r is divisible by any of the terms in $\{\mathrm{Lt}_\sigma(g_1),\ldots,\mathrm{Lt}_\sigma(g_s)\}$.*

(2) *For all i such that $q_i \neq 0$ we have $\mathrm{Lt}_\sigma(q_i g_i) \leq_\sigma \mathrm{Lt}_\sigma(f)$.*

Proof. Since the leading term of h becomes smaller with respect to σ during each execution of (4) or (5), and since σ is a term ordering, the algorithm is finite. Next we note that the equality $f = q_1 g_1 + \cdots + q_s g_s + h + r$ holds throughout the execution of the algorithm. Since the algorithm stops when it reaches $h = 0$, we obtain the desired equality. The fact that the two additional conditions hold follows immediately from the descriptions of steps (4) and (5). $\qquad\square$

The division algorithm produces a polynomial $\mathrm{NR}_{\sigma,G}(f) = r$ which depends on the order of the elements in $G = (g_1,\ldots,g_s)$. This polynomial is called the **normal remainder** of f with respect to G. Besides the Division Algorithm, we need one further ingredient for Buchberger's Algorithm.

Definition 13.1.11. Let σ be a term ordering, let $f_i, f_j \in P \setminus \{0\}$, and let $\mathrm{Lm}_\sigma(f_k) = c_k t_k$ with $c_k \in K$ and $t_k \in \mathbb{T}^n$ for $k \in \{i,j\}$. Then we let $t_{ij} = \mathrm{lcm}(t_i, t_j)/t_i$ and $t_{ji} = \mathrm{lcm}(t_i, t_j)/t_j$ and call the polynomial $S_{ij} = \frac{1}{c_i} t_{ij} f_i - \frac{1}{c_j} t_{ji} f_j$ the **S-polynomial** of f_i and f_j.

The idea behind the definition of S_{ij} is that both $t_{ij} f_i$ and $t_{ji} f_j$ have the same leading term $\mathrm{lcm}(t_i, t_j)$ and that this term cancels in S_{ij}. Therefore the polynomial S_{ij} has a chance to produce a "new" leading term, and this is used in Buchberger's Algorithm as follows.

Theorem 13.1.12 (Buchberger's Algorithm). *Let $P = K[x_1,\ldots,x_n]$, let σ be a term ordering on $\mathbb{T}^n$, and let $f_1,\ldots,f_s \in P$ be non-zero polynomials which generate an ideal $I = \langle f_1,\ldots,f_s \rangle$. Consider the following instructions.*

(1) *Let $G = (f_1,\ldots,f_s)$ and $B = \{(i,j) \mid 1 \leq i < j \leq s\}$.*

(2) *Repeat the following steps until $B = \emptyset$. Then return G and stop.*

(3) *Choose a pair $(i,j) \in B$ and remove it from B.*

(4) *Compute the S-polynomial S_{ij} of f_i and f_j and its normal remainder $S'_{ij} = \mathrm{NR}_{\sigma,G}(S_{ij})$. If $S'_{ij} = 0$ then continue with step (2).*

(5) *Increase s by one, append $f_s = S'_{ij}$ to G, and append $\{(i,s) \mid 1 \leq i \leq s - 1\}$ to B. Then continue with step (2).*

This is an algorithm which computes a σ-Gröbner basis G of the ideal I.

For a proof of the finiteness and correctness of this algorithm, we refer to [KR1], Section 2.5. To get a feeling for the performance of this algorithm, let us apply it in the case of Example 13.1.7.

Example 13.1.13. Let $P = K[x, y]$, let $\sigma = \mathtt{drl}$, and let $f_1 = x^2 - 1$ and $f_2 = xy - 1$. Following Buchberger's Algorithm, we compute $S_{12} = yf_1 - xf_2 = x - y$ and $S'_{12} = S_{12}$ and append $f_3 = S'_{12} = x - y$ to G. At this point we have $B = \{(1, 3), (2, 3)\}$.

Working on the pair $(1, 3)$ yields $S_{13} = f_1 - xf_3 = xy - 1$ and $S'_{13} = 0$. Next we check the pair $(2, 3)$ and get $S_{23} = f_2 - yf_3 = y^2 - 1$ and $S'_{23} = S_{23}$. Thus we append $f_4 = y^2 - 1$ to G and have $B = \{(1, 4), (2, 4), (3, 4)\}$. Working out these three pairs produces $S'_{ij} = 0$ each time, so that the algorithm returns $G = (f_1, f_2, f_3, f_4)$ and stops.

Gröbner bases and Buchberger's Algorithm have numerous applications in computational algebra. We end this section by listing a few of them which will be of use for algebraic cryptography. For detailed proofs of these results, we refer to [KR1].

Proposition 13.1.14 (Macaulay's Basis Theorem). *Let $f_1, \ldots, f_s \in P$ be polynomials which generate an ideal $I = \langle f_1, \ldots, f_s \rangle$. We choose a term ordering σ and use Buchberger's Algorithm to compute a σ-Gröbner basis $G = (g_1, \ldots, g_t)$ of I. Then the residue classes of the terms in the set*

$$\mathbb{O}_\sigma(I) = \mathbb{T}^n \setminus \mathrm{Lt}_\sigma(I) = \{t \in \mathbb{T}^n \mid no\ term\ \mathrm{Lt}_\sigma(g_i)\ divides\ t\}$$

form a K-vector space basis of P/I.

Proposition 13.1.15 (Ideal Membership Test). *Given a polynomial ideal I and a polynomial $f \in P$, we can check as follows whether $f \in I$ holds:*
(1) *Choose a term ordering σ and use Buchberger's Algorithm to compute a σ-Gröbner basis G of I.*
(2) *Using the Division Algorithm, check whether $\mathrm{NR}_{\sigma,G}(f) = 0$.*

13.2 Commutative Gröbner Basis Cryptosystems

The computation of a Gröbner basis using Buchberger's Algorithm has a very high worst-case complexity: EXPSPACE. The idea underlying Gröbner basis cryptosystems is to utilize this fact by constructing a cryptosystem whose security depends on the difficulty of computing a certain Gröbner basis. The first such cryptosystem was suggested by M. Fellows and N. Koblitz in 1994 (see [FK]) and is called the **Polly Cracker Cryptosystem**.

Polly Cracker Key Generation
Let K be a finite field and $P = K[x_1, \ldots, x_n]$ a polynomial ring over K. Suppose that Bob wants to send a message m to Alice. The message space we use is $\mathcal{P} = K$, and the ciphertext space is $\mathcal{C} = P$. The secret key k will be an element of $\mathcal{K} = K^n$. To generate it, we choose a random vector $k = (a_1, \ldots, a_n)$ in K^n.

Next Alice chooses $s \geq 1$ and random polynomials $f_1, \ldots, f_s \in P$. Then she computes the polynomials $g_i = f_i - f_i(a_1, \ldots, a_n)$ for $i = 1, \ldots, s$. The tuple $(g_1, \ldots, g_s) \in P^s$ will serve as Alice's public key.

Encryption and Decryption

To encrypt a message $m \in K$, Bob chooses random polynomials $h_1, \ldots, h_s \in P$ and computes the ciphertext polynomial $c = m + h_1 g_1 + \cdots + h_s g_s$. In order to decrypt a ciphertext unit $c \in P$, Alice evaluates $c(a_1, \ldots, a_n)$ and obtains the original message unit m.

The correctness of this cryptosystem follows from the fact that we have

$$g_i(a_1, \ldots, a_n) = f_i(a_1, \ldots, a_n) - f_i(a_1, \ldots, a_n) = 0,$$

and thus

$$c(a_1, \ldots, a_n) = m + \sum_{i=1}^{n} h_i(a_1, \ldots, a_n) g_i(a_1, \ldots, a_n) = m.$$

How is this cryptosystem related to Gröbner bases? Let us analyse its security. To decrypt a ciphertext unit $c \in P$, an attacker has to find an element $z \in K$ such that $c + z$ is contained in the ideal $I = \langle g_1, \ldots, g_s \rangle$. To do this, he may try to compute a Gröbner basis G of I and then determine z using the equality $z = \mathrm{NR}_{\sigma,G}(c)$. Hence the Polly Cracker cryptosystem is not secure if an attacker is able to compute a Gröbner basis of I.

To make matters worse, the attacker may actually get away with computing a small part of a Gröbner basis of I. In the computation of the normal remainder $\mathrm{NR}_{\sigma,G}(c)$, only finitely many elements of the Gröbner basis G are used. Hence it may suffice to determine a **partial Gröbner basis** in order to decipher a single message.

A further weakness of the Polly Cracker cryptosystem is revealed by a statistical analysis of the ciphertext. In the computation of $c = m + h_1 g_1 + \cdots + h_s g_s$, usually only a small percentage of the terms cancels out completely. Most terms of the products $h_i g_i$ will be contained in the ciphertext. Thus, if t is a term occurring in one of the secret polynomials g_i, many of its multiples will tend to show up in the ciphertext polynomial c. It is clear that a careful statistical analysis of greatest common divisors of the terms in c may reveal individual terms t. Then one can simplify the corresponding polynomial h_i and create a simpler ciphertext for the same message. By repeating this construction, it is frequently possible to decipher c.

In spite of these problems, Koblitz suggested in [Ko2], Ch. 5, several special cases of Polly Cracker cryptosystems whose security seems to be high at first glance.

Example 13.2.1 (The Graph-3-Colouring Cryptosystem). A **graph** is a pair $\Gamma = (V, E)$ consisting of a set of **vertices** $V = \{v_1, \ldots, v_r\}$ and a set of **edges** $E = \{e_1, \ldots, e_s\} \subseteq V \times V$, each of which connects two vertices. A **3-colouring** of a graph $\Gamma = (V, E)$ is

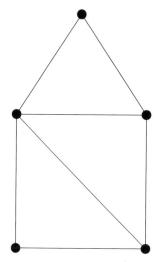

Fig. 13.1: 3-colourable.

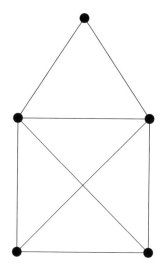

Fig. 13.2: not 3-colourable.

a map $\phi : V \longrightarrow \mathbb{Z}_3$ such that $\phi(v_i) \neq \phi(v_j)$ whenever there is an edge $e_k = (v_i, v_j)$ connecting v_i and v_j. The graph Γ is called **3-colourable** if a 3-colouring exists.

Given a graph Γ, the following two problems are NP-hard:
- *Decision problem:* Is Γ 3-colourable?
- *Search problem:* If Γ is 3-colourable, find a 3-colouring!

Based on these problems, we construct a Polly Cracker cryptosystem as follows. Alice chooses a *suitable* 3-colourable graph $\Gamma = (V, E)$ for which she knows a 3-colouring. (The question which graphs could be considered suitable will be discussed below.) Thus her secret key is a tuple $(a_{ij}) \in V \times \mathbb{Z}_3$ such that

$$a_{ij} = \begin{cases} 1 & \text{if vertex } i \text{ has colour } j, \\ 0 & \text{otherwise.} \end{cases}$$

Alice's public key then consists of the following $4r + 3s$ polynomials in the ring $K[x_{ij} \mid i \in \{1, \dots, r\}, j \in \{0, 1, 2\}]$, where $K = \mathbb{Z}_3 = \{\bar{0}, \bar{1}, \bar{2}\}$, $r = \#V$, and $s = \#E$:
- $f_i = x_{i1} + x_{i2} + x_{i3} - 1$ for $i = 1, \dots, r$
- $g_{ijk} = x_{ij}x_{ik}$ for $i = 1, \dots, r$ and $0 \leq j < k \leq 2$
- $h_{ijk} = x_{ik}x_{jk}$ if $(i, j) \in E$ and $0 \leq j \leq 2$

Based these keys, Alice and Bob use the Polly Cracker cryptosystem.

The following observations show that the security of this cryptosystem relies on the difficulty of solving the 3-colouring search problem. First we check that (a_{ij}) is a common zero of the polynomials in the public key.

(1) For $i = 1, \dots, r$, we have $f(a_{i0}, a_{i1}, a_{i2}) = a_{i0} + a_{i1} + a_{i2} - 1 = 0$, since every vertex has precisely one colour.

(2) For $i = 1, \ldots, r$ and $0 \le j < k \le 2$, we have $g_{ijk}(a_{i0}, a_{i1}, a_{i2}) = a_{ij}a_{ik} = 0$ since every vertex has precisely one colour.

(3) For $i, j \in \{1, \ldots, r\}$ such that $(i, j) \in E$ and for every $k \in \{0, 1, 2\}$, we have $h_{ijk}(a_{i0}, a_{i1}, a_{i2}) = a_{ik}a_{jk} = 0$ because not both vertices v_i and v_j have colour k.

Conversely, let us show that every common zero (c_{ij}) of the polynomials in the public key yields a 3-colouring of Γ.

(4) Since $c_{ij}c_{ik} = 0$ for $j \neq k$, at most one of the numbers $\{c_{i0}, c_{i1}, c_{i2}\}$ is non-zero.

(5) Since $c_{i0} + c_{i1} + c_{i2} = 1$, it follows that precisely one of the numbers $\{c_{i0}, c_{i1}, c_{i2}\}$ is one and the other two are zero.

(6) Since $c_{ik}c_{jk} = 0$ for $(i, j) \in E$, two vertices which are connected by an edge have different colours.

The upshot of this discussion is that the knowledge of a 3-colouring of Γ is sufficient to break the graph 3-colouring cryptosystem. Finding a 3-colouring is equivalent to computing a common zero (a_{ij}) of the set of public polynomials, and therefore equivalent to computing a Gröbner basis $G = \{x_{ij} - a_{ij} \mid i \in \{1, \ldots, r\}, j \in \{0, 1, 2\}\}$ of an ideal containing the public polynomials. If we find a graph for which a 3-colouring is hard to find, a Gröbner basis of this kind will be hard to compute.

So, what is the problem with this construction? First of all, we have to find a *suitable* graph. For most concrete instances of the graph 3-colouring search problem, the actual complexity will be much lower than the worst-case complexity. In fact, in most cases it will be linear in the size of Γ. Therefore we may have to choose very large graphs to make this problem infeasible. However, then we create a very large public key $\{f_i, g_{ijk}, h_{ijk}\}$ and the amount of data that we have to transmit in order to send a message consisting of a single element of $\mathbb{Z}_3$ is huge. In other words, the **data rate** of this cryptosystem will be abysmal. Finally, we still face the problem to choose the graph Γ (and thus the public key polynomials) such that the attacks using partial Gröbner bases and statistical analysis of the terms in the ciphertext become infeasible.

Until now, nobody has been able to find a convincing simultaneous solution to the problems inherent in the Polly Cracker cryptosystem. In order to get a wider choice of constructions, we introduce a far-reaching generalization, namely the general **Commutative Gröbner Basis Cryptosystem (CGBC)**.

CGBC Setup and Key Generation

Let K be a finite field, let $P = K[x_1, \ldots, x_n]$, let σ be a term ordering on $\mathbb{T}^n$, let $I \subset P$ be an ideal, and let $G = \{g_1, \ldots, g_s\}$ be a σ-Gröbner basis of I. The set G will serve as Alice's secret key. To construct a public key, she chooses a set of terms $\mathcal{O}$ in $\mathbb{T}^n \setminus \mathrm{Lt}_\sigma(I)$. The K-vector space $\mathcal{P} = \langle \mathcal{O} \rangle_K$ generated by $\mathcal{O}$ is the set of plaintext units for the CGBC. The set of ciphertext units will be $\mathcal{C} = P$. Finally, Alice chooses *suitable* polynomials $f_1, \ldots, f_t \in I$ which constitute the public key of the CGBC. (The correct interpretation of "suitable" will be discussed below.)

CGBC Encryption and Decryption

Let $m \in \langle \mathbb{O} \rangle_K$ be the plaintext unit that Bob wants to send to Alice. He chooses *suitable* (e.g., sufficiently random) polynomials $h_1, \ldots, h_t \in P$ and computes $c = m + h_1 f_1 + \cdots + h_t f_t$. To decrypt a ciphertext unit $c \in P$, Alice calculates $m = \mathrm{NR}_{\sigma,G}(c)$.

The correctness of this procedure follows from fact that division by a Gröbner basis yields an ideal membership test (see Prop. 13.1.15). For $m' = \mathrm{NR}_{\sigma,G}(c)$ we have $m' \equiv c \pmod{I}$, and therefore $m' - m \equiv 0 \pmod{I}$. Since m and m' are constants and $I \neq P$, they have to be equal.

General commutative Gröbner basis cryptosystems face a number of possible attacks by Eve, the evil attacker. Let us discuss some of them.

(1) Obviously, the cryptosystem can be broken if Eve succeeds in computing *any* Gröbner basis of I. One difficulty may be that she does not know I, but only a smaller ideal $J = \langle f_1, \ldots, f_t \rangle$ for which a Gröbner basis may be hard to come by. On the other hand, it suffices for Eve to find *any* Gröbner basis of *any* ideal containing J.

(2) As for the Polly Cracker cryptosystem, Eve can decipher an individual ciphertext unit c if she knows a sufficiently large partial Gröbner basis of J (or I). The reason is that, in the division steps reducing $c \longrightarrow \mathrm{NR}_{\sigma,G}(c)$, only some of the elements of G are involved.

(3) In the calculation of a ciphertext unit $c = m + h_1 f_1 + \cdots + h_t f_t$, usually not many terms in the polynomials $h_i f_i$ cancel out completely. Unless special care is taken in the choice of the coefficient polynomials h_i, large subsets of the terms in c will have non-trivial greatest common divisors. Using statistical analysis, Eve may determine individual terms in h_i and simplify the ciphertext without changing the underlying plaintext.

(4) In the computation of c, a large amount of cancellation of higher-degree terms has to occur to make the following **Linear Algebra Attack** infeasible: Eve guesses a degree bound for the polynomials h_i and writes them down using indeterminate coefficients. Then the equality $c = m + h_1 f_1 + \cdots + h_t f_t$ yields a linear system of equations for these unknown coefficients. In addition, Eve may proceed to solve this system degree by degree (with respect to the terms in c) and utilize heuristics to exclude the appearance of certain terms in the polynomials h_i, thereby reducing the sizes of the partial linear systems of equations.

(5) Finally, Eve may resort to a chosen-ciphertext-attack. If she is able to decipher illegal ciphertexts consisting of the terms t_i generating $\mathrm{Lt}_\sigma(I)$, she may construct a Gröbner basis $\{\tilde{g}_1, \ldots, \tilde{g}_s\}$ of I via $\tilde{g}_i = t_i - \mathrm{NR}_{\sigma,G}(t_i)$. Hence Alice has to restrict the set of terms $\mathbb{O}$ defining the plaintext unit space such that illegal ciphertexts like t_i are revealed by having terms from $(\mathbb{T}^n \setminus \mathrm{Lt}_\sigma(I)) \setminus \mathbb{O}$ in their decryption. Then her decryption algorithm can reject illegal ciphertexts.

Altogether, it has proven quite difficult to generate secure instances of commutative Gröbner basis cryptosystems, in particular if the need for an acceptable data rate is taken into account. Nevertheless, finding a large enough number of such instances is an attractive goal. We have seen that it is possible to encode NP-hard problems into the task of computing a Gröbner basis. Thus Gröbner basis cryptosystems are candidates for **Post Quantum Cryptography**. Later we study suggestions to overcome the aforementioned difficulties by resorting to non-commutative algebraic structures (such as free associative algebras or certain group rings) which support analogous Gröbner basis theories.

13.3 Algebraic Attacks Using Gröbner Bases

After seeing the difficulties involved in using (commutative) Gröbner bases for constructing new cryptosystems, let us turn the tables and try to use Gröbner bases to break existing cryptosystems. This topic is known as **algebraic attacks** and has gained widespread attention after F. C. Faugère used it to meet the HFE Challenge (see [Fau]). Let us explain the basic idea first.

Every encryption map of a cryptosystem can be represented by a map sending bit-tuples to bit-tuples. Mathematically, we represent it as a map $\epsilon : \mathbb{Z}_2^n \longrightarrow \mathbb{Z}_2^s$ where $\mathcal{P} \subseteq \mathbb{Z}_2^n$ and $\mathcal{C} \subseteq \mathbb{Z}_2^s$. A map like this is *always* polynomial, that is for $K = \mathbb{Z}_2$ there exist polynomials $f_1, \ldots, f_s \in K[x_1, \ldots, x_n]$ such that

$$\epsilon(a_1, \ldots, a_n) = (f_1(a_1, \ldots, a_n), \ldots, f_s(a_1, \ldots, a_n))$$

for all $(a_1, \ldots, a_n) \in K^n$. Of course, these polynomials $f_1, \ldots, f_s$ are not uniquely determined. Suppose that $g_1, \ldots, g_s \in K[x_1, \ldots, x_n]$ are further polynomials with the property that

$$\epsilon(a_1, \ldots, a_n) = (g_1(a_1, \ldots, a_n), \ldots, g_s(a_1, \ldots, a_n))$$

for all $(a_1, \ldots, a_n) \in \mathcal{P}$. Then the differences $f_i - g_i$ are contained in the **vanishing ideal**

$$I(\mathcal{P}) = \{h \in K[x_1, \ldots, x_n] \mid h(a_1, \ldots, a_n) = 0 \text{ for all } (a_1, \ldots, a_n) \in \mathcal{P}\}$$

of $\mathcal{P}$. It is easy to check that $I(\mathcal{P})$ is in fact a polynomial ideal. Below we shall see an effective and efficient algorithm to compute a system of generators of this vanishing ideal. In this setting, we can mount different algebraic attacks using one of the following methods.

Remark 13.3.1 (Algebraic Known-Plaintext-Attack). Suppose that Alice and Bob are using a symmetric cipher and one or more plaintext-ciphertext pairs are known to Eve. She represents the input bits of the encryption map by indeterminates $x_1, \ldots, x_n$, certain bits of intermediate results by indeterminates y_{ij}, and the bits of the unknown key by indeterminates $k_1, \ldots, k_\ell$. Then Eve writes down the encryption map $\epsilon : \mathbb{Z}_2^n \longrightarrow \mathbb{Z}_2^s$ using polynomials $f_1, \ldots, f_s \in \mathbb{Z}_2[x_1, \ldots, x_n, \{y_{ij}\}, k_1, \ldots, k_\ell]$. By substituting the known plaintext-ciphertext pairs, Eve obtains a system of polynomial equations in the indeterminates y_{ij} and $k_1, \ldots, k_\ell$.

In many cases, this polynomial system of equations has a unique solution ($\{a_{ij}\}$, $b_1, \ldots, b_\ell$) defined over $\mathbb{Z}_2$. In the language of Gröbner bases, this means that the ideal generated by those equations has the Gröbner basis $G = \{y_{ij} - a_{ij}\} \cup \{k_1 - b_1, \ldots, k_\ell - b_\ell\}$. If Eve is able to compute that Gröbner basis, she can use the secret key $(b_1, \ldots, b_\ell)$ to decrypt further ciphertext units. She has broken the cryptosystem.

The situation of this remark can for instance occur if Bob encrypts a pdf-file (whose first four bytes are %PDF) and uses no further precautions. In less favourable situations Eve may still try to launch the following attack.

Remark 13.3.2 (Algebraic Ciphertext-Only-Attack). Suppose that Alice and Bob are using a symmetric cipher. Eve intercepts a number of ciphertext units. As in the preceding remark, she represents the encryption map $\epsilon : \mathbb{Z}_2^n \longrightarrow \mathbb{Z}_2^s$ with polynomials

$$f_1^{(m)}, \ldots, f_s^{(m)} \in \mathbb{Z}_2[x_1^{(m)}, \ldots, x_n^{(m)}, y_{ij}^{(m)}, k_1, \ldots, k_\ell]$$

using different sets of indeterminates $x_1^{(m)}, \ldots, x_n^{(m)}, y_{ij}^{(m)}$ for every intercepted ciphertext unit, but the same indeterminates $k_1, \ldots, k_\ell$ for the key bits. Then Eve tries to solve the resulting polynomial system for $k_1, \ldots, k_\ell$ and thereby break the cryptosystem.

It is clear that similar algebraic attacks can be mounted against public key ciphers and stream ciphers. For the sake of brevity, we describe only the first case in more detail.

Remark 13.3.3 (Algebraic Attacks to Public Key Systems). Suppose that Alice and Bob use a public key cryptosystem. Since Eve knows the public key, she can describe the encryption map $\epsilon : \mathbb{Z}_2^n \longrightarrow \mathbb{Z}_2^s$ using polynomials $f_1, \ldots, f_s \in \mathbb{Z}_2[x_1, \ldots, x_n]$ which do not involve any indeterminates related to the secret key. Hence, if Eve want to attack a single ciphertext unit, it suffices to solve the polynomial system $f_i(x_1, \ldots, x_n) = c_i$ where $i = 1, \ldots, s$ and c_i is the i-th bit of the ciphertext.

Eve may also decide to attack the public key. For this purpose she describe the decryption map $\delta : \mathbb{Z}_2^s \longrightarrow \mathbb{Z}_2^n$ by polynomials $g_j(y_1, \ldots, y_s, k_1, \ldots, k_\ell)$ where the indeterminates $y_1, \ldots, y_s$ represent the bits of the ciphertext and the indeterminates $k_1, \ldots, k_\ell$ represent the bits of the secret key. Then Eve uses the publicly known encryption map to create as many plaintext - ciphertext pairs as she needs and sets up arbitrarily many polynomial equations in the indeterminates $k_1, \ldots, k_\ell$ whose only solution over $\mathbb{Z}_2$ is the secret key.

Let us apply this former method in an extremely simple case.

Example 13.3.4. Suppose that Alice and Bob are using the RSA cryptosystem with the public keys $n = 15$, $e = 5$ and the secret key $d = 5$. (Notice that $\phi(15) = 8$ and $de \equiv 1 (\mathrm{mod}\ 8)$.) The plaintext and ciphertext units are elements of $(\mathbb{Z}/15\mathbb{Z})^\times$ and will be represented by elements $a_0 + 2a_1 + 4a_2 + 8a_3$ with $a_i \in \{0, 1\}$.

Then the encryption map $\epsilon(a_0, a_1, a_2, a_3) = (c_0, c_1, c_2, c_3)$ is determined by $c_0 + 2c_1 + 4c_2 + 8c_3 = (a_0 + 2a_1 + 4a_2 + 8a_3)^5$. In view of the relations $a_i^2 = a_i$ for

$a_i \in \mathbb{Z}_2$, this yields the polynomial representation

$$c_0 = a_0 a_1 a_2 a_3 + a_0 a_1 a_2 + a_0 a_2 + a_0 a_3 + a_2 a_3 + a_0 + a_3$$
$$c_1 = a_0 a_1 a_2 a_3 + a_0 a_1 a_2 + a_0 a_1 a_3 + a_0 a_2 a_3 + a_0 a_1 + a_1 + a_2 + a_3$$
$$c_2 = a_1 a_2 a_3 + a_0 a_1 + a_1 a_2 + a_1 a_3 + a_1 + a_2$$
$$c_3 = a_0 a_1 a_2 a_3 + a_0 a_1 a_2 + a_1 a_2 a_3 + a_0 a_1 + a_0 a_2 + a_0 a_3 + a_2 a_3 + a_3.$$

For instance, if Eve intercepts the ciphertext bits $(c_0, c_1, c_2, c_3) = (1, 1, 0, 0)$, she can now compute a Gröbner basis of the ideal $\langle f_i(a_0, a_1, a_2, a_3) - c_i \mid i = 0, 1, 2, 3 \rangle$ where the polynomials f_i correspond to the right-hand sides above. The result is $G = \{a_0 - 1, a_1 - 1, a_2, a_3\}$. This corresponds to the fact that the plaintext $a_0 + 2a_1 + 4a_2 + 8a_3 = 3$ encrypts to the given ciphertext.

At the heart of these algebraic attacks lies the problem of solving a large system of polynomial equations over a finite field K, where we usually have $K = \mathbb{Z}_2$. Let $f_1, \ldots, f_s \in K[x_1, \ldots, x_n]$ be the polynomials which define the polynomial system

$$(S) \begin{cases} f_1(x_1, \ldots, x_n) = 0 \\ \quad\vdots \\ f_s(x_1, \ldots, x_n) = 0. \end{cases}$$

Each solution of the polynomial system (S) satisfies every equation of the form $g_1 f_1 + \cdots + g_s f_s = 0$ where $g_1, \ldots, g_s \in K[x_1, \ldots, x_n]$ are arbitrary polynomials. In other words, each solution of (S) is a **zero** of the polynomial ideal $I = \langle f_1, \ldots, f_s \rangle$. The set of all zeros of I is called the **zero set** of I and is denoted by $\mathcal{Z}(I)$.

Notice that we are always looking for solutions of the given polynomial systems which are defined over K, where $K = \mathbb{F}_q$ is a finite field. Therefore we my append the equations $x_i^q - x_i = 0$ to the system in question to make sure that the computed solution is indeed an element of K (and not of a larger finite field). These equations are also called the **base field equations** and the ideal $\langle x_1^q - x_1, \ldots, x_n^q - x_n \rangle$ is called the **base field ideal**.

Before we address particular methods for solving large polynomial systems over finite fields, we want to point out algorithms which can be used to describe the encryption map via polynomials. One way is of course to trace the explicit description of the encryption map and to represent each step by polynomials. For certain steps (such as S-boxes) or certain encryption maps, we can try to consider them as *black box algorithms* and apply the following algorithm for multivariate polynomial interpolation.

Theorem 13.3.5 (The Buchberger-Möller Algorithm). *Let K be a finite field and $\mathbb{X} = \{(p_1, k_1), \ldots, (p_s, k_s)\}$ a finite set of points in $K^n \times K^\ell$. Assume that the elements $p_i \in K^n$ are plaintext units and the elements $k_i \in K^\ell$ are keys of a given cryptosystem. For $i = 1, \ldots, s$, let $(c_{i1}, \ldots, c_{im}) \in K^m$ be the ciphertext unit obtained by encrypting p_i using the key k_i.*

Let Q be the polynomial ring $Q = K[x_1, \ldots, x_n, y_1, \ldots, y_\ell]$ and let σ be a term ordering on Q. Consider the following sequence of instructions.

(1) *Let $g = \emptyset$, $\mathcal{O} = \emptyset$, $S = \emptyset$, $L = \{1\}$, and $M = (m_{ij})$ a matrix having initially s columns and 0 rows.*

(2) *If $L = \emptyset$ then continue with step (6). Otherwise, let $t = \min_\sigma(L)$.*

(3) *Compute* eval$(t) = (t(p_1, k_1), \ldots, t(p_s, k_s))$ *and row reduce this vector against the rows of M. The result is of the form*

$$(v_1, \ldots, v_s) = \text{eval}(t) - \sum_i a_i \cdot (m_{i1}, \ldots, m_{is})$$

with $a_i \in K$.

(4) *If $(v_1, \ldots, v_s) = 0$ then append the polynomial $t - \sum_i a_i s_i$ to G where s_i is the i-th element of S. Delete all multiples of t in L and continue with step (2).*

(5) *Now let $(v_1, \ldots, v_s) \neq 0$. Append $(v_1, \ldots, v_s)$ as a new row to M, append $t - \sum_i a_i s_i$ to S, append t to $\mathcal{O}$, and append to L all elements of the set $\{x_1 t, \ldots, x_n t, y_1 t, \ldots, y_\ell t\}$ which are not in L or in $\text{Lt}_\sigma(G)$. Then continue with step (2).*

(6) *Using row operations, transform M into a diagonal matrix. Mimick these row operations on the elements of S.*

(7) *For $j = 1, \ldots, m$, let $f_j = \sum_i c_{ij} s_i$ where s_i is the i-th element of S. Return $(f_1, \ldots, f_m)$ and G and stop.*

*This is an algorithm which computes a tuple of polynomials $(f_1, \ldots, f_m) \in Q^m$ such that $c_{ij} = f_j(p_i, k_i)$ for $i = 1, \ldots, s$ and $j = 1, \ldots, m$, as well as a set of polynomials $G \subset Q$ which is a σ-Gröbner basis of the **vanishing ideal***

$$I(\mathbb{X}) = \{g \in Q \mid g(p_i, k_i) = 0 \text{ for } i = 1, \ldots, s\}.$$

A proof of this theorem is contained in [KR2], Section 6.3. Given a cryptosystem whose encryption map we can apply to individual plaintext units and keys, the Buchberger-Möller algorithm can be used to construct a polynomial map which correctly represents the encryption map for many (or even all) inputs. A further, common way to apply the theorem is to determine polynomial representations of S-Boxes in symmetric ciphers.

13.3.1 The Gröbner Basis Attack

Above we showed that breaking many types of cryptosystems can be reduced to solving large systems of polynomial equations over finite fields. For a long time, this had been considered an utterly impossible task. The situation changed when J.-C. Faugère broke the HFE Challenge in [Fau] using a Gröbner basis computation. Let us outline the general method.

Let K be a finite field, let $P = K[x_1, \ldots, x_n]$, and let $f_1, \ldots, f_s \in P$ be polynomials defining a system of equations

$$(S) \begin{cases} f_1(x_1, \ldots, x_n) = 0 \\ \vdots \\ f_s(x_1, \ldots, x_n) = 0 \, . \end{cases}$$

We form the polynomial ideal $I = \langle f_1, \ldots, f_s, x_1^q - x_1, \ldots, x_n^q - x_n \rangle$ where $q = \#K$. Our task is to find the zero set

$$\mathcal{Z}(I) = \{(a_1, \ldots, a_n) \in K^n \mid f(a_1, \ldots, a_n) = 0 \text{ for all } f \in I\}.$$

To solve this task, it is sufficient to compute any Gröbner basis of the ideal I. Then there is an efficient algorithm, called the **FGLM Algorithm** (see [FGLM] or [DE], Section 2.2.4), to compute the **reduced Gröbner basis** of I with respect to the lexicographic term ordering $\sigma = \text{Lex}$. Using a linear coordinate transformation, we can then bring I into **normal x_n-position**, that is we can make sure that the last coordinates of the points of $\mathcal{Z}(I)$ are pairwise distinct (see [KR1], Section 3.7). Finally, we are in the setting of the **Shape Lemma** (cf. [KR1], Theorem 3.7.25) which says that the reduced Lex-Gröbner basis G of I has the following shape:

$$G = \{x_1 - g_1(x_n), \ \ldots, \ x_{n-1} - g_{n-1}(x_n), \ g_n(x_n)\}$$

By factoring $g_n(x_n)$, it is then easy to read off its zeros, and substituting these into the other polynomials of G reveals the zeros of I.

Using the following remark, many of these steps can be simplified in practice.

Remark 13.3.6. When we carry out an algebraic attack and search for the secret key, the polynomial system (S) has frequently a unique solution. In this case every reduced Gröbner basis G of I is of the form $G = \{x_1 - a_1, \ldots, x_n - a_n\}$ where $(a_1, \ldots, a_n) \in K^n$ is the unique solution of (S).

Our next example represents the first important case of a successful Gröbner basis attack.

Example 13.3.7 (The HFE Challenge). Let K be a finite field having $q = 2^n$ elements. The encryption function of the HFE cryptosystem can be described as a composition $\epsilon = S \circ F \circ T : K \longrightarrow K$ where S, T are invertible linear maps and F can be identified (via an identification $\mathbb{F}_{2^n} \cong \mathbb{Z}_2^n$) with a polynomial map $f : \mathbb{Z}_2^n \longrightarrow \mathbb{Z}_2^n$ given by $f(x_1, \ldots, x_n) = (q_1(x_1, \ldots, x_n), \ldots, q_n(x_1, \ldots, x_n))$ with *quadratic* polynomials $q_i \in \mathbb{Z}_2[x_1, \ldots, x_n]$.

Then the secret key of HFE is the triple (S, F, T) and the public key is the tuple of polynomials $(p_1, \ldots, p_n)$ with $p_i \in \mathbb{Z}_2[x_1, \ldots, x_n]$ representing the encryption map $S \circ F \circ T$ as above. To break this system, it is clearly sufficient to solve the set of polynomial equations $p_i(x_1, \ldots, x_n) = c_i$ for $i = 1, \ldots, n$, where c_i is the i-th bit of the ciphertext.

In [Fau], J.-C. Faugère carried out this Gröbner basis attack via an improved version of Buchberger's Algorithm, called the **F5-Algorithm**, to break the first *HFE Challenge* with $n = 80$.

13.3.2 The Integer Programming Attack

In this subsection we will restrict our attention to algebraic attacks based on polynomial systems defined over $\mathbb{Z}_2$. Although the generalization to other finite base fields is straightforward, we want to concentrate on the fundamental principles in the most important case. The task of solving a polynomial system (S) with $f_1, \ldots, f_s \in \mathbb{Z}_2[x_1, \ldots, x_n]$ can be rephrased as follows: Find a tuple $(a_1, \ldots, a_n) \in \{0, 1\}^n$ such that

$$F_1(a_1, \ldots, a_n) \equiv 0 \pmod 2$$

$$\vdots$$

$$F_s(a_1, \ldots, a_n) \equiv 0 \pmod 2$$

where $F_i \in \mathbb{Z}[x_1, \ldots, x_n]$ is the canonical representative of f_i. Thus we are looking for an integer solution $(a_1, \ldots, a_n)$ of this system which satisfies $0 \le a_i \le 1$. This formulation suggests to linearise the system and to apply an Integer Programming (IP) algorithm for finding a solution satisfying the stated bounds. The following proposition turns this idea into an effective algorithm.

Proposition 13.3.8 (The Integer Programming Attack). *Let $f_1, \ldots, f_s \in P = \mathbb{Z}_2[x_1, \ldots, x_n]$. Then the following instructions define an algorithm which computes a tuple $(a_1, \ldots, a_n) \in \{0, 1\}^n$ which defines a zero of the polynomial ideal $I = \langle f_1, \ldots, f_m, x_1^2 + x_1, \ldots, x_n^2 + x_n \rangle$.*

(1) *Reduce $f_1, \ldots, f_s$ modulo the field equations, that is, make their terms squarefree. For $i = 1, \ldots, s$, let T_i be the set of terms of degree ≥ 2 in f_i and s_i the number of terms in f_i.*

(2) *For $i = 1, \ldots, s$, introduce a new indeterminate k_i and write down the linear inequality $K_i : k_i \le \lfloor s_i/2 \rfloor$.*

(3) *For every $t_j \in T_i$, introduce a new indeterminate y_{ij}. For $i = 1, \ldots, s$, write $f_i = \sum_j t_j + \ell_i$ where the sum extends over all j such that $t_j \in T_i$ and where $\ell_i \in P_{\le 1}$. Form the linear equation $F_i : \sum_j y_{ij} + \ell_i - 2 k_i = 0$.*

(4) *For $i \in \{1, \ldots, s\}$ and $t_j \in T_i$, write $t_j = x_{j_1} \cdots x_{j_r}$ with $1 \le j_1 < \cdots < j_r \le n$. Form the linear inequalities $Y_{ij} : y_{ij} - x_i \le 0$ and $Z_{ij} : -y_{ij} + x_{j_1} + \cdots + x_{j_r} - r + 1 \le 0$.*

(5) *For all $i \in \{1, \ldots, s\}$, let $X_i : x_i \le 1$.*

(6) *Choose a linear polynomial $C \in \mathbb{Q}[x_i, y_{ij}, k_i]$ and use an IP solver to find a tuple of natural numbers (a_i, b_{ij}, c_i) which solves the system of linear equations and inequalities $\{K_i, F_i, Y_{ij}, Z_{ij}, X_i\}$ and minimizes C.*

(7) *Return $(a_1, \ldots, a_n)$ and stop.*

Proof. Since we are looking for natural numbers a_i for which X_i holds, we have $a_i \in \{0, 1\}$. Similarly, we have $b_{ij} \in \{0, 1\}$ by X_i and Y_{ij}. Moreover, if $t_j \in T_i$ and if one of the numbers $a_{j_1}, \ldots, a_{j_r}$ is zero then Y_{ij} implies $b_{ij} = 0$. On the other hand, if $a_{j_1} = \cdots = a_{j_r} = 1$ then Z_{ij} implies $b_{ij} \geq 1$. Altogether, this means that b_{ij} equals $a_{j_1} \cdots a_{j_r}$, the value of t_j at $(a_1, \ldots, a_n)$.

Next it follows from F_i that $f_i(a_1, \ldots, a_n) = 2\, k_i$ is an even number, and K_i is nothing but the trivial bound for k_i implied by the number of terms of f_i. In this way the solutions of the IP problem correspond uniquely to the tuples $(a_1, \ldots, a_n) \in \{0, 1\}^n$ which satisfy the above reformulation of the given polynomial system. $\qquad\square$

In this proposition we have not taken any advantage of the possibility to choose the *cost function C*. Obviously, the number of additional indeterminates y_{ij} (and k_i) we have to introduce depends on the sparsity of the system (S). For systems with few quadratic or higher degree terms, even a straightforward, non-optimized implementation yields satisfactory results, as our next example shows.

Example 13.3.9. Given the CTC ("Courtois Toy Cipher") cryptosystem introduced in [Cou] and a plaintext–ciphertext pair, we construct an overdetermined algebraic system of equations in terms of the indeterminates representing key bits and certain intermediate quantities. The task is to solve the system for the key bits. The size of the system depends mainly on two parameters: the number b of simultaneous S-boxes and the number N of encryption rounds used.

For instance, in the case $b = 3$ and $N = 4$ we have to solve a polynomial system having 153 indeterminates, 285 equations, and 180 non-linear terms. Using a standard PC, the public domain IP solver GLPK, and the proposition, this system can be solved in less than a minute.

13.3.3 The SAT Attack

In this subsection we continue to work over the field $K = \mathbb{Z}_2$. In order to solve the polynomial system (S), we proceed as follows:

(1) Linearise the system by introducing a new indeterminate for each term occurring in one of the polynomials.
(2) Having written a polynomial as a sum of indeterminates, introduce new indeterminates to cut it after a certain number of terms. (This number is called the *cutting number.*)
(3) Convert the reduced sums into their logical equivalents using a CNF conversion.

Here CNF means **conjunctive normal form** and represents a set of propositional logic clauses of a certain type. Such sets of CNF clauses can be solved very efficiently using so-called **SAT solvers**. Therefore the transformation to a problem in logic is currently the fastest and most powerful way to solve polynomial systems over $\mathbb{Z}_2$.

To explain this technique in more detail, we start by elaborating the transformation of a polynomial equation into its logical equivalent. Let $M = \{X_1,\ldots,X_n\}$ be a set of *boolean variables* (atomic formulas), and let $\widehat{M}$ be the set of all (propositional) logical formulas that can be constructed from them, that is all formulas involving the operations $\neg$, $\wedge$, and $\vee$.

Definition 13.3.10. Let $f \in \mathbb{Z}_2[x_1,\ldots,x_n]$. A logical formula $F \in \widehat{M}$ is called a **logical representation** of f if $\phi_a(F) = f(a_1,\ldots,a_n) + 1$ for every $a = (a_1,\ldots,a_n) \in \mathbb{Z}_2^n$. Here ϕ_a denotes the boolean value of F at the tuple of boolean values a where $1 = \texttt{true}$ and $0 = \texttt{false}$.

The following lemma provides the central step for the logical conversion process.

Lemma 13.3.11. *Let $f \in \mathbb{Z}_2[x_1,\ldots,x_n,y]$ be of the form $f = \ell_1\cdots\ell_s + y$, where $1 \le s \le n$ and $\ell_i \in \{x_i,\, x_i + 1\}$ for $i = 1,\ldots,s$. We define formulas $L_i = X_i$ if $\ell_i = x_i$ and $L_i = \neg X_i$ if $\ell_i = x_i + 1$. Then*

$$F = (\neg Y \vee L_1) \wedge \cdots \wedge (\neg Y \vee L_s) \wedge (Y \vee \neg L_1 \vee \cdots \vee \neg L_s)$$

*is a logical representation of f. Notice that F is in **conjunctive normal form (CNF)** and has $s + 1$ clauses.*

Proof. Let $a = (a_1,\ldots,a_n,b) \in \mathbb{Z}_2^{n+1}$. By induction on s, we show $\phi_a(F) = f(a) + 1$. In the case $s = 1$ we have $f = x_1 + y + c$ with $c \in \{0,1\}$ and $F = (\neg Y \vee L_1) \wedge (Y \vee \neg L_1)$ where $L_1 = X_1$ if $c = 0$ and $L_1 = \neg X_1$ if $c = 1$. The claim $\phi_a(F) = f(a) + 1$ follows easily with the help of a truth table.

Now we prove the inductive step, assuming that the claim has been shown for $s-1$ factors ℓ_i, that is for $f' = \ell_1\cdots\ell_{s-1}$ and the corresponding formula F'. To begin with, we assume that $\ell_s = x_s$ and distinguish two sub-cases.
(1) If $a_s = 0$, we have $\phi_a(F) = \phi_a(\neg Y \vee L_1) \wedge \cdots \wedge (\neg Y \vee L_{s-1}) \wedge \neg Y = \phi_b(\neg Y)$ and $f(a) = b$. This shows $\phi_a(F) = f(a) + 1$.
(2) If $a_s = 1$, we have $f(a) = f'(a)$. Using $\phi_a(L_s) = 1$, we obtain

$$\phi_a(F) = \phi_a(\neg Y \vee L_1) \wedge \cdots \wedge (\neg Y \vee L_{s-1}) \wedge (Y \vee \neg L_1 \vee \cdots \vee \neg L_{s-1}) = \phi_a(F').$$

Hence the inductive hypothesis yields $\phi_a(F) = \phi_a(F') = f'(a) + 1 = f(a) + 1$.

In the case $\ell_s = x_s + 1$, the proof proceeds in exactly the same way. $\qquad\square$

Using this lemma, we define a **standard conversion strategy** for $f \in \mathbb{Z}_2[x_1,\ldots,x_n]$ as follows. Choose a **cutting number** $\ell \ge 3$. For each non-linear term t appearing in f, introduce a new indeterminate y and a new boolean variable Y. Substitute y for t in f and append the clause corresponding to $t + y$ in the lemma to the set of clauses.

After f has been linearised, choose a polynomial g consisting of $\le \ell - 1$ terms of f, introduce a new indeterminate z, replace f by $f - g + z$ and collect the polynomial $g + z$ in a set L. Repeat this step until all of f has been cut to short sums of terms.

Finally, use the fact that the logical representation of $g + z$ is $\neg G \Leftrightarrow Z$ repeatedly to transform L into a set of CNF clauses.

Let us apply this transformation to an actual example polynomial.

Example 13.3.12. Let $f = x_1 x_2 + x_2 x_3 + x_1 + x_2 + 1 \in \mathbb{Z}_2[x_1, x_2, x_3]$. We convert f into a set of CNF clauses as follows. First we let $y_1 = x_1 x_2$ and start with the corresponding clauses $L = \{\neg Y_1 \vee X_1, \ \neg Y_1 \vee X_2, \ Y_1 \vee \neg X_1 \vee \neg X_2\}$. Then we set $y_2 = x_2 x_3$ and append the analogous three clauses to L.

Now we have $f = y_1 + y_2 + x_1 + x_2 + 1$ and choose $\ell = 3$. Letting $g_1 = z_1 + y_1 + y_2$, we have corresponding clauses $G_1 = \{\neg Z_1 \vee Y_1 \vee Y_2, \ Z_1 \vee \neg Y_1 \vee Y_2, \ Z_1 \vee Y_1 \vee \neg Y_2, \ \neg Z_1 \vee \neg Y_1 \vee \neg Y_2\}$. Similarly, the polynomials $g_2 = z_2 + x_1 + x_2$ leads to a set of four clauses G_2 and $f = z_1 + z_2 + 1$ corresponds to $F = \{\neg Z_1 \vee Z_2, \ Z_1 \vee \neg Z_2\}$.

Altogether, the standard CNF conversion of f consists of the 16 clauses in $L \cup G_1 \cup G_2 \cup F$.

The application of state-of-the-art SAT solvers now allows us to solve very big polynomial systems arising from algebraic attacks. Here is a case in point.

Example 13.3.13. In the CTC family of cryptosystems (see [Cou]), we choose $b = 6$ simultaneous SBoxes and $N = 6$ encryption rounds. A straightforward algebraic attack yields a polynomial system consisting of 612 quadratic equations in 468 indeterminates over the field $\mathbb{Z}_2$. The above transformation strategy converts this system to a set of 5065 propositional logical clauses in 937 boolean variables. Standard SAT solvers such as CryptoMiniSat solve this SAT instance in a few seconds on a standard PC.

With this example we end our excursion into algebraic attacks. For the interested reader, we suggest to consult [Kre] for further variants.

13.4 Exercises

13.1 Prove that the orderings $\leq_{\text{lex}}$ and $\leq_{\text{drl}}$ defined in Example 13.1.3 are indeed term orderings.

13.2 Define an ordering relation $\leq_{\text{rlex}}$ on $\mathbb{T}^n$ by letting $t = x_1^{\alpha_1} \cdots x_n^{\alpha_n} \leq_{\text{rlex}} t' = x_1^{\beta_1} \cdots x_n^{\beta_n}$ if and only if the last non-zero component of $(\alpha_1 - \beta_1, \ldots, \alpha_n - \beta_n)$ is negative. Show that this **reverse lexicographic ordering** is a multiplicative ordering relation, but not a well-ordering. How are the indeterminates ordered by $\leq_{\text{rlex}}$?

13.3 Let σ be a term ordering on $\mathbb{T}^n$, and let $f \in K[x_1, \ldots, x_n] \setminus \{0\}$. Prove that the leading term ideal of the principal ideal $\langle f \rangle$ is generated by $\text{Lt}_\sigma(f)$.

13.4 Using the term ordering $\sigma = \text{drl}$, show that the following sets of polynomials are σ-Gröbner bases of the ideals they generate.

(a) $G_1 = \{x^2 - 1, xy - 1, x - y, y^2 - 1\}$ in $K[x, y]$ (see Example 13.1.9)

(b) $G_2 = \{y^2 - x, z^3 - x\}$ in $K[x, y, z]$

(c) $G_3 = \{x^2 - y^2, xy^2 - z^3, y^4 - xz^3\}$ in $K[x, y, z]$

13.5 Generalize Example 13.2.1 to a Graph-p-Colouring Cryptosystem, where $p \geq 2$ is a prime. Describe a suitable polynomial ring and the polynomials which form Alice's public key.

13.6 Let K be a field which contains three solutions of $x^3 = 1$, i.e., three cubic roots of unity. Suppose that these cubic roots of unity represent three colours. Define a suitable ideal in a polynomial ring $K[x_1, \ldots, x_n]$ such that 3-colourings of a graph $\Gamma = (V, E)$ correspond uniquely to zeros of this ideal.

13.7 Given a graph $\Gamma = (V, E)$, a **perfect code** in Γ is a subset $S \subseteq V$ such that every vertex of Γ is in S or joined to precisely one element of S by an edge. For every $v \in V$, let N_v be the **neighbourhood** of v, i.e., the set consisting of v and all vertices of Γ which are joined to v by an edge. The problem of finding a perfect code in Γ is known to be NP-complete.

(a) Let $V = \{v_1, \ldots, v_r\}$, and let $P = \mathbb{Z}_2[x_1, \ldots, x_n]$. For $i = 1, \ldots, r$, we let $f_i = 1 - \sum_{j \in N_{v_i}} x_j$, and if $\text{dist}_\Gamma(v_j, v_k) \leq 2$, we let $g_{jk} = x_j x_k$. Show that the polynomials f_i, g_{jk} generate an ideal whose zeros correspond uniquely to perfect codes in Γ.

(b) Using (a), construct an instance of the Polly Cracker Cryptosystem whose security depends on the Graph Perfect Code search problem in Γ.

13.8 Suppose that Alice and Bob are using the RSA cryptosystem with the public keys $n = 77$ and $e = 11$, and the secret key $d = 11$. We represent an element $c_0 + 2c_1 + 4c_2 + 8c_3 + 16c_4 + 32c_5 + 64c_6$ in $\mathbb{Z}_{77}$ with $c_i \in \{0, 1\}$ by the tuple $(c_0, \ldots, c_6)$ in $\mathbb{Z}_2^7$.

(a) Compute polynomials in $\mathbb{Z}_2[x_0, \ldots, x_6]$ which represent the encryption map $\epsilon : \mathbb{Z}_2^7 \longrightarrow \mathbb{Z}_2^7$.

(b) Decrypt the ciphertext $(0, 1, 1, 1, 0, 1, 0)$ by solving the corresponding polynomial system via a Gröbner basis calculation.

14 Non-Commutative Gröbner Basis Methods

14.1 Non-Commutative Gröbner Bases

Since we encountered some problems when we tried to use commutative polynomials for constructing secure Gröbner basis cryptosystems, it is natural to examine the possibility to use non-commutative algebraic structures. Given a finite set of **letters** $X = \{x_1, \ldots, x_n\}$, a **word** in X is an element of the form $w = x_{i_1} x_{i_2} \cdots x_{i_k}$ with $i_j \in \{1, \ldots, n\}$. Here we denote the **empty word** by 1 and the set of all words by $\langle X \rangle$. It is clear that concatenation makes $\langle X \rangle$ into a monoid with neutral element 1. We call it the **free monoid** on X.

Definition 14.1.1. Let K be a field and $X = \{x_1, \ldots, x_n\}$. The K-vector space with basis $\langle X \rangle$ can be made into a K-algebra by extending the multiplication of words K-linearly. In other words, given $f = c_1 w_1 + \cdots + c_s w_s$ and $g = c_1' w_1' + \cdots + c_t' w_t'$ with $c_i, c_j' \in K$ and $w_i, w_j' \in \langle X \rangle$, we let $f \cdot g = \sum_{i,j} c_i c_j' w_i w_j'$. This defines a K-algebra which is denoted by $K\langle X \rangle$ and is called the **free associative K-algebra** on X or the **non-commutative polynomial ring** in the indeterminates $x_1, \ldots, x_n$.

Our goal in this section is to develop a Gröbner basis theory for $K\langle X \rangle$ which is analogous to the Gröbner basis theory for the commutative polynomial ring. We will see that most definitions and results carry over easily, but Buchberger's Algorithm turns into an enumerating procedure.

Let us start with some basic definitions.

Definition 14.1.2. Let K be a field and $X = \{x_1, \ldots, x_n\}$ a set of letters.
(1) Given a word $w = x_{i_1} \cdots x_{i_k} \in \langle X \rangle$ with $i_1, \ldots, i_k \in \{1, \ldots, n\}$, the number $\ell(w) = k$ is called the **length** of w. The length of the empty word is defined to be zero.
(2) Given two words $w, w' \in \langle X \rangle$, the word w' is called a **subword** of w if w is of the form $w = uw'u'$ with words $u, u' \in \langle X \rangle$.
(3) Given a non-zero non-commutative polynomial $f = c_1 w_1 + \cdots + c_s w_s$ in $K\langle X \rangle$ with $c_i \in K \setminus \{0\}$ and $w_i \in \langle X \rangle$, the elements c_i are called the **coefficients** of the words w_i in f, and the non-negative integer $\deg(f) = \max\{\ell(w_i) \mid i = 1, \ldots, s\}$ is called the **degree** of f.

In order to compute effectively with non-commutative polynomials, we need to represent them in a unique way. Thus we need to order the words in a suitable way.

Definition 14.1.3. A **word ordering** σ on $\langle X \rangle$ is a complete ordering relation which satisfies the following two additional conditions.
(1) The ordering σ is compatible with multiplication, that is if two words $w, w' \in \langle X \rangle$ satisfy an inequality $w \leq_\sigma w'$ and if $u, u' \in \langle X \rangle$ are further words, then we have $uwu' \leq_\sigma uw'u'$.

(2) The ordering σ is a well-ordering, that is every descending chain of words $w_1 \geq_\sigma w_2 \geq_\sigma \cdots$ becomes eventually stationary.

Notice that a word ordering σ has the additional property that $w \geq_\sigma 1$ for all $w \in \langle X \rangle$, because $1 >_\sigma w$ implies $1 >_\sigma w >_\sigma w^2 >_\sigma \cdots$, in contradiction to the well-ordering property.

The most straightforward candidate for a word ordering seems to be the **lexicographic word ordering** lex defined by $w = x_{i_1} \cdots x_{i_k} \leq_{\text{lex}} w' = x_{j_1} \cdots x_{j_m}$ if and only if $w' = ww''$ with $w'' \in \langle X \rangle$ or $i_1 < j_1$ or $i_1 = j_1, i_2 < j_2$, etc. However, this ordering is neither compatible with multiplication nor a well-ordering, as the inequalities $x_2^2 >_{\text{lex}} x_2$ and $x_2 x_1 >_{\text{lex}} x_2^2 x_1 >_{\text{lex}} \cdots$ show. To construct a true word ordering, we have to modify it as follows.

Example 14.1.4. The **length-lexicographic word ordering** llex on $\langle X \rangle$ is defined as follows. Given two words $w = x_{i_1} \cdots x_{i_k}$ and $w' = x_{j_1} \cdots x_{j_m}$, we let $w \leq_{\text{llex}} w'$ if and only if $\ell(w) < \ell(w')$ or if both words have the same length and $w \leq_{\text{lex}} w'$.

It is easy to check that llex is in fact a word ordering. For instance, it satisfies $x_1 >_{\text{llex}} x_2$ and $x_2^2 >_{\text{llex}} x_2$ and $x_1 x_2 >_{\text{llex}} x_2 x_1$.

Another method to construct word orderings is to consider the given words as commutative terms in $K[x_1, \ldots, x_n]$, compare them using a term ordering, and break ties using lex. Further word orderings will be discussed in the next section.

In non-commutative polynomial rings there are several kinds of ideals. Let K be a field and $X = \{x_1, \ldots, x_n\}$ a set of letters.

Definition 14.1.5. Let I be a subset of $K\langle X \rangle$.
(1) The set I is called a **left ideal** in $K\langle X \rangle$ if I is an additive subgroup of $K\langle X \rangle$ and if we have $K\langle X \rangle \cdot I \subseteq I$.
(2) The set I is called a **right ideal** in $K\langle X \rangle$ if I is an additive subgroup of $K\langle X \rangle$ and if we have $I \cdot K\langle X \rangle \subseteq I$.
(3) The set I is called a **two-sided ideal** in $K\langle X \rangle$, or simply an **ideal** in $K\langle X \rangle$, if I is both a left and a right ideal in $K\langle X \rangle$.

For instance, the set of all polynomials in $K\langle X \rangle$ whose **constant coefficient** (that is the coefficient of the word 1) is zero forms a two-sided ideal in $K\langle X \rangle$. A general method to construct two-sided ideals can be obtained as follows.

Definition 14.1.6. Let $S \subseteq K\langle X \rangle$ be a subset.
(1) The set
$$\langle S \rangle = \{f_1 s_1 g_1 + \cdots + f_r s_r g_r \mid r \geq 0, \ f_i, g_i \in K\langle X \rangle, \ s_i \in S\}$$
is a two-sided ideal in $K\langle X \rangle$. It is called the ideal **generated by** S. In this case the set S is called a **system of generators** of $\langle S \rangle$.
(2) An ideal I in $K\langle X \rangle$ is called **finitely generated** if there exists a finite subset S of $K\langle X \rangle$ such that $I = \langle S \rangle$.

Examples of finitely generated ideals in $K\langle X\rangle$ are the *zero ideal* $\langle 0\rangle = \{0\}$, the *unit ideal* $K\langle X\rangle$, *principal ideals* $\langle f\rangle = \{gfh \mid g, h \in K\langle X\rangle\}$, and the *irrelevant ideal* $\langle x_1, \dots, x_n\rangle$ which consists of all non-commutative polynomials having constant coefficient zero.

Not every ideal in $K\langle X\rangle$ is finitely generated, as our next example shows.

Example 14.1.7. Let $X = \{x, y\}$, and let I be the two-sided ideal in $K\langle X\rangle = K\langle x, y\rangle$ generated by $S = \{xy^i x \mid i \geq 1\}$. Then no generator $xy^i x$ of I is contained in the two-sided ideal $\langle xyx, xy^2 x, \dots, xy^{i-1} x\rangle$. Since every finite set of generators of I can be represented using finitely many elements of S, it follows that I is not finitely generated.

Given a word ordering σ, the following definitions correspond to the analogous definitions for the commutative case.

Definition 14.1.8. Let σ be a word ordering on $\langle X\rangle$, and let $f = c_1 w_1 + \cdots + c_s w_s \in K\langle X\rangle \setminus \{0\}$ be a non-zero non-commutative polynomial, where $c_i \in K \setminus \{0\}$ and where $w_i \in \langle X\rangle$ are words satisfying $w_1 >_\sigma w_2 >_\sigma \cdots >_\sigma w_s$.

(1) The word $\mathrm{Lw}_\sigma(f) = w_1$ is called the **leading word** of f.
(2) The element $\mathrm{Lc}_\sigma(f) = c_1$ is called the **leading coefficient** of f.
(3) We let $\mathrm{Lm}_\sigma(f) = \mathrm{Lc}_\sigma(f) \cdot \mathrm{Lw}_\sigma(f)$ and call it the **leading monomial** of f.
(4) Given a two-sided ideal I in $K\langle X\rangle$, the two-sided ideal

$$\mathrm{Lw}_\sigma(I) = \langle \mathrm{Lw}_\sigma(f) \mid f \in I \setminus \{0\}\rangle$$

is called the **leading word ideal** of I.
(5) Given a two-sided ideal I in $K\langle X\rangle$, we also let $\mathcal{O}_\sigma(I) = \{w \in \langle X\rangle \mid w \notin \mathrm{Lw}_\sigma(I)\}$.

Note that we did not define the leading word and the leading coefficient of the zero polynomial. Leading words of polynomials satisfy a few simple rules.

Remark 14.1.9. Let σ be a word ordering on $\langle X\rangle$, let $w, w' \in \langle X\rangle$, and let $f, f' \in K\langle X\rangle \setminus \{0\}$.

(1) If $f + f' \neq 0$ then we have $\mathrm{Lw}_\sigma(f + f') \leq_\sigma \max_\sigma \{\mathrm{Lw}_\sigma(f), \mathrm{Lw}_\sigma(f')\}$.
(2) We have $\mathrm{Lw}_\sigma(wfw') = w\, \mathrm{Lw}_\sigma(f)\, w'$.
(3) We have $\mathrm{Lw}_\sigma(ff') = \mathrm{Lw}_\sigma(f)\, \mathrm{Lw}_\sigma(f')$.

Our next example demonstrates that the leading word ideal of a two-sided ideal need not be finitely generated, even if the ideal is finitely generated. In particular, just as in the commutative case, the leading words of the polynomials in a system of generators of I do, in general, not generate the leading word ideal of I.

Example 14.1.10. Let I be the principal ideal in $K\langle x, y\rangle$ generated by $f = x^2 - xy$, and let σ be a word ordering on $\langle x, y\rangle$ such that $x >_\sigma y$. Then for all $i \geq 1$, the polynomials $g_i = xy^i x - xy^{i+1}$ are contained in I, as the equality $g_{i+1} = xy^i f + g_i(y - x)$ and induction show. In particular, it follows that $J = \langle xy^i x \mid i \geq 0\rangle$ is contained in $\mathrm{Lw}_\sigma(I)$. In fact, it is not difficult to verify that J equals $\mathrm{Lw}_\sigma(I)$. (For instance, we could apply the Buchberger Procedure below to the set $\{f, g_1, g_2, \dots\}$ and verify that it is a σ-Gröbner basis of I.)

Thus the ideal I is generated by a single polynomial, but $\mathrm{Lw}_\sigma(I)$ is not finitely generated.

Given a two-sided ideal I, the set $\mathbb{O}_\sigma(I)$ is an **order ideal** in $\langle X \rangle$, that is, it is closed under the formation of subwords. This set plays a prominent role in the following non-commutative version of Macaulay's Basis Theorem.

Proposition 14.1.11 (Macaulay's Basis Theorem). *Let σ be a word ordering on $\langle X \rangle$, and let I be a two-sided in $K\langle X \rangle$. Then the residue classes of the words in $\mathbb{O}_\sigma(I)$ form a K-basis of $K\langle X \rangle / I$.*

Proof. First we show that these residue classes generate $K\langle X \rangle / I$. For this it suffices to show that $B = \langle \mathbb{O}_\sigma(I) \rangle_K + I$ equals $K\langle X \rangle$. Suppose it does not. Then there exists a non-zero polynomial $f \in K\langle X \rangle \setminus B$ having a minimal leading word with respect to σ.

If this leading word is contained in $\mathrm{Lw}_\sigma(I)$, there exists a polynomial $g \in I$ such that $\mathrm{Lw}_\sigma(f) = w\,\mathrm{Lw}_\sigma(g)\,w'$ for some $w, w' \in \langle X \rangle$. But then the polynomial $h = f - (\mathrm{Lc}_\sigma(f)/\mathrm{Lc}_\sigma(g))\,w\,g\,w'$ continues to be contained in $\langle X \rangle \setminus B$ and has a smaller leading word than f, a contradiction.

It remains to consider that case that $\mathrm{Lw}_\sigma(f)$ is not in $\mathrm{Lw}_\sigma(I)$. Hence this word is in $\mathbb{O}_\sigma(I)$ and $g = f - \mathrm{Lc}_\sigma(f)\,\mathrm{Lw}_\sigma(f)$ is contained in $K\langle X \rangle \setminus B$ and has a smaller leading word than f, contradicting the choice of f.

Finally, we prove linear independence. Suppose that there exists a polynomial $f = c_1 w_1 + \cdots + c_s w_s \in I \setminus \{0\}$ such that $c_i \in K$ and $w_i \in \mathbb{O}_\sigma(I)$ for $i = 1, \ldots, s$. Then one of the words w_i is the leading word of f and contained in both $\mathbb{O}_\sigma(I)$ and $\mathrm{Lw}_\sigma(I)$, a contradiction again. $\qquad\square$

Now we are ready to introduce the non-commutative version of Gröbner bases.

Definition 14.1.12. Let σ be a word ordering on $\langle X \rangle$, and let I be a two-sided ideal in $K\langle X \rangle$. A set of polynomials $G \subseteq I$ is called a σ-**Gröbner basis** of I if we have $\mathrm{Lw}_\sigma(I) = \langle \mathrm{Lw}_\sigma(g) \mid g \in G \setminus \{0\} \rangle$.

As we have seen above, non-commutative Gröbner bases may be infinite. Thus there can be no algorithm computing them in general. Nevertheless, we will see that there is an **enumerative procedure**. This is a procedure which computes the elements of a Gröbner basis one by one and has the property that the union of all computed non-commutative polynomials is a Gröbner basis. An important ingredient for this procedure is the non-commutative analogue of the Division Algorithm.

Proposition 14.1.13 (Non-Commutative Division Algorithm). *Let $f, g_1, \ldots, g_s \in K\langle X \rangle \setminus \{0\}$, and let σ be a word ordering on $\langle X \rangle$. Consider the following instructions.*

(1) *Let $q_1 = \cdots = q_s = 0$, $r = 0$ and $h = f$.*

(2) *Repeat the following steps until $h = 0$. Then return $(q_1, \ldots, q_s)$ and r and stop.*

(3) *Repeat the following step as often as possible.*

(4) *Find the smallest $i \in \{1, \ldots, s\}$ such that $\mathrm{Lw}_\sigma(g_i)$ is a subword of $\mathrm{Lw}_\sigma(h)$. If such an i exists, write $\mathrm{Lw}_\sigma(h) = w\,\mathrm{Lw}_\sigma(g_i)w'$ with $w, w' \in \langle X \rangle$, append the triple $(\mathrm{Lc}_\sigma(h)/\mathrm{Lc}_\sigma(g_i), w, w')$ to q_i and replace h by $h - (\mathrm{Lc}_\sigma(h)/\mathrm{Lc}_\sigma(g_i))\,wg_iw'$.*
(5) *Replace r by $r + \mathrm{Lm}_\sigma(h)$ and h by $h - \mathrm{Lm}_\sigma(h)$.*
This is an algorithm which computes tuples $q_1, \ldots, q_s$ of triples $(c_{ij}, w_{ij}, w'_{ij})$ and $r \in K\langle X \rangle$ such that $f = \sum_{i=1}^s \sum_j c_{ij}\,w_{ij}\,g_i\,w'_{ij} + r$ and such that the following conditions hold.
(1) *No word of r is a multiple of any of the words in $\{\mathrm{Lw}_\sigma(g_1), \ldots, \mathrm{Lw}_\sigma(g_s)\}$.*
(2) *For all i, j we have $w_{ij}\,\mathrm{Lw}_\sigma(g_i)\,w'_{ij} \leq_\sigma \mathrm{Lw}_\sigma(f)$.*

The proof of this Division Algorithm is completely analogous to the proof of Proposition 13.1.10. As above, we call the polynomial r returned as part of the output of the Division Algorithm the **normal remainder** of f with respect to $G = (g_1, \ldots, g_s)$ and denote it by $\mathrm{NR}_{\sigma, G}(f)$.

The non-commutative analogues of critical pairs and S-polynomials are defined as follows.

Definition 14.1.14. Let $G = (g_1, \ldots, g_s)$ be a tuple of non-zero elements of $K\langle X \rangle$.
(1) For all $i, j \in \{1, \ldots, s\}$, a quadruple $(w_i, w'_i; w_j, w'_j)$ of words in $\langle X \rangle$ is called an **obstruction** of g_i and g_j if we have $(1/\mathrm{Lc}_\sigma(g_i))\,w_i\,\mathrm{Lw}_\sigma(g_i)\,w'_i = (1/\mathrm{Lc}_\sigma(g_j))\,w_j \cdot \mathrm{Lw}_\sigma(g_j)w'_j$.
(2) For $i \in \{1, \ldots, s\}$, an obstruction of g_i and g_i is called a **self-obstruction** of g_i.
(3) For $i, j \in \{1, \ldots, s\}$, the set of all obstructions of g_i and g_j is denoted by $\mathrm{Obs}(i, j)$.
(4) For every obstruction $\omega = (w_i, w'_i; w_j, w'_j)$ in $\mathrm{Obs}(i, j)$, the non-commutative polynomial $S_{ij}(\omega) = (1/\mathrm{Lc}_\sigma(g_i))\,w_ig_iw'_i - (1/\mathrm{Lc}_\sigma(g_j))\,w_jg_jw'_j$ is called the **S-polynomial** of ω.

Using these definitions, we can now formulate **Buchberger's Procedure** for enumerating non-commutative Gröbner bases. (Sometimes this is also called **Mora's Algorithm**, although it is clearly no algorithm.)

Theorem 14.1.15 (Buchberger's Procedure). *Let σ be a word ordering on $\langle X \rangle$, and let $f_1, \ldots, f_s \in K\langle X \rangle$ be non-zero polynomials which generate a two-sided ideal $I = \langle f_1, \ldots, f_s \rangle$. Consider the following instructions.*
(1) *Let $G = (f_1, \ldots, f_s)$, and let B be the union of all sets $\mathrm{Obs}(i, j)$ such that $1 \leq i \leq j \leq s$.*
(2) *Repeat the following steps until $B = \emptyset$. Then return G and stop.*
(3) *Using a **fair strategy**, choose an obstruction $\omega = (w_i, w'_i; w_j, w'_j) \in B$ and remove it from B. (By a fair strategy we mean a strategy which ensures that every obstruction is eventually selected.)*
(4) *Compute the S-polynomial $S_{ij}(\omega)$ and its normal remainder $S'_{ij}(\omega) = \mathrm{NR}_{\sigma, G}(S_{ij})$. If $S'_{ij}(\omega) = 0$ then continue with step (2).*
(5) *Increase s by one, append $f_s = S'_{ij}(\omega)$ to G, and append all sets $\mathrm{Obs}(i, s)$ such that $1 \leq i \leq s - 1$ to B. Then continue with step (2).*

This is a procedure which enumerates a σ-Gröbner basis G of the ideal I. If the ideal I has a finite σ-Gröbner basis, the procedure will stop after finitely many steps and return a finite σ-Gröbner basis of I.

A proof of this theorem using the current notation is, for instance, contained in [Xiu]. The proof of the preceding theorem is based on the following characterization of non-commutative Gröbner bases.

Proposition 14.1.16 (Buchberger's Criterion). *Let σ be a word ordering on $\langle X \rangle$, let $f_1, \ldots, f_s \in K\langle X \rangle$ be non-zero polynomials which generate a two-sided ideal $I = \langle f_1, \ldots, f_s \rangle$, and let $G = (f_1, \ldots, f_s)$. Then the following conditions are equivalent.*
(1) *The tuple G is a σ-Gröbner basis of I.*
(2) *For every obstruction ω of G, we have $\mathrm{NR}_{\sigma, G}(S_{ij}(\omega)) = 0$.*

Again a proof using the current notation can be found in [Xiu]. Notice that Bucherger's Criterion implies that we can check in finitely many steps whether a given finite set of non-commutative polynomials is a Gröbner basis.

Let us apply the Buchberger Procedure in a concrete example.

Example 14.1.17. In the non-commutative polynomial ring $\mathbb{Z}_2\langle x_1, \ldots, x_6 \rangle$, we consider the two-sided ideal $I = \langle f_1, f_2 \rangle$ generated by $f_1 = x_3(x_1 x_2)^3 + x_4(x_1 x_2)^2 + x_3 + x_4$ and $f_2 = (x_2 x_1)^3 x_5 + (x_2 x_1)^2 x_5 + x_5 + x_6$. Using the word ordering $\sigma = \mathtt{llex}$, we have $\mathrm{Lw}_\sigma(f_1) = x_3(x_1 x_2)^3$ and $\mathrm{Lw}_\sigma(f_2) = (x_2 x_1)^3 x_5$. Let us follow the steps of the Buchberger Procedure.

1. Let $G = (g_1, g_2)$ where $g_1 = f_1$ and $g_2 = f_2$. Let $B = \mathrm{Obs}(1, 1) \cup \mathrm{Obs}(2, 2) \cup \mathrm{Obs}(1, 2)$. The sets $\mathrm{Obs}(1, 1)$ and $\mathrm{Obs}(2, 2)$ contain only obstructions **without overlap**, i.e., obstructions derived from leading words $\mathrm{Lw}_\sigma(g_i)$, $\mathrm{Lw}_\sigma(g_j)$ without a word w such that $\mathrm{Lw}_\sigma(g_i)$ ends in w and $\mathrm{Lw}_\sigma(g_j)$ starts with w. The set $\mathrm{Obs}(1, 2)$ contains obstructions of three types:
   (i) obstructions $\omega_1 = (1, x_1 x_5; x_3 x_1, 1)$, $\omega_2 = (1, x_1 x_2 x_1 x_5; x_3 x_1 x_2 x_1, 1)$, and $\omega_3 = (1, (x_1 x_2)^2 x_1 x_5; x_3(x_1 x_2)^2 x_1 x_5, 1)$
   (ii) all obstructions $(1, w(x_2 x_1)^3 x_5; x_3(x_1 x_2)^3 w, 1)$ with a word $w \in \langle X \rangle$
   (iii) all obstructions $((x_1 x_2)^3 x_5 w, 1; 1, w x_3(x_1 x_2)^3)$ with a word $w \in \langle X \rangle$
4. For all obstructions ω in $\mathrm{Obs}(1, 1) \cup \mathrm{Obs}(2, 2)$, we get $\mathrm{NR}_{\sigma, G}(S_{ii}(\omega)) = 0$.
4. For the obstruction ω_1, we get $\mathrm{NR}_{\sigma, G}(S_{12}(\omega_1)) = x_3 x_1 x_6 + x_4 x_1 x_5$.
5. We append $g_3 = x_3 x_1 x_6 + x_4 x_1 x_5$ to G and update B.
4. For ω_2, we get $\mathrm{NR}_{\sigma, G}(S_{12}(\omega_2)) = x_3 x_1 x_2 x_1 x_6 + x_4 x_1 x_2 x_1 x_5$.
5. We append $g_4 = x_3 x_1 x_2 x_1 x_6 + x_4 x_1 x_2 x_1 x_5$ to G and update B.
4. For ω_3, we get $\mathrm{NR}_{\sigma, G}(S_{12}(\omega 3)) = x_3(x_1 x_2)^2 x_1 x_6 + x_4(x_1 x_2)^2 x_1 x_5$.
5. We append $g_5 = x_3(x_1 x_2)^2 x_1 x_6 + x_4(x_1 x_2)^2 x_1 x_5$ to G and update B.
4. For the obstructions ω of type (ii) and (iii), we get $\mathrm{NR}_{\sigma, G}(S_{12}(\omega)) = 0$.
4. For all obstructions ω in $\mathrm{Obs}(i, j)$ with $j \in \{3, 4, 5\}$ and $1 \leq i \leq j$, we get $\mathrm{NR}_{\sigma, G}(S_{ij}(\omega)) = 0$.
2. The procedure stops and returns $G = (g_1, \ldots, g_5)$.

Altogether, the tuple $G = (g_1, \ldots, g_5)$ is a σ-Gröbner basis of I.

One important application of the Buchberger Procedure is the following **Ideal Membership Test** for two-sided ideals in $K\langle X \rangle$:

Corollary 14.1.18. *Given a σ-Gröbner basis G of a two-sided ideal I in $K\langle X \rangle$ and $f \in K\langle X \rangle$, we have $f \in I$ if and only if $\mathrm{NR}_{\sigma,G}(f) = 0$.*

Proof. Clearly, if $\mathrm{NR}_{\sigma,G}(f) = 0$ then we can collect the reductions and arrive at a representation of f as an element of the ideal generated by G, i.e., as an element of I. Conversely, if we have $f \in I$, then the normal remainder $\mathrm{NR}_{\sigma,G}(f)$ is also contained in I. It follows that it has to be zero, since otherwise its leading term would be in $\mathcal{O}_\sigma(I) = \langle X \rangle \setminus \mathrm{Lw}_\sigma(I)$. □

The Buchberger Procedure can be applied to some problems for finitely presented groups introduced in Section 9.8. For this purpose, we need to introduce the following ring.

Definition 14.1.19. Let G be a group and K a field. Then the K-vector space $K[G] = \bigoplus_{g \in G} K \cdot g$ has a natural ring structure given by the K-linear extension of the multiplication in G. The resulting ring is called the **group ring** of G over K.

In other words, for two elements $\sum_{g \in G} a_g\, g$ and $\sum_{g \in G} b_g\, g$ of $K[G]$, we let

$$\left(\sum_{g \in G} a_g\, g \right) \cdot \left(\sum_{g \in G} b_g\, g \right) = \sum_{g,h \in G} a_g b_h\, (g \cdot h)$$

where only finitely many coefficients $a_g b_h$ are non-zero.

If $G = \langle x_1, \ldots, x_n; r_1 = \cdots = r_m = 1 \rangle$ is a finitely presented group, we can represent the group ring over K by

$$K[G] = K\langle x_1, \ldots, x_n, y_1, \ldots, y_n \rangle / \langle x_i y_i - 1, y_i x_i - 1, r_j - 1 \mid i = 1 \ldots n, j = 1 \ldots m \rangle.$$

Here the indeterminates y_i represent the inverses of the residue classes of the elements x_i and the relators r_i have to be written as words in x_i, y_i.

Remark 14.1.20. Let $G = \langle x_1, \ldots, x_n; r_1 = \cdots = r_m = 1 \rangle$ be a finitely presented group, and let w be a word in the letters $\{x_1, \ldots, x_n, y_1, \ldots, y_n\}$, where y_i represents x_i^{-1}. The **word problem** in G asks us to decide effectively whether w represents the identity element of G.

The following instructions provide a **semi-decision procedure** for the word problem in G which is based on Buchberger's Procedure. Here "semi-decision" means that the procedure will terminate and give the correct answer if w represents the identity element of G. However, if the correct answer is "no", the procedure terminates and gives the correct answer only if the ideal defining $K[G]$ has a finite Gröbner basis with respect to the chosen word ordering. If the Gröbner basis is infinite, the procedure will run forever and never produce an answer.

(1) Consider the non-commutative polynomial ring $K\langle x_1, \ldots, x_n, y_1, \ldots, y_n \rangle$ and the two-sided ideal I defining $K[G]$ given above. Let H be the stated tuple of generators of I. Choose a word ordering σ.

(2) Compute $\mathrm{NR}_{\sigma,H}(w)$. If the result is zero, return YES and stop.

(3) Run one iteration of the Buchberger procedure, starting with the system of generators H. Afterwards, update H to the resulting partial Gröbner basis.

(4) If we have $B = \emptyset$ in the Buchberger Procedure, i.e., if the computed tuple H is indeed a Gröbner basis of I and if $\mathrm{NR}_{\sigma,H}(w) \neq 0$, return NO and stop. Otherwise, continue with step 1.

Recall that the Buchberger Procedure enumerates a σ-Gröbner basis of I. If w represents the identity element, the normal remainder of w with respect to this Gröbner basis is zero. In this reduction to zero only finitely many Gröbner basis elements are involved. Therefore, if we repeat steps 1 and 2 often enough, all necessary Gröbner basis elements will have been found and the procedure stops with the correct answer. Moreover, if the Buchberger Procedure produces a finite Gröbner basis, the answer is correct by the above Ideal Membership Test.

14.2 Elimination and its Applications

In the following we continue to use the setting of the last section. We have a field K, a set of letters $X = \{x_1, \ldots, x_n\}$, and the non-commutative polynomial ring $K\langle X \rangle$. Elimination is an important technique in computer algebra and will be essential for Gröbner basis cryptography. The following definition provides the necessary terminology.

Definition 14.2.1. Let $L \subset X$ be a subset of the given set of letters, and let $\widehat{X} = X \setminus L$.

(1) A word ordering σ on $\langle X \rangle$ is called an **elimination ordering** for L, if every polynomial $f \in K\langle X \rangle \setminus \{0\}$ such that $\mathrm{Lw}_\sigma(f) \in \langle \widehat{X} \rangle$ satisfies $f \in K\langle \widehat{X} \rangle$.

(2) Given a two-sided ideal $I \subseteq K\langle X \rangle$, the ideal $\widehat{I} = I \cap K\langle \widehat{X} \rangle$ is called the **elimination ideal** of I with respect to L.

It is clear that $\widehat{I}$ is a two-sided ideal in $K\langle \widehat{X} \rangle$. It consists of all non-commutative polynomials in I which do not involve the letters from L. Let us show that elimination orderings exist.

Example 14.2.2. Let $\ell \in \{1, \ldots, n\}$, and let $L = \{x_1, \ldots, x_\ell\}$. We define a word ordering elim on $\langle X \rangle$ as follows. Given a word $w \in \langle X \rangle$ and a letter x_i, the number $\deg_{x_i}(w)$ is the number of occurrences of the letter x_i in w. Given two words $w_1, w_2 \in \langle X \rangle$, we let $w_1 \leq_{\mathrm{elim}} w_2$ if and only if there exists an index $j \in \{1, \ldots, n\}$ such that $\deg_{x_i}(w_1) = \deg_{x_i}(w_2)$ for $i < j$ and $\deg_{x_j}(w_1) < \deg_{x_j}(w_2)$, or if $\deg_{x_i}(w_1) = \deg_{x_i}(w_2)$ for $i = 1, \ldots, n$ and $w_1 \leq_{\mathrm{lex}} w_2$.

It is easy to check that elim is an elimination ordering for L, independent of the actual value of ℓ.

For instance, in $X = \langle x_1, x_2 \rangle$ we have $x_1 >_{\texttt{elim}} x_2^2$ and $x_1 x_2^2 <_{\texttt{elim}} x_2^3 x_1$ and $x_1 x_2^2 >_{\texttt{elim}} x_2^2 x_1$.

In the following we let $L \subset \{1, \ldots, n\}$ and $\widehat{X} = X \setminus L$. The property of σ being an elimination ordering for L can be rephrased by saying that if the letters of L do not occur in a word $w \in \langle X \rangle$, they do not occur in any word $w' \in \langle X \rangle$ such that $w' \leq_\sigma w$.

The following lemma is easily verified and will be used in the main theorem below.

Lemma 14.2.3. *Let σ be a word ordering on $\langle X \rangle$, and let $\widehat{\sigma}$ be the restriction of σ to $\langle \widehat{X} \rangle$. Then also $\widehat{\sigma}$ is a word ordering.*

Now we are ready to state and prove the main theorem of this section.

Theorem 14.2.4 (Computation of Elimination Ideals). *Let $L \subset X$, let σ be an elimination ordering for L, and let $\widehat{\sigma}$ be the restriction of σ to $\langle \widehat{X} \rangle$. Given a two-sided ideal $I \subseteq K\langle X \rangle$ and a σ-Gröbner basis G of I, the set $\widehat{G} = G \cap K\langle \widehat{X} \rangle$ is a $\widehat{\sigma}$-Gröbner basis of the elimination ideal $\widehat{I} = I \cap K\langle \widehat{X} \rangle$.*

Proof. It is clear that $\widehat{G}$ is contained in $\widehat{I}$. We have to show that the leading words of the polynomials in $\widehat{G}$ generate the leading word ideal $\mathrm{Lw}_{\widehat{\sigma}}(\widehat{I})$. For $f \in \widehat{I} \setminus \{0\}$, we have $\mathrm{Lw}_\sigma(f) = \mathrm{Lw}_{\widehat{\sigma}}(f) \in \langle \widehat{X} \rangle$. Since G is a σ-Gröbner basis of I, we find words $w_1, w_2 \in \langle X \rangle$ and $g \in G$ such that $\mathrm{Lw}_\sigma(f) = w_1 \mathrm{Lw}_\sigma(g) w_2$. As this word is contained in $\langle \widehat{X} \rangle$, it follows that $w_1, w_2, \mathrm{Lw}_\sigma(g) \in \langle \widehat{X} \rangle$. From the hypothesis that σ is an elimination ordering, we get that $g \in K\langle \widehat{X} \rangle$, and therefore $g \in \widehat{G}$. This was to be shown. $\qquad\square$

A first application of this theorem is the following method for computing the intersection of two-sided ideals in $K\langle X \rangle$.

Proposition 14.2.5. *Let $f_1, \ldots, f_s, g_1, \ldots, g_t \in K\langle X \rangle$, let $I = \langle f_1, \ldots, f_s \rangle$ and $J = \langle g_1, \ldots, g_t \rangle$ be the two-sided ideals generated by these non-commutative polynomials, and let y be a further letter. Then we have*

$$I \cap J = \langle y f_1, \ldots, y f_s, (1-y) g_1, \ldots, (1-y) g_t, y x_1 - x_1 y, \ldots, y x_n - x_n y \rangle \cap K\langle X \rangle.$$

In particular, a system of generators of the intersection ideal $I \cap J$ can be enumerated using the Buchberger Procedure.

Proof. Given $h \in I \cap J$, we can write $h = p_1 f_1 p_1' + \cdots + p_s f_s p_s'$ and $h = q_1 g_1 q_1' + \cdots + q_t g_t q_t'$ with $p_i, p_i', q_j, q_j' \in K\langle X \rangle$. Then we have

$$h = yh + (1-y)h = \sum_{i=1}^{s} y p_i f_i p_i' + \sum_{j=1}^{t} (1-y) q_j g_j q_j'$$

$$= \sum_{i=1}^{s} p_i y f_i p_i' + \sum_{j=1}^{t} q_j (1-y) g_j q_j' + r$$

with a polynomial $r \in \langle y x_1 - x_1 y, \ldots, y x_n - x_n y \rangle$. Hence h is contained in the right-hand side of the claimed formula.

Conversely, let h be contained in this right-hand side. Then we have $p_i, p_i', q_j, q_j' \in K\langle X, y\rangle$ such that $h = \sum_{i=1}^{s} p_i y f_i p_i' + \sum_{j=1}^{t} q_j(1-y)g_j q_j' + r$ with $r \in \langle yx_i - x_i y \mid i = 1, \dots, n\rangle$. Substituting $y \mapsto 1$ in this representation shows $h \in I$, and substituting $y \mapsto 0$ shows $h \in J$. ☐

It is clear that this proposition can be extended to intersections of finitely many two-sided ideals in $K\langle X\rangle$ in a straightforward way. The main application of elimination is the possibility to compute kernels and images of K-algebra homomorphisms. For this purpose we use the following setting.

Let $Y = \{y_1, \dots, y_m\}$ be a further set of letters, and let $I \subset K\langle X\rangle$ and $J \subset K\langle Y\rangle$ be two-sided ideals. We consider a homomorphism of K-algebras

$$\phi : K\langle X\rangle/I \longrightarrow K\langle Y\rangle/J$$

given by $\phi(\bar{x}_i) = \bar{f}_i$ with $f_i \in K\langle Y\rangle$ for $i = 1, \dots, n$.

Proposition 14.2.6 (Kernels of Algebra Homomorphisms). *The kernel of ϕ satisfies*

$$\mathrm{Ker}(\phi) = ((\langle x_1 - f_1, \dots, x_n - f_n\rangle + J \cdot K\langle X, Y\rangle) \cap K\langle X\rangle) + I.$$

In particular, it can be enumerated by applying the Buchberger Procedure to compute this elimination ideal.

Proof. Given $g \in K\langle X\rangle$ such that $\bar{g} \in \mathrm{Ker}(\phi)$, we have $g(f_1, \dots, f_n) \in J$. By replacing f_i by $(f_i - x_i) + x_i$ in this equality and multiplying out, we obtain $g = h + g(f_1, \dots, f_n) \in \langle x_1 - f_1, \dots, x_n - f_n\rangle + J$ where $h \in \langle x_1 - f_1, \dots, x_n - f_n\rangle$. Hence $\bar{g}$ is contained in the right-hand side.

Conversely, let $g \in K\langle X\rangle$ be such that $\bar{g}$ is contained in the right-hand side. Then there exist $p_i, p_i' \in K\langle X, Y\rangle$ such that $g = \sum_{i=1}^{n} p_i(x_i - f_i)p_i' + h$ with $h \in J \cdot K\langle X, Y\rangle$. Substituting $x_i \mapsto f_i$ in this equality shows $g(f_1, \dots, f_n) \in J$, and therefore $\bar{g} \in \mathrm{Ker}(\Phi)$. ☐

One application of this procedure is that we have a semi-decision procedure for checking whether an element $\bar{f}$ of a finitely generated algebra $K\langle X\rangle/I$ is **algebraic** or **transcendental** over K. Namely, we can apply the preceding proposition to the K-algebra homomorphism

$$\phi : K[z] \longrightarrow K\langle X\rangle/I \quad \text{given by } z \mapsto \bar{f}.$$

If the procedure finds a non-trivial element in $\mathrm{Ker}(\phi)$, we can also compute the **minimal polynomial** of $\bar{f}$, i.e., the monic polynomial which generates $\mathrm{Ker}(\phi)$.

Another application of the preceding proposition is a semi-decision procedure for checking if an element of a monoid or group has a finite order. This can be done as follows.

Remark 14.2.7. Let $G = \langle x_1, \dots, x_n; r_1 = \dots = r_m = 1\rangle$ be a finitely presented monoid having the cancellation property or a monoid presentation of a finitely presented group. Then the monoid ring over $\mathbb{Z}_2$ satisfies $\mathbb{Z}_2[G] = \mathbb{Z}_2\langle X\rangle/I$ where $I = \langle r_1 - 1, \ldots, r_m - 1\rangle$. Let $w \in \langle X\rangle$ be a word representing an element $\bar{w}$ of G.

To check whether the order of $\bar{m}$ is finite, we choose an additional letter y and form the ideal $J = \langle y - w, r_1 - 1, \ldots, r_m - 1\rangle$ in $\mathbb{Z}_2\langle X, y\rangle$. Then we let σ be an elimination ordering for X and start the computation of a σ-Gröbner basis G of J.

(1) If we have $G \cap \mathbb{Z}_2[y] \neq \emptyset$ at some point during the computation, the element $\bar{w}$ is algebraic over $\mathbb{Z}_2$. Since the ideal J is generated by binomials, its Gröbner basis consists of binomials. Thus the minimal polynomial of $\bar{w}$ is of the form $z^k + z^\ell$ with $1 \leq k < \ell$. Since G has the cancellation property, it follows that $\bar{w}^{\ell-k} = 1$. Notice that in this case we can compute the order of $\bar{w}$.

(2) If the computation finishes with a Gröbner basis G with $G \cap \mathbb{Z}_2[y] = \emptyset$ then we have $\mathrm{Ker}(\phi) = \{0\}$ and the element $\bar{w}$ has infinite order.

The next proposition provides a semi-decision procedure for checking if an element is in the image of an algebra homomorphism, and yields a preimage if the answer is positive.

Proposition 14.2.8 (Images of Algebra Homomorphisms). *Let $\phi : K\langle X\rangle/I \longrightarrow K\langle Y\rangle/J$ be a K-algebra homomorphism with $\phi(\bar{x}_i) = \bar{f}_i$ as above. Let $D \subseteq K\langle X, Y\rangle$ be the* **diagonal ideal**

$$D = \langle x_1 - f_1, \ldots, x_n - f_n\rangle + J \cdot K\langle X, Y\rangle,$$

let σ be an elimination ordering for Y, and let G be a σ-Gröbner basis of D.
(1) *For $g \in K\langle Y\rangle$, the element $\bar{g}$ is contained in the image of ϕ if and only if $\mathrm{NR}_{\sigma,G}(g) \in K\langle X\rangle$, i.e., if and only if this normal remainder does not involve any letter from Y.*
(2) *Given $g \in K\langle Y\rangle$ such that $g' = \mathrm{NR}_{\sigma,G}(g)$ is contained in $K\langle X\rangle$, we have $\bar{g} = \phi(\bar{g}')$.*

Proof. First we let $g \in K\langle Y\rangle$ such that $g' = \mathrm{NR}_{\sigma,G}(g) \in K\langle X\rangle$. Then there exist elements $p_i, p_i' \in K\langle X, Y\rangle$ and $q \in J \cdot K\langle X, Y\rangle$ such that $g - g'(f_1, \ldots, f_m) = \sum_{i=1}^n p_i(x_i - f_i)p_i' + q$. Substituting $x_i \mapsto f_i$ in this equality yields $g - g'(f_1, \ldots, f_n) \in J$, and therefore $\bar{g} = \phi(\bar{g}')$.

Conversely, let $g \in K\langle Y\rangle$ be such that $\bar{g}$ is contained in the image of ϕ. Then there exists a polynomial $h \in K\langle X\rangle$ such that $g + J = h(f_1, \ldots, f_n) + J$. Since we have $h + D = h(f_1, \ldots, f_n) + D$, we get $g - h \in D$, and hence $\mathrm{NR}_{\sigma,G}(g) = \mathrm{NR}_{\sigma,G}(h)$. Now the facts that $h \in K\langle X\rangle$ and that σ is an elimination ordering for X imply $\mathrm{NR}_{\sigma,G}(g) \in K\langle X\rangle$. $\square$

Based on this proposition, the Buchberger Procedure yields a semi-decision procedure for the **Subalgebra Membership Problem**: given a finitely generated K-subalgebra $S = K\langle \bar{f}_1, \ldots, \bar{f}_n\rangle$ of a K-algebra $K\langle Y\rangle/J$, decide whether a given residue class $\bar{g}$ is contained in S. In the YES case, i.e., if we have $\bar{g} \in S$, this will be detected in finitely many steps and the procedure returns an explicit representation of $\bar{g}$ in terms of the generators $\{\bar{f}_1, \ldots, \bar{f}_n\}$ of the subalgebra S. Otherwise, the procedure may or may not stop.

In group theory, the proposition allows us to semi-decide subgroup membership as follows.

Remark 14.2.9. Let $G = \langle x_1, \ldots, x_n; r_1 = \cdots = r_m = 1 \rangle$ be a finitely presented group, and let $H \subseteq G$ be the subgroup generated by the residue classes of some words $h_1, \ldots, h_k \in \langle X \rangle$. The **subgroup membership problem** (or the **generalized word problem**) asks us to decide whether a given word $w \in \langle X \rangle$ represents an element of H. The following instructions define a semi-decision procedure for this problem.

(1) Let $\{y_1, \ldots, y_n\}$ be further letters. Consider the presentation

$$\mathbb{Z}_2[G] = \mathbb{Z}_2\langle x_1, \ldots, x_n, x_1', \ldots, x_n' \rangle / \langle r_i - 1, \; x_j y_j - 1, y_j x_j - 1 \rangle_{\substack{i=1,\ldots,m \\ j=1,\ldots,n}}.$$

(2) Using the subalgebra membership test, check whether we have $\bar{w} - 1 \in \mathbb{Z}_2\langle \bar{h}_1 - 1, \ldots, \bar{h}_k - 1 \rangle$ inside $\mathbb{Z}_2\langle X, Y \rangle / I$, where I is the defining ideal for $\mathbb{Z}_2[G]$ given above.

To prove correctness of this method, we still have to check that $\bar{w} \in H$ is equivalent to $\bar{w} - 1 \in \mathbb{Z}_2\langle \bar{h}_1 - 1, \ldots, \bar{h}_k - 1 \rangle$. Let $\bar{w} = \bar{h}_{i_1} \cdots \bar{h}_{i_s}$. We proceed by induction on s and find that

$$\bar{w} - 1 = (\bar{h}_{i_1} \cdots \bar{h}_{i_{s-1}} - 1)\bar{h}_{i_s} + (\bar{h}_{i_s} - 1) \in \mathbb{Z}_2\langle \bar{h}_1 - 1, \ldots, \bar{h}_k - 1 \rangle.$$

For a proof of the reverse implication we refer to [MR], Theorem 9.

To use Gröbner bases for other hard problems underlying non-commutative cryptography (such as the problems explained in Section 9.8), we need to generalize Gröbner basis theory from two-sided ideals in $K\langle X \rangle$ to two-sided submodules of a free module over this ring. In particular, we need the concept and calculation of syzygies. These topics are introduced in the next section.

14.3 Gröbner Bases of $K\langle X \rangle$-Modules

Let K be a field, let $X = \{x_1, \ldots, x_n\}$ be a set of letters, and let $K\langle X \rangle$ be the non-commutative polynomial ring over K in the indeterminates $x_1, \ldots, x_n$. The **enveloping algebra** $K\langle X \rangle^{\mathrm{env}} = K\langle X \rangle \otimes_K K\langle X \rangle$ is a two-sided module over $K\langle X \rangle$ in the obvious way: for $f_1, f_2, g_1, g_2 \in K\langle X \rangle$, we let $f_1(g_1 \otimes g_2)f_2 = f_1 g_1 \otimes g_2 f_2$.

In the category of all two-sided $K\langle X \rangle$-modules, the enveloping algebra $K\langle X \rangle^{\mathrm{env}}$ plays the role of a free module of rank 1: given any two-sided $K\langle X \rangle$-module M and an element $v \in M$, there is a unique homomorphism of two-sided $K\langle X \rangle$-modules $\psi: K\langle X \rangle^{\mathrm{env}} \longrightarrow M$ such that $\psi(1 \otimes 1) = v$. Consequently, for every $r \geq 1$, the two-sided $K\langle X \rangle$-module $F_r = \bigoplus_{i=1}^r K\langle X \rangle^{\mathrm{env}}$ is a free module of rank r in this category. For $i = 1, \ldots, r$, we let $e_i = (0, \ldots, 0, 1 \otimes 1, 0, \ldots, 0)$ where $1 \otimes 1$ occurs in the i-th position.

In this section we extend the Gröbner basis theory for two-sided ideals in $K\langle X \rangle$ to a Gröbner basis theory for two-sided submodules of F_r. The main motivation for this is the following notion.

Definition 14.3.1. Let M be a two-sided $K\langle X \rangle$-module, let $v_1, \ldots, v_r \in M$, and let $\psi: F_r \longrightarrow M$ be the uniquely determined homomorphism of two-sided $K\langle X \rangle$-modules satisfying $\psi(e_i) = v_i$ for $i = 1, \ldots, r$. Then the two-sided $K\langle X \rangle$-submodule

$$\mathrm{Syz}(v_1, \ldots, v_r) = \mathrm{Ker}(\psi) = \left\{ \sum_{i=1}^{r} f_i e_i g_i \in F_r \mid \sum_{i=1}^{r} f_i v_i g_i = 0 \right\}$$

is called the **(two-sided) syzygy module** of $(v_1, \ldots, v_r)$.

The computation of syzygy modules is a fundamental technique in computational algebra and is used frequently in non-commutative cryptography. Let us start the development of Gröbner basis theory in this setting.

Definition 14.3.2. A **term** in F_r is an element of the form $w e_i w'$ where $w, w' \in \langle X \rangle$ and $1 \le i \le r$. The set of all terms in F_r is denoted by $\mathbb{T}(F_r)$.

A **module term ordering** on $\mathbb{T}(F_r)$ is a total ordering τ such that $t_1 \le_\tau t_2$ implies $w_1 t_1 w_1' \le_\tau w_2 t_2 w_2'$ for all $t_1, t_2 \in \mathbb{T}(F_r)$ and all $w_1, w_1', w_2, w_2' \in \langle X \rangle$ and such that τ is a well-ordering.

Starting from a word ordering on $\langle X \rangle$, we can define module term orderings as follows.

Example 14.3.3. Let To be a word ordering on $\langle X \rangle$.
(1) For terms $w_1 e_i w_1', w_2 e_j w_2' \in \mathbb{T}(F_r)$ such that $w_1, w_1', w_2, w_2' \in \langle X \rangle$ and $i, j \in \{1, \ldots, r\}$, we let

$$w_1 e_i w_1' \ge_{\mathrm{ToPos}} w_2 e_j w_2' \iff w_1 w_1' >_{\mathrm{To}} w_2 w_2' \text{ or}$$
$$(w_1 w_1' = w_2 w_2' \text{ and } w_1 >_{\mathrm{To}} w_2) \text{ or}$$
$$(w_1 = w_2 \text{ and } i \le j).$$

This defines a module term ordering ToPos on $\mathbb{T}(F_r)$.
(2) For terms $w_1 e_i w_1', w_2 e_j w_2' \in \mathbb{T}(F_r)$ such that $w_1, w_1', w_2, w_2' \in \langle X \rangle$ and $i, j \in \{1, \ldots, r\}$, we let

$$w_1 e_i w_1' \ge_{\mathrm{PosTo}} w_2 e_j w_2' \iff i < j \text{ or}$$
$$(i = j \text{ and } w_1 w_1' >_{\mathrm{To}} w_2 w_2') \text{ or}$$
$$(i = j \text{ and } w_1 w_1' = w_2 w_2' \text{ and } w_1 \ge_{\mathrm{To}} w_2).$$

Again this defines a module term ordering PosTo on $\mathbb{T}(F_r)$.

Given a module term ordering, the following terminology generalizes the usual definitions of leading terms etc.

Definition 14.3.4. Let τ be a module term ordering on $\mathbb{T}(F_r)$.
(1) Given a vector $v \in F_r \setminus \{0\}$, there exists a unique representation $v = c_1 t_1 + \cdots + c_s t_s$ with $c_1, \ldots, c_s \in K \setminus \{0\}$ and $t_1, \ldots, t_s \in \mathbb{T}(F_r)$ satisfying $t_1 >_\tau \cdots >_\tau t_s$. The term $\mathrm{Lt}_\tau(v) = t_1$ is called the **leading term** of v with respect to τ. The element $\mathrm{Lc}_\tau(v) = c_1$ is called its **leading coefficient**.
(2) For a two-sided submodule $M \subseteq F_r$, the two-sided submodule $\mathrm{Lt}_\tau(M) = \langle \mathrm{Lt}_\tau(v) \mid v \in M \setminus \{0\} \rangle$ of F_r is called the **leading term module** of M.

(3) A subset G of a two-sided submodule M of F_r is called a τ-**Gröbner basis** of M if the leading term module $\mathrm{Lt}_\tau(M)$ is generated by the leading terms $\mathrm{Lt}_\tau(f)$ such that $f \in G$.

Based on these definitions, many standard results of Gröbner basis theory generalize in a straightforward way. Let us mention three of them. For detailed proofs we refer the reader to [BK], Section 2.

Proposition 14.3.5 (Macaulay's Basis Theorem). *Let M be a two-sided submodule of F_r. Then the residue classes of the elements in $\mathbb{T}(F_r) \setminus \mathrm{Lt}_\tau(M)$ form a K-vector space basis of F^r/M.*

Proposition 14.3.6 (The Division Algorithm). *Let $s \geq 1$, and let $m, f_1, \ldots, f_s \in F_r \setminus \{0\}$. Consider the following sequence of instructions.*

(1) *For $i = 1, \ldots, s$ let $k_i = 1$, $g_{i1} = g'_{i1} = 0$, $p = 0$ and $v = m$.*
(2) *Find the smallest $i \in \{1, \ldots, s\}$ such that $\mathrm{Lt}_\tau(v) = w\, \mathrm{Lt}_\tau(f_i)w'$ for some $w, w' \in \langle X \rangle$. If such an i exists, increase s by 1, set $g_{ik_i} = \frac{\mathrm{Lc}_\tau(v)}{\mathrm{Lc}_\tau(f_i)}w$, $g'_{ik_i} = w'$ and replace v by $v - \frac{\mathrm{Lc}_\tau(v)}{\mathrm{Lc}_\tau(f_i)}wf_iw'$. If now $v \neq 0$, continue with step 2. Otherwise, continue with step 4.*
(3) *Replace p by $p + \mathrm{Lc}_\tau(v) \cdot \mathrm{Lt}_\tau(v)$ and v by $v - \mathrm{Lc}_\tau(v) \cdot \mathrm{Lt}_\tau(v)$. If now $v \neq 0$, continue with step 2.*
(4) *Return the tuple $((g_{11}, g'_{11}), \ldots, (g_{1k_1}, g'_{1k_1}), \ldots, (g_{s1}, g'_{s1}), \ldots, (g_{sk_s}, g'_{sk_s}))$ and the vector $p \in F_r$.*

This is an algorithm which returns elements $((g_{11}, g'_{11}), \ldots, (g_{sk_s}, g'_{sk_s}))$ and p such that the following conditions are satisfied.

(1) *We have $m = \sum_{i=1}^{s} \sum_{j=1}^{k_i} g_{ij}f_ig'_{ij} + p$.*
(2) *No element of $\mathrm{Supp}(p)$ is contained in $\langle \mathrm{Lt}_\tau(f_1), \ldots, \mathrm{Lt}_\tau(f_s) \rangle$.*
(3) *If $g_{ij} \neq 0 \neq g'_{ij}$ for some $i \in \{1, \ldots, s\}$ and $j \in \{1, \ldots, k_i\}$ then we have $\mathrm{Lt}_\tau(g_{ij}f_ig'_{ij}) \leq_\tau \mathrm{Lt}_\tau(m)$.*
(4) *For all $i \in \{1, \ldots, s\}$ and $j \in \{1, \ldots, k_i\}$ we have*

$$g_{ij}\, \mathrm{Lt}_\tau(f_i)g'_{ij} \notin \langle \mathrm{Lt}_\tau(f_1), \ldots, \mathrm{Lt}_\tau(f_{i-1}) \rangle.$$

(5) *The elements $((g_{11}, g'_{11}), \ldots, (g_{sk_s}, g'_{sk_s}))$ and p are uniquely determined by the preceding conditions (1)–(4).*

Definition 14.3.7. Let $G = (g_1, \ldots, g_s)$ be a tuple of elements $g_i \in F_r$.

(1) A pair (i, j) with $i, j \in \{1, \ldots, s\}$ and $i < j$ is called a **critical pair** of G if there exist words $w_i, w'_i, w_j, w'_j \in \langle X \rangle$ such that $w_i\, \mathrm{Lt}_\tau(g_i)w'_i = w_j\, \mathrm{Lt}_\tau(g_j)w'_j$, such that w_i and w_j have no common prefix, and such that w'_i and w'_j have no common suffix.
(2) For every critical pair (i, j) of G, the element

$$S_{ij} = \frac{1}{\mathrm{Lc}_\tau(g_i)} w_ig_iw'_i - \frac{1}{\mathrm{Lc}_\tau(g_j)} w_jg_jw'_j$$

is called the **S-vector** of g_i and g_j.

Proposition 14.3.8 (Buchberger's Criterion). *Let* $G = \{g_i \mid i \in I\}$ *be a (countable) set of elements in* F_r *which generate a two-sided submodule* $M = \langle G \rangle$ *of* F_r, *and let B be the set of critical pairs between elements of G. Then the set G is a* τ-*Gröbner basis of M if and only if* $\mathrm{NR}_{\tau,G}(S_{ij}) = 0$ *for all* $(i, j) \in B$.

Finally, we are ready to formulate the module analogue of Buchberger's Procedure.

Theorem 14.3.9 (Buchberger's Procedure for Modules). *Let* $G = \{g_1, \ldots, g_s\}$ *be a finite set of elements in* $F_r \setminus \{0\}$ *which generates a two-sided* $K\langle X \rangle$-*submodule* $M = \langle G \rangle$ *of* F_r. *Consider the following sequence of instructions.*

(1) *Let* $H = G$, *let B be the set of critical pairs of H, and let* $s' = s$.

(2) *If* $B = \emptyset$, *return G and stop. Otherwise choose a pair* $(i, j) \in B$ *using a fair strategy and delete it from B.*

(3) *Compute the S-vector* S_{ij} *and its normal remainder* $\mathrm{NR}_{\tau,H}(S_{ij})$. *If the result is zero, continue with step 2.*

(4) *Increase* s' *by one. Append* $g_{s'} = \mathrm{NR}_{\tau,H}(S_{ij})$ *to H, and append the elements of* $\{(i, s') \mid 1 \le i < s' \text{ and } (i, s') \text{ is a critical pair}\}$ *to B. Continue with step 2.*

This is a procedure which enumerates a tuple H of vectors which form a τ-*Gröbner basis of M. If M has a finite* τ-*Gröbner basis, the procedure stops after finitely many steps and the vectors of the resulting tuple H form a finite* τ-*Gröbner basis of M.*

A proof using the current notation and terminology is contained in [BK], Section 2. Moreover, it is shown there that Gröbner bases for two-sided submodules of F_r generalize Gröbner bases for two-sided ideals in $K\langle X \rangle$ in the following way: under the epimorphism $\pi : F_1 \longrightarrow K\langle X \rangle$ given by $e_1 \mapsto 1$, two-sided submodules of F_1 containing $\langle x_1 e_1 - e_1 x_1, \ldots, x_n e_1 - e_1 x_n \rangle$ correspond uniquely to two-sided ideals in $K\langle X \rangle$, and their Pos–σ Gröbner bases correspond to σ-Gröbner bases of their images in the obvious way.

It is clear that the Buchberger Procedure for Modules results in a semi-decision procedure for submodule membership. Since this in completely analogous to the ideal case, we leave it to the interested reader. Besides the analogues of the applications of Gröbner bases introduced in Section 4.2, we have one new application which is particular to modules and which is explained next.

In the following we let L be a subset of $\{1, \ldots, r\}$, and we let $\widehat{F}_r$ denote the free two-sided $K\langle X \rangle$-module generated by $\{e_i \mid i \in \{1, \ldots, r\} \setminus L\}$.

Definition 14.3.10. A module term ordering τ on $\mathbb{T}(F_r)$ is called a **component elimination ordering** for L if every element $m \in F_r \setminus \{0\}$ such that $\mathrm{Lt}_\tau(m) \in \widehat{F}_r$ is contained in $\widehat{F}_r$.

Let $M \subseteq F_r$ be a two-sided submodule. The two-sided submodule $M \cap \widehat{F}_r$ of $\widehat{F}_r$ is called the **component elimination module** of M with respect to L.

The following example shows that component elimination orderings exist.

Example 14.3.11. Let $i \in \{1,\ldots,r\}$, and let $L = \{1,\ldots,i\}$. If σ is a term ordering on $\langle X \rangle$ then the module ordering $\tau = \text{Pos-}\sigma$ is a component elimination ordering for L. Namely, let $m \in F_r \setminus \{0\}$ be such that $\text{Lt}_\tau(M) = w_1 e_j w_1' \in \hat{F}_r$. Then every term $t = w_2 e_k w_2' \in \text{Supp}(m)$ satisfies $t \leq_\tau \text{Lt}_\tau(m)$. This implies $k \geq j$, and we conclude that $t \in \hat{F}_r$ and $m \in \hat{F}_r$.

The following proposition shows how one can compute component elimination modules. In fact, it yields a Gröbner basis with respect to the restriction to $\hat{F}_r$ of the given component elimination ordering.

Proposition 14.3.12 (Computing Component Elimination Modules). *Let M be a two-sided submodule of F_r, let $L \subseteq \{1,\ldots,r\}$, and let τ be a component elimination ordering for L. Furthermore, let G be a τ-Gröbner basis o M, and let $\hat{\tau}$ be the restriction of τ to $\mathbb{T}(\hat{F}_r)$. Then the set $\hat{G} = G \cap \hat{F}_r$ is a $\hat{\tau}$-Gröbner basis of $M \cap \hat{F}_r$.*

Proof. Let $m \in (M \cap \hat{F}_r) \setminus \{0\}$. Then we have $\text{Lt}_{\hat{\tau}}(m) = \text{Lt}_\tau(m) \in \text{Lt}_\tau(M)$ because $\hat{\tau}$ is the restriction of τ. Since G is a τ-Gröbner basis of M, there exists an element $g \in G$ such that $\text{Lt}_{\hat{\tau}}(m) = w \, \text{Lt}_\tau(g) w'$ for some $w, w' \in \langle X \rangle$. But then we have $\text{Lt}_\tau(g) \in \hat{F}_r$, and the assumption that τ is a component elimination ordering for L yields $g \in \hat{F}_r$, i.e., $g \in \hat{G} = G \cap \hat{F}_r$. Now the fact that $\text{Lt}_\tau(g) = \text{Lt}_{\hat{\tau}}(g)$ concludes the proof. $\square$

Module component elimination is the key ingredient for the computation of syzygies in a non-commutative setting. Recall that the two-sided syzygy module $\text{Syz}(G)$ of a tuple $G = (g_1,\ldots,g_s)$ of elements of F_r was defined as the kernel of the homomorphism $\lambda : F_s \longrightarrow F_r$ of two-sided $K\langle X \rangle$-modules which is given by $\varepsilon_i \mapsto g_i$ for $i = 1,\ldots,s$. The computation of two-sided syzygy modules is based on the following proposition.

Proposition 14.3.13. *Let F_{r+s} be the free two-sided $K\langle X \rangle$-module of rank $r + s$, and let $\{e_1,\ldots,e_{r+s}\}$ be its canonical basis. Let $G = (g_1,\ldots,g_s)$ be a tuple of elements of $F_r \setminus \{0\}$, let $\hat{F}_{r+s} = \langle e_{r+1},\ldots,e_{r+s} \rangle$, and for every $m \in F_r$ let $\overline{m}$ denote the corresponding element in F_{r+s} under the canonical injection $e_i \mapsto e_i$. Let U be the two-sided submodule of F_{r+s} generated by $\{\overline{g}_1 - e_{r+1},\ldots,\overline{g}_s - e_{r+s}\}$. Then we have*

$$U \cap \hat{F}_{r+s} \cong \text{Syz}(G).$$

In particular, we can enumerate the syzygy module $\text{Syz}(G)$ using module component elimination.

Proof. Let us consider the homomorphism $\psi : \hat{F}_{r+s} \longrightarrow F_s$ given by $e_{r+i} \mapsto \varepsilon_i$ for $i = 1,\ldots,s$. By restricting ψ to $U \cap \hat{F}_{r+s}$, we obtain an injective homomorphism φ. Therefore it suffices to prove that the image of φ is $\text{Syz}(G)$. Let $m = \sum_{i=1}^{s} \sum_{j \in \mathbb{N}} c_{ij} w_{ij} \varepsilon_i w_{ij}'$ be an element of $\text{Syz}(G)$ with $c_{ij} \in K$ and $w_{ij}, w_{ij}' \in \langle X \rangle$ for $i = 1,\ldots,s$. Notice that for every $j \in \mathbb{N}$, all but finitely many of the elements c_{ij} are zero. Then the element $\overline{m} = \sum_{i=1}^{s} \sum_{j \in \mathbb{N}} c_{ij} w_{ij} e_{r+i} w_{ij}' = \sum_{i=1}^{s} \sum_{j \in \mathbb{N}} c_{ij} w_{ij} \overline{g}_i w_{ij}' - \sum_{i=1}^{s} \sum_{j \in \mathbb{N}} c_{ij} w_{ij} (\overline{g} - e_{r+i}) w_{ij}'$ is contained in $U \cap \hat{F}_{r+s}$, and it satisfies $\varphi(\overline{m}) = m$.

Now suppose that $U \cap \widehat{F}_{r+s}$ contains $\overline{m} = \sum_{i=1}^{s} \sum_{j \in \mathbb{N}} c_{ij} w_{ij} e_{r+i} w'_{ij}$. Then we have $\lambda(\varphi(\overline{m})) = \sum_{i=1}^{s} \sum_{j \in \mathbb{N}} c_{ij} w_{ij} \overline{g}_i w'_{ij} = \sum_{i=1}^{s} \sum_{j \in \mathbb{N}} c_{ij} w_{ij} (\overline{g} - e_{r+i}) w'_{ij} + \sum_{i=1}^{s} \sum_{j \in \mathbb{N}} c_{ij} w_{ij} e_{r+i} w'_{ij} \in U$. Moreover, none of the generators $e_{r+1}, \ldots, e_{r+s}$ appears in the representation of $\lambda(\varphi(\overline{m}))$. Since U is generated by the elements $\{\overline{g}_1 - e_{r+1}, \ldots, \overline{g}_s - e_{r+s}\}$, this implies $\lambda(\varphi(\overline{m})) = 0$. Hence we get $\varphi(\overline{m}) \in \mathrm{Syz}(G)$. $\square$

Using this result, it is easy to formulate a procedure for the computation of a two-sided syzygy module.

Corollary 14.3.14 (Computing Two-Sided Syzygy Modules). *Let $g_1, \ldots, g_s \subseteq F_r \setminus \{0\}$, and let $G = (g_1, \ldots, g_s)$. Let $\varphi : \widehat{F}_{r+s} \longrightarrow F_s$ be the homomorphism defined by $e_{r+i} \mapsto \varepsilon_i$ for $i = 1, \ldots, s$, and for every $m \in F_r$ let $\overline{m}$ denote the corresponding element in F_{r+s}. Consider the following sequence of instructions.*

(1) *Choose a component elimination ordering τ for $L = \{1, \ldots, r\}$ on $\mathbb{T}(F_{r+s})$.*
(2) *Compute a τ-Gröbner basis H of the two-sided submodule $U = \langle \overline{g}_1 - e_{r+1}, \ldots, \overline{g}_s - e_{r+s} \rangle$ of F_{r+s}.*
(3) *Compute $\widehat{H} = H \cap \widehat{F}_{r+s}$. Return $\varphi(\widehat{H})$ and stop.*
This is a procedure which enumerates a $\widehat{\tau}$-Gröbner basis of the two-sided syzygy module $\mathrm{Syz}(G)$, where $\widehat{\tau}$ is the restriction of τ to $\mathbb{T}(\widehat{F}_{r+s})$.

Proof. By the above proposition, the two-sided module $U \cap \widehat{F}_{r+s}$ is isomorphic to $\mathrm{Syz}(G)$. Since $U \cap \widehat{F}_{r+s}$ is also the component elimination module of U with respect to L, Proposition 14.3.12 implies the claim. $\square$

Using the computation of two-sided syzygies, we arrive at the following Gröbner basis algorithm for solving the **Conjugator Search Problem (CSP)** in certain finitely presented groups. This problem has been used as a basis for group-based cryptosystems in previous chapters.

Problem CSP: Given a group G and two elements $g, h \in G$ which are known to be conjugated to each other (i.e., such that there exists an element $a \in G$ for which $ag = ha$), find a conjugator (i.e., find such an element a).

In the sequel we make the following assumptions.
(1) The group G is finitely presented: $G = \langle x_1, \ldots, x_n; r_1 = \cdots = r_m = 1 \rangle$
(2) There exists a word ordering σ on $\langle X \rangle$ such that $R = \{r_1 - 1, \ldots, r_m - 1\}$ is a σ-Gröbner basis of the defining ideal in $\mathbb{Z}_2 \langle X \rangle$ of the group ring $\mathbb{Z}_2[G]$.

Thus we can use R to represent the residue class of a word $w \in \langle X \rangle$ in G uniquely by its normal remainder $\mathrm{NR}_{\sigma, R}(w)$. In particular, we can use R to solve the word problem in G effectively. In this setting, the following algorithm solves CSP in G. We outline its proof which is based on the results of this section. For the full details, we refer to [BK], Section 5.

Proposition 14.3.15 (The Conjugator Search Algorithm). *In the setting described above, let* $w, w' \in \langle X \rangle$ *be two words representing conjugated elements of the group G. Consider the following sequence of instructions.*

(1) *Let* F_6 *be the free two-sided module of rank 6 over* $K \langle X \rangle$. *In* F_6 *form the two-sided submodule*

$$U = \langle e_1 w - e_3, \, e_1 w' - e_4, \, e_2 - e_3 - e_5, \, e_2 + e_4 + e_6,$$
$$e_1(w_1 - w'_1), \ldots, e_1(w_t - w'_t), e_2(w_1 - w'_1), \ldots, e_2(w_t - w'_t),$$
$$x_1 e_1 - e_1 x_1, \ldots, x_n e_1 - e_1 x_n, x_1 e_2 - e_2 x_1, \ldots, x_n e_2 - e_2 x_n \rangle.$$

(2) *Choose the following module term ordering* τ *on* $\mathbb{T}(F_6)$: *for* $t_1, t'_1, t_2, t'_2 \in \langle X \rangle$, *let*

$$t_1 e_i t'_1 >_\tau t_2 e_j t'_2 \iff i < j \text{ or}$$
$$(i = j \text{ and } t'_1 >_\sigma t'_2) \text{ or}$$
$$(i = j \text{ and } t'_1 = t'_2 \text{ and } t_1 >_\sigma t_2).$$

Compute an interreduced two-sided τ-*Gröbner basis H of U.*

(3) *In H there exist elements whose leading term is of the form* $t_i e_5$ *where* $t_i \in \langle X \rangle$ *is the normal remainder with respect to R. Return the words* t_i *and stop.*

This is an algorithm which solves the Conjugator Search Problem in G.

Proof. It is clear that the module term ordering τ defined in step 2 is an elimination ordering for both $L = \{1, 2\}$ and $L' = \{1, 2, 3, 4\}$. Hence the elements of $H \cap \langle e_5, e_6 \rangle$ form a Gröbner basis of the intersection of $\mathrm{Syz}(w, w')$ and $\mathrm{Syz}(1, -1)$.

By assumption, there exists a word $a \in \langle X \rangle$ representing a conjugator such that $a w = w' a$. Therefore $a e_5 - e_6 a$ represents a syzygy in $\mathrm{Syz}(w, w')$ and is contained in U. In particular, there exists an element $h \in H$ whose leading term is of the form $\mathrm{Lt}_\tau(h) = t e_5$ with $t \in \langle X \rangle$. Since the elements $(w_i - w'_i) e_5$ are all contained in U, we may assume that the word t is a normal remainder with respect to R.

Observe that we can consider the computation of $\mathrm{Syz}(w, w') \cap \mathrm{Syz}(1, -1)$ as the composition of the computation of the two individual syzygy moduls combined with the computation of the intersection of two submodules. For all three Gröbner basis computations, we start with a system of generators consisting of binomials. Hence also the computed Gröbner bases consist of binomials. Consequently, the element $h \in H$ found above is of the form $h = t e_5 + b e_5 c$ or $h = t e_5 + b' e_6 c'$. In the first case, the definition of τ yields $c = 1$ and $t >_\sigma b$. This contradicts the fact that we assumed t to be a normal remainder with respect to R. Therefore only $h = t e_5 + b' e_6 c'$ is possible. Here $u \in \mathrm{Syz}_{K[G]}(1, -1)$ yields $t = b' c'$. Hence the element $a = (b')^{-1} t$ satisfies $a \varepsilon_5 - \varepsilon_6 a = (b')^{-1} u \in \mathrm{Syz}_{K[G]}(w, w')$. Thus the word $a \in \langle X \rangle$ represents the desired conjugator.

Let us recall that the computation of the Gröbner basis necessary in step (2) is an enumerating procedure. After a new Gröbner basis element has been found and fully interreduced, we can check whether it has the shape required by step (3). Since we assume that w and w' are conjugates, a suitable element u will be discovered eventually, i.e., our instructions can be performed in such a manner that they define an algorithm. □

Up to now, we have used non-commutative Gröbner bases mainly to deal with the computational problems mentioned in Section 9.8 effectively. In the final section of this chapter, we use them to construct cryptosystems whose security relies on the difficulty of computing certain Gröbner bases.

14.4 Non-Commutative Gröbner Basis Cryptosystems

As mentioned previously, the worst-case bounds for the complexity of computing a Gröbner basis are very high. This has led many researchers to suggest cryptosystems whose security is based on the difficulty of computing a certain Gröbner basis. Initial attempts at doing this via commutative polynomials met with a lot of skepticism and mixed success. However, generalizing the approach to submodules of two-sided free modules over the non-commutative polynomial ring provides sufficient freedom to define cryptosystems which are resistant to all known attacks. In fact, in this section we show that moving from polynomial rings to modules over polynomial rings has the added advantage that we can include the operation of a group (or a ring) on a finite (or countably infinite) set in the encryption algorithm. In this way, we can show that all public key cryptosystems discussed in the earlier chapters of this book are special cases of Gröbner basis cryptosystems.

To define the most general kind of Gröbner basis cryptosystem, we use a setting which is slightly more general than that of the preceding sections. Instead of working over the non-commutative polynomial ring $K\langle X \rangle$, we work over a monoid ring $K[M]$ where M is a finitely presented monoid defined by a convergent rewriting system, i.e., such that the defining ideal of $K[M]$ has a finite Gröbner basis. Of course, the non-commutative polynomial ring is then nothing but the special case of a free monoid M. More precisely, we consider the following setting.

Let $X = \{x_1, \ldots, x_n\}$ be a finite alphabet, and let $M = \langle x_1, \ldots, x_n; \ell_1 = r_1, \ldots, \ell_m = r_m \rangle$ be a finitely presented monoid. Then the monoid ring $K[M]$ satisfies $K[M] = K\langle X \rangle / I_M$, where I_M is the two-sided ideal generated by $R = \{\ell_1 - r_1, \ldots, \ell_m - r_m\}$. We assume that there exists a word ordering σ on $\langle X \rangle$ such that R is a σ-Gröbner basis of I_M and $\ell_i >_\sigma r_i$ for $i = 1, \ldots, m$. Consequently, we can represent every element of M by a word $w \in \langle X \rangle$, and there is a unique representation by a word $\mathrm{NR}_{\sigma,R}(w)$ which is in **normal form**, i.e., the normal remainder w.r.t. R of some other word representing the same monoid element.

Furthermore, we let $r \geq 1$ and $F_r = (K[M] \otimes K[M])^r$ the free two-sided $K[M]$-module of rank r. The standard basis of F_r is denoted by $\{e_1, \ldots, e_r\}$. Let τ be a module term ordering on $\mathbb{T}(F_r)$, and for a two-sided submodule U of F_r we let $\mathcal{O}_\tau(U) = \mathbb{T}(F_r) \setminus \mathrm{Lt}_\tau(U)$.

Definition 14.4.1. A **(general, two-sided) Gröbner basis cryptosystem** consists of the following data.

(1) **Public Information:** the monoid presentation of M, the free two-sided module F_r, a subset S of the set $\mathbb{O}_\tau(U)$, and finitely many elements $u_1, \ldots, u_s \in U$.

(2) **Secret Information:** a τ-Gröbner basis G of a two-sided submodule U of F_r.

(3) **Encryption Procedure:** A plaintext unit is an element $m \in \langle S \rangle_K$. The corresponding ciphertext unit is $c = m + f_1 u_1 g_1 + \cdots + f_s u_s g_s$ with suitably (e.g. randomly) chosen elements $f_i, g_j \in K[M]$.

(4) **Decryption Procedure:** Compute $m = \mathrm{NR}_{\tau,G}(c)$.

The correctness of this cryptosystem follows from the fact that every element of F_r has a unique normal remainder with respect to G and this normal remainder is contained in $\langle \mathbb{O}_\tau(U) \rangle_K$. Notice that the correct choice of the elements $f_i, g_j \in K[M]$ in the encryption procedure is crucial for the security of the cryptosystem. It depends on the concrete setting in which this cryptographic primitive is used. For some settings this issue will be discussed later in this section. A similar definition of Gröbner basis cryptosystems is used in [AKr], but it relies on one-sided rather than two-sided modules. However, most remarks and constructions carry over from this case to the two-sided case without problems.

Next we collect some easy remarks about properties and variations of Gröbner basis cryptosystems.

Remark 14.4.2. Let a Gröbner basis cryptosystem be given as above.

(1) If an attacker can compute G, he can break the cryptosystem. In general, the computation of Gröbner bases is EXPSPACE-hard.

(2) The attacker knows $u_1, \ldots, u_s$ and S, but not a system of generators of U. We can make his task difficult by choosing $u_1, \ldots, u_s$ such that a Gröbner basis of $\langle u_1, \ldots, u_s \rangle$ is hard to compute.

(3) The advantage of using modules (rather than two-sided ideals in $K\langle X \rangle$) is that one can encode hard combinatorial or number theoretic problems in the action of the terms on the canonical basis vectors (see the examples below).

(4) The free module F_r is not required to be finitely generated. Any concrete calculation will involve only finitely many components.

(5) We shall also consider the following **variant:** for the ciphertext unit we construct a pair $c = (f_0, m f_0 + f_1 u_1 g_1 + \cdots + f_s u_s g_s)$ where $f_0 \in K[M]$ is a further randomly chosen element. Then the decoding procedure consists of computing $\mathrm{NR}_{\tau,G}(m f_0 + f_1 u_1 g_1 + \cdots + f_s u_s g_s) = \mathrm{NR}_{\tau,G}(m f_0)$ and "dividing" by f_0 to obtain m. In this way we achieve some additional data hiding: the summand $m f_0$ on the right hand-side resembles the other summands. However, there is no general method for performing the "division" $\mathrm{NR}_{\tau,G}(m f_0) \mapsto m$. We have to provide an explicit procedure in every individual example.

In the following we give some examples of Gröbner basis cryptosystems. In particular, we show that many classical cryptosystems can be realized as special cases.

Example 14.4.3. Let $K = \mathbb{F}_q$ be a finite field, where $q = p^e$ with a prime number p and $e > 0$. Let M be the monoid $M = \mathbb{N}^n = \langle x_1, \ldots, x_n; x_i x_j = x_j x_i = 1 \text{ for all } i, j \rangle$. We use the module F_1 and submodules containing $\langle x_i e_1 - e_1 x_i \rangle$, i.e., ideals in the commutative polynomial ring $K[M] = K[x_1, \ldots, x_n]$. Choose a point $(a_1, \ldots, a_n) \in K^n$. Let $U = \langle x_1 - a_1, \ldots, x_n - a_n \rangle$ and choose elements $u_1, \ldots, u_s \in U$ such that $u_i(a_1, \ldots, a_n) = 0$ for $i = 1, \ldots, s$. Consider the following Gröbner basis cryptosystem.

(1) **Public information:** The polynomial ring $K[x_1, \ldots, x_n]$, a term ordering τ on $\mathbb{T}^n$, the set $\mathbb{O}_\tau(U) = \{1\}$, and the polynomials $u_1, \ldots, u_s$.
(2) **Secret key:** The point $(a_1, \ldots, a_n) \in K^n$ corresponding to the Gröbner basis $G = \{x_1 - a_1, \ldots, x_n - a_n\}$ of the ideal U.
(3) **Encryption procedure:** A plaintext unit $m \in K$ is encrypted as the polynomial $c = m + u_1 f_1 + \cdots + u_s f_s$ with randomly chosen polynomials $f_1, \ldots, f_s \in K[M]$.
(4) **Decryption procedure:** Compute $m = \mathrm{NR}_{\tau, G}(c) = c(a_1, \ldots, a_n)$.

Clearly, this is Neal Koblitz' Polly Cracker Cryptosystem from Section 13.2. Unfortunately, it allows many dangerous attacks, as we have seen in Section 13.3.

A number of improvements of Koblitz' original approach have been proposed. Many of them fit our scheme.

Example 14.4.4. In the setting of the preceding example, choose a second commutative polynomial ring $Q = K[y_1, \ldots, y_m]$ and polynomials $g_1, \ldots, g_m$ in $K[M]$. In this way there is a K-algebra homomorphism $\phi : Q \longrightarrow K[M]$ given by $\phi(y_i) = g_i$ for $i = 1, \ldots, m$. Choose a point $(\xi_1, \ldots, \xi_n) \in K^n$ and elements $f_1, \ldots, f_s \in Q$ such that $\phi(f_1), \ldots, \phi(f_s) \in U = \langle x_1 - \xi_1, \ldots, x_n - \xi_n \rangle$. Now construct the following Gröbner basis cryptosystem.

(1) **Public information:** The rings $K[M]$ and Q, the homomorphism ϕ, the term $\mathbb{O}_\tau(U) = \{1\}$, and the polynomials $f_1, \ldots, f_s \in Q$.
(2) **Secret key:** The point $(\xi_1, \ldots, \xi_n) \in K^n$, or equivalently, the Gröbner basis $\{x_1 - \xi_1, \ldots, x_n - \xi_n\}$ of the ideal U in $K[M]$.
(3) **Encryption procedure:** We proceed in a similar way to the variant above. A plaintext unit is an element $m \in K$. We choose random polynomials $h \in \langle f_1, \ldots, f_s \rangle$, a polynomial $h' \in \mathrm{Ker}(\phi)$, and a random exponent $\kappa \in \mathbb{N}^n$. Then we send $c = (y^\kappa, my^\kappa + h + h')$ where $y = (y_1, \ldots, y_m)$. In other words, an attacker knows the pair $(\phi(y)^\kappa, m\phi(y)^\kappa + \phi(h))$.
(4) **Decryption procedure:** Compute $\bar{v} = [m\phi(y)^\kappa + \phi(h)](\xi_1, \ldots, \xi_n) = m\phi(y)^\kappa \cdot (\xi_1, \ldots, \xi_n)$ and obtain $m = \bar{v}/[\phi(y)^\kappa(\xi_1, \ldots, \xi_n)]$.

This is Le van Ly's **Polly Two Cryptosystem** (see [vLy]). Compared to Polly Cracker, it has the advantage that the usual linear algebra attacks do not work. It appears that an attacker has no choice but to compute a (possibly hard) Gröbner basis. Supposedly hard concrete instances of this cryptosystem have been suggested, but were not able to withstand side-channel attacks.

Now we show that the RSA Cryptosystem is in fact a Gröbner Basis Cryptosystem.

Example 14.4.5. Let $K = \mathbb{Z}_2$, let $X = \{x, y\}$, and let $M = \mathbb{N}^2 = \langle x, y; xy = yx \rangle$. Then $K[M] = K[x, y]$ is the commutative polynomial ring in two indeterminates. Moreover, let $p, q \gg 0$ be two distinct prime numbers, let $n = pq$, and let $\Pi = \mathbb{Z}_n^*$ be the set of residue classes prime to n. We use the free module $F_n = \bigoplus_{i=0}^{n-1} K[x, y] e_i$ and the term ordering $\tau = \texttt{deglex-pos}$. Choose a number $\epsilon \in \mathbb{Z}_{(p-1)(q-1)}^*$ and compute the inverse d of ϵ in $\mathbb{Z}_{(p-1)(q-1)}^*$.

(1) **Public information:** The module F_n (and thus the number n), the set $\mathcal{O}_\tau(U) = \{e_0, \ldots, e_{n-1}\}$, the number ϵ, and the vectors $\{u_1, \ldots, u_s\} = \{e_i x - e_{i^\epsilon \bmod n} \mid i = 0, \ldots, n - 1\} \cup \{e_i xy - e_i \mid i = 0, \ldots, n - 1\}$.

(2) **Secret key:** The secret key consists of the primes p and q and the number d. Equivalently, the secret key is the τ-Gröbner basis $G = \{u_1, \ldots, u_s\} \cup \{e_i y - e_{i^d \bmod n}; i = 0, \ldots, n - 1\}$ of $U = \langle G \rangle$.

(3) **Encryption procedure:** A plaintext unit is a vector $e_m \in \mathcal{O}_\tau(U)$. To encrypt it, we form $e_m + (xye_m - e_m) - (xe_m - ye_{m^\epsilon \bmod n}) \in e_m + U$ to obtain the ciphertext unit $c = ye_{m^\epsilon \bmod n}$.

(4) **Decryption procedure:** Compute $\mathrm{NR}_{\tau, G}(ye_{m^\epsilon \bmod n}) = e_{m^{\epsilon d} \bmod n} = e_m$.

It is easy to see that this is the Gröbner basis version of the RSA cryptographic primitive studied in Section 7.5. If an attacker is able to factor n, he can break the system. This is equivalent to being able to find d. In the Gröbner basis version, the problem the attacker faces is that he does not know the Gröbner basis elements $ye_i - e_{i^d \bmod n}$.

Also discrete logarithms can be used in Gröbner basis cryptosystems.

Example 14.4.6. Let $K = \mathbb{Z}_2$, let $X = \{x\}$, and let $M = \langle X \rangle \cong \mathbb{N}$. Then $K[M] = K[x]$ is the univariate polynomial ring over K. Moreover, let $p \gg 0$ be a prime number. We use the $K[x]$-module $F_{2p-2} = \bigoplus_{i=1}^{p-1} K[x]\epsilon_i \oplus \bigoplus_{j=1}^{p-1} K[x]e_j$ where ϵ_i, e_j are the standard basis vectors. Let g be a generator of the multiplicative group $\mathbb{Z}_p^*$, and let $\tau = \texttt{deg-pos}$ with $\epsilon_i >_\tau e_j$ for all $i, j = 1, \ldots, p - 1$. Choose a number $a \in \{1, \ldots, p - 1\}$ and compute $b = g^a \bmod p$. Now we introduce the following Gröbner basis cryptosystem.

(1) **Public information:** The module F_{2p-2}, i.e., the prime p, the set $\mathcal{O}_\tau(U) = \{e_1, e_2, \ldots, e_{p-1}\}$, the number b, and the vectors $\{u_1, \ldots, u_s\} = \{\epsilon_1 - e_1\} \cup \{x\epsilon_i - \epsilon_{gi} \mid i = 1, \ldots, p - 1\} \cup \{xe_j - e_{bj} \mid j = 1, \ldots, p - 1\}$ where all indices are computed modulo p.

(2) **Secret key:** The number $a \in \{1, \ldots, p - 1\}$, or equivalently, the τ-Gröbner basis $G = \{u_1, \ldots, u_s\} \cup \{\epsilon_i - e_{i^a} \mid i = 1, \ldots, p - 1\}$ of $U = \langle G \rangle$.

(3) **Encryption procedure:** A plaintext unit is of the form $e_1 + e_m$ with a number $m \in \{0, \ldots, p - 1\}$. Using the variant, we randomly choose a number $k \in \{0, \ldots, p - 1\}$, form $(e_1 + e_m)x^k$ and send the ciphertext unit $c = \epsilon_{g^k} + e_{mb^k} \in x^k(e_1 + e_m) + \langle u_1, \ldots, u_s \rangle$.

(4) **Decryption procedure:** We compute $\text{NR}_{\tau,G}(c) = e_{b^k} + e_{mb^k}$. Since $e_{b^k} + e_{mb^k}$ reduces via G to $x^k(e_1 + e_m)$, we have to "divide" this vector by x^k. To this end, it suffices to compute $m = (mb^k)/(b^k)$ in K and to form $e_1 + e_m$.

Clearly, this is the Gröbner basis version of the ElGamal Cryptosystem of Section 8.1. It can be broken if the attacker is able to compute the discrete logarithm a of $b = g^a$ or k of g^k. In the Gröbner basis version, an attacker can only reduce using ϵ_{g^k} via u_i to $x^k \epsilon_1$ and then via u_1 to $x^k e_1$. This takes $k \gg 0$ reduction steps. If one knows a, one can get rid of the vector ϵ_{g^k} by using just one reduction step $\epsilon_{g^k} \longrightarrow e_{g^{ka}} = e_{b^k}$.

The next example is based on non-commutative polynomials. In order to prevent linear algebra attacks, T. Rai suggested in his doctoral thesis [Rai] to construct Gröbner basis cryptosystems utilizing two-sided ideals in $K\langle X \rangle$.

Example 14.4.7. Let K be a (finite) field, let $X = \{x_1, \ldots, x_n\}$, and let $M = \langle X \rangle$. Then $K[M] = K\langle X \rangle$ is the non-commutative polynomial ring. We choose a two-sided ideal $I \subseteq K[M]$ for which we know a finite (two-sided) Gröbner basis $G = \{g_1, \ldots, g_t\}$ with respect to some word ordering τ.

(1) **Public information:** The ring $K[M]$, the set $\mathbb{O}_\tau(U)$, and a finite subset $\{u_1, \ldots, u_s\} \in I$ such that computing a Gröbner basis of $\langle u_1, \ldots, u_s \rangle$ is infeasible.

(2) **Secret key:** The τ-Gröbner basis G of I.

(3) **Encryption procedure:** A plaintext unit m is an element in $\langle \mathbb{O}_\tau(U) \rangle_K$. The corresponding ciphertext is $c = m + f_1 u_1 g_1 + \cdots + f_s u_s g_s$ where the non-commutative polynomials f_i, g_i are suitably chosen so that in the computation of c many leading term cancellations occur (see [Rai], Section 4.1).

(4) **Decryption procedure:** Compute $m = \text{NR}_{\tau,G}(c)$ using the Gröbner basis G.

In [Rai] several concrete instances of this cryptosystem are proposed. They offer good resistence to linear algebra attacks because using indeterminate coefficients for the polynomials f_i and g_j leads to systems of quadratic equations in these coefficients which cannot be solved using linear algebra. However, one has to take great care to make these instances secure against attackers who are able to compute partial Gröbner bases (see [Rai], Chapter 4).

Our approach also includes the group based cryptosystems studied in Chapter 11. The following Gröbner basis cryptosystem relies on the difficulty of solving the conjugator search problem in certain groups.

Example 14.4.8. Let K be a field, and let $M = \langle x_1, \ldots, x_n; r_1 = \cdots = r_m = 1 \rangle$ be a finitely presented group. We use the free two-sided $K[M]$-module $F_r = \bigoplus_{\bar{w} \in M} K[M]\, e_{\bar{w}}\, K[M] \oplus \bigoplus_{\bar{w} \in M} K[M]\, e_{\bar{w}}\, K[M]$ (which is possibly of infinite rank). Moreover, let $\tau = \texttt{llex-pos}$ be such that $\epsilon_{\bar{w}} >_\tau e_{\bar{u}}$ for all $w, u \in M$. Choose $a, g \in M$ and compute $g' = a^{-1}ga$. Now consider the following Gröbner basis cryptosystem.

(1) **Public information:** The module F_r, the elements $g, g' \in M$, a set $B \subseteq \{b \in M; ba = ab\}$, the set $\mathbb{O}_\tau(U) = \{e_{\bar{w}}; \bar{w} \in M\}$, and the vectors $\{u_\lambda; \lambda \in \Lambda\} = \{\epsilon_i h - \epsilon_{h^{-1}ih} \mid i, h \in M\} \cup \{\epsilon_g - e_{g'}\} \cup \{e_j k - e_{k^{-1}jk} \mid j, k \in M\}$.

(2) **Secret key:** The element $a \in M$, or equivalently, the τ-Gröbner basis $G = \{u_\lambda \mid \lambda \in \Lambda\} \cup \{e_i - e_{a^{-1}ia} \mid i \in M\}$ of the two-sided submodule $U = \langle G \rangle$ of F_r.

(3) **Encryption procedure:** Randomly choose an element $b \in B$. A plaintext unit $m \in M$ is written in the form $\epsilon_g + e_{g'\tilde{m}}$, where $\tilde{m} = bmb^{-1}$. Then we use the elements u_λ to obtain the ciphertext unit $c = \epsilon_{b^{-1}gb} + e_{b^{-1}g'\tilde{m}b}$.

(4) **Decryption procedure:** Find $\mathrm{NR}_{\tau,G}(c) = e_{a^{-1}g''a} + e_{b^{-1}g'bm} = e_{b^{-1}g'b} + e_{b^{-1}g'bm}$ first, where $g'' = b^{-1}gb$. Then determine m using the equality $m = (b^{-1}g'b)^{-1} \cdot (b^{-1}g'bm)$.

As one can readily check, this is a Gröbner basis version of an ElGamal like cryptosystem based on a group with a "hard" Diffie-Hellman conjugacy problem, i.e., the problem to find $a^{-1}b^{-1}gba$ given g, $a^{-1}ga$ and $b^{-1}gb$ where a and b commute. One can solve this problem if, given g and $g' = a^{-1}ga$, one can find a_1, a_2 such that $a_1ga_2 = g'$ and a_1, a_2 commute with the elements from B. The advantage of knowing the Gröbner basis is that one can pass from $\epsilon_{g''}$ to the corresponding e_i without going through the reduction steps $\epsilon_g \longrightarrow e_{g'}$. The computation of that Gröbner basis is equivalent to finding a.

To perform the encryption step explicitly, one has to perform the following simple computations in the group: Conjugate g' with b to obtain $b^{-1}g'b$ and multiply with the plaintext m. Conjugate g with b to obtain $b^{-1}gb$.

If we want to decrypt the ciphertext $\epsilon_{b^{-1}gb} + e_{b^{-1}g'\tilde{m}b}$ knowing the secret a, an explicit decryption amounts to performing the following: conjugate $b^{-1}gb$ with a to obtain the Gröbner basis element $\epsilon_{b^{-1}gb} - e_{a^{-1}b^{-1}gba}$, reduce c via this element in one step to $\mathrm{NR}_{\tau,G}(c) = e_{a^{-1}b^{-1}gba} + e_{b^{-1}a^{-1}gabm}$ and obtain m by multiplying the inverse of the first index with the second index.

So, all computations performed to encrypt and decrypt are actually computations in the group M.

In Chapter 11 braid groups have been suggested for this kind of cryptosystem. However, we mentioned that there is a polynomial time algorithm to solve the Diffie-Hellman conjugacy problem in braid groups. If one chooses reasonable parameters, this algorithm is not feasible today, but it seems that the braid group based version of this cryptosystem may not be secure in the future.

In the final part of this section we collect some remarks about the security of Gröbner basis cryptosystems.

Remark 14.4.9 (Linear Algebra Attacks). Several types of linear algebra attacks have been proposed which apply to special Gröbner basis cryptosystems.

(1) The basic type is the attack proposed in the original paper [FK]. In the equation $c = m + f_1u_1g_1 + \cdots + f_su_sg_s$, the attacker regards the coefficients of f_i, g_j as unknowns and tries to solve the resulting system of equations. In our setup, it is possible to make this attack infeasible:

(a) In general, the system of equations will be quadratic rather than linear.
(b) Now suppose that the system of equations is linear in a particular setup. By choosing a large set $\mathcal{O}_\tau(U)$, we can make the plaintext m "similar" to the cyphertext c, so that the resulting linear system of equations involves $|\mathcal{O}_\tau(U)|$ coefficients. By using a module of large rank, we can make the solution of this linear system infeasible. Moreover, since we are working over a monoid or group ring, many products $(e_i t) t'$ with $e_i t \in \mathrm{Supp}(u_j)$ and $t' \in \mathrm{Supp}(f_j)$ are going to yield the same term, so that the corresponding coefficients cannot be recovered.
(2) The "intelligent" linear algebra attack suggested by H. W. Lenstra described in [Bar] is based on the idea that in the equation $c = m + f_1 u_1 g_1 + \cdots + f_s u_s g_s$ one can guess the terms t occurring in the support of $u_1, \ldots, u_s$ if $t \cdot \mathrm{Supp}(f_i)$ intersects $\mathrm{Supp}(u)$, and that the list of all such terms is not too large. As before, in our approach this attack can be repelled in several ways, for instance by choosing a large set $\mathcal{O}_\tau(U)$, by working over group rings, or by using a free module of large rank. In each case sufficient cancellation happens during the computation of the cyphertext.

Remark 14.4.10 (The Attack Using Characteristic Terms). If a representation $c = m + f_1 u_1 g_1 + \cdots + f_s u_s g_s$ is such that there are terms in c which do not belong to $\mathcal{O}_\tau(U)$ and therefore not to $\mathrm{Supp}(m)$, then it is sometimes possible to reveal individual messages by performing suitable linear algebra on the coefficients of c and the elements f_i, g_j. In particular, this works if there exist "characteristic terms", i.e., terms which occur in just one of the elements f_i, g_j. By recognizing multiples of these terms in the ciphertext one can then reconstruct a constant message unit. As before, this attack rests on the fact that plaintext units are small, i.e., that $\mathcal{O}_\tau(U)$ is small. Furthermore, if several products $t \cdot t'$ with $t \in \mathrm{Supp}(u_i)$ and $t' \in \mathrm{Supp}(f_i) \cup \mathrm{Supp}(g_j)$ contribute to one coefficient of c, this attack becomes infeasible. Thus the defensive measures described above apply.

Remark 14.4.11 (Chosen Ciphertext Attacks). In the proposed cryptosystems the receiver has no method to detect invalid ciphertexts. In addition, since decryption is K-linear, the chosen ciphertext attacks described in [Ko2] are possible. However, by using suitable hash functions the system can be made secure in the following way: The sender appends a suitable random value to his message, computes the hash value of the result, and transmits the ciphertext of the message together with the ciphertext of the random value and the hash value. If the receiver detects an invalid ciphertext, he refuses to decrypt or returns a meaningless or random decryption result.

Altogether, it appears that there are sufficiently many defenses for Gröbner Basis Cryptosystems to the known attacks. In fact, breaking them in general would entail breaking essentially all cryptosystems in practical use today.

14.5 Exercises

14.1 Let $X = \{x_1, \ldots, x_n\}$, and let σ be a complete ordering relation on $\langle X \rangle$ which is compatible with multiplication. Prove that the following conditions are equivalent.

(a) The relation σ is a well-ordering, i.e., every non-empty subset of $\langle X \rangle$ has a minimal element.

(b) Every descending chain $w_1 \geq_\sigma w_2 \geq_\sigma \cdots$ in $\langle X \rangle$ is eventually stationary.

(c) For every $w \in \langle X \rangle$, we have $w \geq_\sigma 1$.

14.2 Show that the length-lexicographic word ordering `llex` defined in Example 14.1.4 is indeed a word ordering.

14.3 Let K be a field, let $I = \langle x^2 - xy \rangle \subset K\langle x, y \rangle$, and let σ be a word ordering on $\langle x, y \rangle$ such that $x >_\sigma y$. Prove that $\mathrm{Lw}_\sigma(I) = \langle xy^i x \mid i \geq 0 \rangle$ (see Example 14.1.10).

14.4 Prove the Non-Commutative Division Algorithm 14.1.13. Proceed as in the proof of the commutative case 13.1.10.

14.5 Let $P = \mathbb{Z}_2\langle x_1, \ldots, x_6 \rangle$, let $\sigma = $ `llex`, and let $G = (g_1, \ldots, g_5)$, where the non-commutative polynomials g_i are defined as in Example 14.1.17. Using Buchberger's Criterion 14.1.16, prove that G is a σ-Gröbner basis of the two-sided ideal $\langle G \rangle$.

14.6 Consider the **dihedral group** $D_3 = \langle x, y; x^3 = y^2 = (xy)^2 = 1 \rangle$ and its group ring $\mathbb{Z}_2[D_3] = \mathbb{Z}_2\langle x, y \rangle / I$, where $I = \langle x^3 + 1, y^2 + 1, (xy)^2 + 1 \rangle$.

(a) Show that I has the `llex`-Gröbner basis $G = \{y^2 + 1, yxy + x^2, yx^2 + xy, xyx + y, x^2y + yx, x^3 + 1\}$.

(b) Find the shortest word representing the inverse of $\overline{x^2 y}$ in D_3.

14.7 Let $X = \{x_1, \ldots, x_n\}$ and $\ell \in \{1, \ldots, n\}$. Show that the ordering `elim` on $\langle X \rangle$ defined in Example 14.2.2 is a word ordering and an elimination ordering for $L = \{x_1, \ldots, x_\ell\}$.

14.8 Work out the proof of Lemma 14.2.3.

14.9 Consider the **infinite dihedral group** $D_\infty = \langle x, y; y^2 = (xy)^2 = 1 \rangle$.

(a) Using the method of Remark 14.2.7, prove that the element $\bar{x} \in D_\infty$ has infinite order.

(b) Similarly, compute the order of the element $\overline{yxy}$ in the group $D_3 = \langle x, y; x^3 = y^2 = (xy)^2 = 1 \rangle$.

14.10 Let $G = \langle x_1, \ldots, x_n; r_1 = \cdots = r_m = 1 \rangle$ be a finitely presented group, let $Y = \{y_1, \ldots, y_n\}$ be a set of further letters such that y_i represents x_i^{-1}, let $h_1, \ldots, h_k \in \langle X, Y \rangle$ be words whose residue classes generate a subgroup $H = \langle \bar{h}_1, \ldots, \bar{h}_k \rangle$ of G, and let S be the subalgebra $S = \mathbb{Z}_2\langle \bar{h}_1 - 1, \ldots, \bar{h}_k - 1 \rangle$ of $\mathbb{Z}_2[G]$. Prove that, for a word $w \in \langle X, Y \rangle$, the following conditions are equivalent.

(a) $\bar{w} \in H$

(b) $\bar{w} - 1 \in S$

14.11 Consider the free $\mathbb{Q}\langle x, y \rangle$-module F_2 of rank 2 and the module term ordering $\tau = $ `Pos-llex` on $\mathbb{T}(F_2)$. Let M be the $\mathbb{Q}\langle x, y \rangle$-submodule of F_2 generated by $g_1 = $

$x_2 x_1 e_1 x_2 + e_2$ and $g_2 = e_1 x_2^2 + x_1 e_1$. Using Buchberger's Procedure for Modules 14.3.9, enumerate a τ-Gröbner basis of M. Show that $G = \{g_1, g_2, x_2 x_1^{k-1} e_1 + (-1)^k e_2 x_2^{2k-5} \mid k \geq 3\}$ is a τ-Gröbner basis of M.

14.12 Formulate and prove a semi-decision procedure for the submodule membership problem in finitely generated $K\langle X \rangle$-submodules of F_r.

14.13 Let K be a field, let $X = \{x_1, \ldots, x_n\}$, and let $M = \langle v_1, \ldots, v_s \rangle$ and $N = \langle u_1, \ldots, u_t \rangle$ be two $K\langle X \rangle$-submodules of the two-sided free module F_r. Consider the inclusions $\iota_1 : F_r \hookrightarrow F_{2r}$ given by $\iota_1(e_i) = e_i$ and $\iota_2 : F_r \hookrightarrow F_{2r}$ given by $\iota_2(e_i) = e_{r+i}$. Moreover, let

$$V = \langle \iota_1(v_i) + \iota_2(v_i) \mid i = 1, \ldots, s \rangle + \langle \iota_2(u_j) \mid j = 1, \ldots, t \rangle.$$

Prove that $M \cap N = V \cap \langle e_1, \ldots, e_r \rangle$ and conclude that a system of generators of $M \cap N$ can be enumerated via Buchberger's Procedure for Modules.

14.14 Let K be a field. Consider the following special case of the Gröbner basis cryptosystem described in Example 14.4.7.

Secret Key: $h = z - c \in K\langle x, y, z \rangle$ with $c \in K$.

Public Information: $f = x - a$, $g = y - b$, $u_1 = fhg + gh$, and $u_2 = ghf + hg$ in $K\langle x, y, z \rangle$ with $a, b \in K$.

Show that this instance is not secure, because the secret key can be derived from the public information.

14.15 Let K be a field. Consider the following special case of the Gröbner basis cryptosystem described in Example 14.4.7.

Secret Key: $f_1 = z - a, f_2 = w - b$ in $K\langle x, y, z, w \rangle$ with $a, b \in K$.

Public Information: $u_1 = x f_1 y + x f_2 y + y f_1 + y f_2 + f_1 + f_2$ and $u_2 = y f_1 x + y f_2 x + f_1 y + f_2 y + f_1 + f_2$.

Show that the secret key cannot be read off the public information. Then explain how this instance can be broken by computing a partial Gröbner basis of $U = \langle u_1, u_2 \rangle$.

14.16 Let K be a field. Consider the following special case of the Gröbner basis cryptosystem described in Example 14.4.7. In $K\langle x_1, \ldots, x_6 \rangle$, consider the words $w_1 = x_1 x_3 x_5 x_4 x_2 x_6$, $w_2 = x_1 x_5 x_2 x_4 x_3 x_6$, and $w_3 = x_1 x_2 x_3 x_4 x_5 x_6$.

Secret Key: $h = w_3 + c_1 x_1 + \cdots + c_6 x_6 + c_0$, where $c_0, \ldots, c_6 \in K \setminus \{0\}$ are arbitrary constants.

Public Information: $f = w_1 + a_1 x_1 + \cdots + a_6 x_6 + a_0$ $g = w_2 + b_1 x_1 + \cdots + b_6 x_6 + b_0$, where $a_i, b_j \in K \setminus \{0\}$ are arbitrary constants, $u_1 = fhg + gh$, and $u_2 = ghf + hg$.

(a) Prove that the private key can be deduced from the public information as follows: first show that $\mathrm{Lw}_\sigma(h) = w_3$, $\mathrm{Lw}_\sigma(f) = w_1$, and $\mathrm{Lw}_\sigma(g) = w_2$ for a suitable word ordering σ. Then find the coefficients $c_0, \ldots, c_6$ from the coefficients of u_1 and u_2.

(b) Modify this instance such that the attack described in (a) fails.

15 Lattice-Based Cryptography

15.1 Lattice-Based Cryptography

Lattice-based cryptography refers to asymmetric cryptographic protocols based on the use of lattices. These types of protocols are of interest because they can be efficiently constructed with good key spaces and in some cases there are true security proofs tying them to provably hard problems. There has been a great deal of research on lattice methods and in this final chapter we give an overview of this area.

While lattice-based cryptography has been studied for several decades, there has been renewed interest in lattice-based techniques as prospects for a real quantum computer improve. Unlike more widely used and known public key cryptography such as the RSA or Diffie-Hellman cryptosystems which are easily attacked by a quantum computer, some lattice-based cryptosystems appear likely to be resistant to attack by both classical and quantum computers.

Lattices were first studied by mathematicians Joseph Louis Lagrange and Carl Friedrich Gauss, and many results on lattices fall into an area of mathematics called the **geometry of numbers**. In 1996, Miklós Ajtai showed, in a seminal result, the possibility to use lattices in a cryptographic primitive. He and Dwork then developed this into a workable encryption system.

In this book we have examined many aspects of mathematical cryptography, using various types of algebraic platforms, in order to obtain diverse and secure cryptographic protocols. The classical methods for public key cryptography, Diffie-Hellman and RSA, are number theory based and could be called commutative cryptography. They are based on supposedly hard number theoretic problems; the discrete log problem for Diffie-Hellman and the difficulty of factoring large integers for RSA. However, there are no formal complexity proofs that these are hard (in a complexity sense) and therefore no real security proofs for these cryptographic methods. Despite this, up to this point there have been no successful general attacks against them (see Chapter 6). In 1997, Shor developed a factoring algorithm, now called **Shor's quantum factoring algorithm**, that can be implemented on a quantum computer that successfully attacks both Diffie-Hellman and RSA. We refer to Shor's article [Sho] for a discussion of quantum computers and quantum algorithms. If and when a workable and efficient quantum computer is constructed, Shor's algorithm will make both Diffie-Hellman and RSA insecure.

Elliptic curve cryptography adapted the basic methods of Diffie-Hellman and El-Gamal, to the environment of elliptic curves and elliptic curve groups, which as in the number theory cases were abelian. Elliptic curve techniques were efficiently implementable and reduced the size of the necessary key space. However, these methods are also susceptible to quantum attacks.

The general group-based techniques, that we discussed in Chapters 9, 10 and 11 and the Gröbner basis methods discussed in Chapters 13 and 14, were attempts to develop new and efficient and perhaps more secure cryptographic platforms, which have the potential to withstand quantum attacks.

Lattices are discrete additive subgroups of $\mathbb{R}^n$ with many nice properties. These will be introduced and discussed in Section 15.3. There are several hard decision problems on lattices, such as the closest vector problem, and the shortest vector problem (see Section 15.4). In 1996 Ajtai showed how to use these hard problems to construct one-way functions with trapdoors, and lattice-based cryptography was born.

15.2 General Cryptoprimitives and Cryptographic Platforms

In public key cryptography, a **cryptographic primitive**, or **cryptoprimitive**, is a basic cryptographic algorithm that can be used as a building block to develop a cryptographic protocol. For example, in the RSA method, the basic idea is that if p, q are distinct odd primes and $(e, \phi(pq)) = 1$ and $ed \equiv 1 \pmod{\phi(pq)}$ then for any $M \in \mathbb{Z}_{pq}$ we have

$$M^{ed} = M \text{ in } \mathbb{Z}_{pq}$$

(see Chapter 7). From this various protocols can be built.

Recall that a **cryptographic platform** is a mathematical object, usually an algebraic object, upon which a cryptoprimitive can be developed. The cryptoprimitive, combined with the cryptographic platform, can then be used to develop a cryptosystem or other cryptographic protocols. For a public key, or asymmetric cryptosystem, there is generally a hard problem based on the platform that is used to develop a one-way function that becomes the cryptoprimitive. Further this hard problem should have a trapdoor, that is, an easy way to invert, and hence decode, if a secret key is known.

As we discussed in the chapters on group-based and Gröbner basis based cryptosystems there are several features that are needed by a mathematical system to possess in order to be a good platform for a public key cryptosystem.

First there must be a relatively easy and efficiently computable method to encrypt within the system. Further if the system is algebraic, so that elements and hence messages, can be combined, there should be good diffusion. By this we mean that we cannot within the system determine easily g_1 or g_2 if we know $x = g_1 g_2$.

Next there should exist a hard problem concerning the system that lends itself to the formation of a trapdoor function, that is, a one-way function that has a trapdoor.

Finally, if possible, there should be a security proof, that is, a proof tying the security of the cryptosystem to a known hard problem in the complexity sense.

Briefly, we again look at three of the cryptosystems that we have already examined. To begin, let us consider the RSA method. The basic cryptographic platform is the modular ring $\mathbb{Z}_n$ where $n = pq$ for p, q large odd primes. It is easy to have plaintext messages as either an element in $\mathbb{Z}_n$ or as a block using $\mathbb{Z}_n$ (see Chapter 7).

The basic cryptoprimitive for RSA depends on the fact that if e is a randomly chosen integer with $1 < e < \phi(n)$ with $(e, \phi(pq)) = 1$ and $ed \equiv 1 \pmod{\phi(pq)}$ then for any $M \in \mathbb{Z}_{pq}$ we have (see Chapter 7)

$$M^{ed} = M \text{ in } \mathbb{Z}_{pq}.$$

The basic encryption is then to encrypt $P \in \mathbb{Z}_n$ by P^e. From this, the various RSA cryptosystems and protocols can be developed.

The hard problem, upon which this is based, is the factorization problem: given n and knowing $n = pq$ with p, q distinct large primes to find p, q. Although this is assumed hard, and no general method has been found to solve it or to break RSA, there is no security proof in the complexity sense.

Now consider Diffie-Hellman and the related ElGamal method. As in RSA, the cryptographic platform is $\mathbb{Z}_n$ for some large n. The basic goal is to determine a key q with $1 < q < n$. The basic hard problem is the discrete log problem: knowing g and g^h in $\mathbb{Z}_n$ to find h.

Elliptic curve methods shifted the cryptographic platform to elliptic curve groups but still used the discrete log problem as the background for the cryptoprimitive. The elliptic curve methods were efficient and significantly reduced the size of the necessary key space.

Although the above methods and variations of these methods are still the basis for the most widely used public key cryptographic protocols there are two problems, one theoretical and one quite practical. First there are no security proofs. No-one is really certain of how difficult or easy is the factorization problem or the discrete log problem. Certain instances of both can be solved (see Chapter 6) so care must be taken in choosing the parameters for either RSA or Diffie-Hellman.

Second, and perhaps more serious, is the susceptibility of these systems to attacks based on Shor's quantum factorization algorithm. This algorithm, given a truly workable quantum computer, breaks these methods. Although at present the best quantum computer can only factor small products of primes, such as 15, there is a general belief that a workable quantum computer will eventually be built.

It was for these two reasons that non-abelian group based cryptosystems were developed. For the Ko-Lee method the cryptographic platform is a non-abelian group G with an efficiently computable normal form (see Chapters 9 and 10) and two large mutually commuting subgroups. The supposed hard problem is the conjugacy problem. For an actual implementation the braid groups, B_n, were suggested (see Chapter 11). As discussed in Chapter 11, for small values of the parameter n, there have been some successful attacks on braid-group based cryptosystems.

In 1996, Ajtai [Aj1] showed how to construct a hash function and hence a cryptoprimitive based on a hard problem in a lattice in $\mathbb{R}^n$. From this, several cryptosystems were built that were efficient and resistant to attacks based on quantum factoring.

Therefore lattice-based cryptographic constructions provide an alternative, if and when quantum algorithms can really be implemented. Many lattice-based protocols are quite efficient, and can compete with the standard alternatives. Further, in general they are relatively simple to implement and are all believed to be secure against attacks using conventional or quantum computers.

In terms of security, lattice-based cryptographic constructions can be divided into two types. The first includes practical proposals, which are generally efficient, but lack a true supporting security proof. The second type admits true theoretical security proofs based on the worst-case hardness of lattice problems, but only a few of them are efficient enough to be used in practice.

In the next section we discuss lattices and their properties and then introduce the hard problems that are necessary for cryptography.

15.3 Lattices and Their Properties

We consider $\mathbb{R}^n$ as Euclidean n-space. This is an n-dimensional real space with the usual Euclidean inner product and norm. Hence the elements of $\mathbb{R}^n$ are n-tuples of real numbers $(x_1, x_2, \ldots, x_n)$ and if $\mathbf{u} = (x_1, \ldots, x_n)$, $\mathbf{v} = (y_1, \ldots, y_n)$ then

$$\mathbf{u} \cdot \mathbf{v} = x_1 y_1 + \cdots + x_n y_n$$

while the norm or magnitude of the vector $\mathbf{u}$ is

$$|\mathbf{u}| = \sqrt{x_1^2 + \cdots + x_n^2}.$$

The distance between $\mathbf{u}$ and $\mathbf{v}$ is given by

$$d(\mathbf{u}, \mathbf{v}) = |\mathbf{u} - \mathbf{v}| = \sqrt{\sum_{i=1}^{n} (x_i - y_i)^2}$$

while the angle between $\mathbf{u}$ and $\mathbf{v}$ is

$$\theta = \cos^{-1}\left(\frac{\mathbf{u} \cdot \mathbf{v}}{|\mathbf{u}||\mathbf{v}|}\right).$$

If $\mathbf{u} \cdot \mathbf{v} = 0$ then $\mathbf{u}$ and $\mathbf{v}$ are **orthogonal** or **perpendicular**.

Let $\mathbf{v}_1, \ldots, \mathbf{v}_k$ be an independent set of vectors in $\mathbb{R}^n$ and $V = \langle \mathbf{v}_1, \ldots, \mathbf{v}_k \rangle$, the vector subspace of $\mathbb{R}^n$ spanned by these vectors. The **lattice** L generated by $\mathbf{v}_1, \ldots, \mathbf{v}_k$ consists of all integer linear combinations of $\mathbf{v}_1, \ldots, \mathbf{v}_k$, that is,

$$L = \{n_1 \mathbf{v}_1 + \cdots + n_k \mathbf{v}_k \mid n_i \in \mathbb{Z}\}.$$

A **lattice** in $\mathbb{R}^n$ is a subset L of $\mathbb{R}^n$ which is the lattice generated by $\mathbf{v}_1, \ldots, \mathbf{v}_k$ for some set of independent vectors in $\mathbb{R}^n$. The set $\mathbf{v}_1, \ldots, \mathbf{v}_k$ is called a **lattice basis** for L. If each $\mathbf{v}_i \in \mathbb{Z}^n$ for all $i = 1, \ldots, k$, that is, if each component of each vector in a lattice basis is an integer, then L is a subset of $\mathbb{Z}^n$ and is called an **integral lattice**.

A lattice L is clearly an additive subgroup of $\mathbb{R}^n$. Further, since the elements in L are integral linear combinations of a lattice basis there is a positive constant $c > 0$ such that $|\mathbf{u} - \mathbf{v}| > c$ for all different elements $\mathbf{u}, \mathbf{v} \in L$. This is sufficient to characterize lattices. A subset $S \subset \mathbb{R}^n$ is **discrete** if there exists a positive constant $c > 0$ such that $|\mathbf{u} - \mathbf{v}| > c$ for all different elements $\mathbf{u}, \mathbf{v} \in S$. Then:

Theorem 15.3.1. *A subset $S \subset \mathbb{R}^n$ is a lattice if and only if S is a discrete additive subgroup of $\mathbb{R}^n$.*

Given a basis $\mathbf{v}_1, \ldots, \mathbf{v}_k$ for a vector subspace V in $\mathbb{R}^n$, it is important to know what are equivalent bases for V. In particular we have the result that $\mathbf{u}_1, \ldots, \mathbf{u}_k$ is also a basis if and only if there exists an invertible $k \times k$ real matrix T such that

$$T \begin{pmatrix} \mathbf{v}_1 \\ \vdots \\ \mathbf{v}_k \end{pmatrix} = \begin{pmatrix} \mathbf{u}_1 \\ \vdots \\ \mathbf{u}_k \end{pmatrix}.$$

For lattices the same result is almost true. Suppose that $\mathbf{v}_1, \ldots, \mathbf{v}_k$ is a lattice basis for L. Then $\mathbf{u}_1, \ldots, \mathbf{u}_k$ is also a lattice basis if and only if there exists an invertible **integral** matrix T (invertible over $\mathbb{Z}$) with

$$T \begin{pmatrix} \mathbf{v}_1 \\ \vdots \\ \mathbf{v}_k \end{pmatrix} = \begin{pmatrix} \mathbf{u}_1 \\ \vdots \\ \mathbf{u}_k \end{pmatrix}.$$

Since T is invertible over $\mathbb{Z}$ we must have $\det(T) = \pm 1$.

If L is an n-dimensional lattice in $\mathbb{R}^n$ with lattice basis $\{\mathbf{v}_1, \ldots, \mathbf{v}_n\}$ then the **fundamental domain** for L is the subset of $\mathbb{R}^n$ given by

$$\mathcal{F} = \{t_1 \mathbf{v}_1 + \cdots + t_n \mathbf{v}_n \mid 0 \le t_i < 1, i = 1, \ldots, n\}.$$

The translates of the fundamental domain will cover or **tile** all of $\mathbb{R}^n$. More precisely:

Theorem 15.3.2. *Let L be an n-dimensional lattice within $\mathbb{R}^n$ and $\mathcal{F}$ its fundamental domain. Then for every vector $\mathbf{w} \in \mathbb{R}^n$ there exists unique vectors $\mathbf{u} \in \mathcal{F}$ and $\mathbf{v} \in L$ such that $\mathbf{w} = \mathbf{u} + \mathbf{v}$. Hence the **translates** of the fundamental domain*

$$\mathbf{v} + \mathcal{F} = \{\mathbf{u} + \mathbf{v} \mid \mathbf{u} \in \mathcal{F}\}$$

over all $\mathbf{u} \in V$ covers precisely all of $\mathbb{R}^n$.

The n-dimensional volume of a lattice's fundamental domain is an invariant that plays an important role in studying the lattice. We look at this soon. But first we consider special bases for vector subspaces of $\mathbb{R}^n$.

If V is a vector subspace of $\mathbb{R}^n$ then a basis $\mathbf{v}_1, \ldots, \mathbf{v}_k$ of V is an **orthogonal basis** if $\mathbf{v}_i \cdot \mathbf{v}_j = 0$ if $i \neq j$. They are an **orthonormal basis** if each $\mathbf{v}_i$ has magnitude 1, that is, if $|\mathbf{v}_i| = 1$ for all $i = 1, \ldots, \mathbf{v}_k$.

Given a finite-dimensional subspace V of $\mathbb{R}^n$, it is always possible to find an orthonormal basis. This is a consequence of the **Gram-Schmidt orthogonalization procedure**. This is a procedure to change any independent set into an orthogonal set that spans the same subspace. Once we have an orthogonal basis then dividing each basis vector by its magnitude provides an orthonormal basis.

Theorem 15.3.3 (Gram-Schmidt Orthogonalization Procedure). *Let V be a subspace of $\mathbb{R}^n$. If $\{\mathbf{u}_1, \ldots, \mathbf{u}_m\}$ is an independent set of V then $\{\mathbf{v}_1, \ldots, \mathbf{v}_m\}$ is an orthogonal set where the $\mathbf{v}_i$ are defined inductively by*

$$\mathbf{v}_1 = \mathbf{u}_1 \text{ and } \mathbf{v}_{k+1} = \mathbf{u}_{k+1} - \sum_{i=1}^{k} \frac{\mathbf{u}_{k+1} \cdot \mathbf{v}_i}{\mathbf{v}_i \cdot \mathbf{v}_i} \mathbf{v}_i.$$

*The orthogonal set $\{\mathbf{v}_1, \ldots, \mathbf{v}_m\}$ is called the **Gram-Schmidt orthogonalization** or **GSO** of $\{\mathbf{u}_1, \ldots, \mathbf{u}_m\}$.*

Corollary 15.3.4. *Any finite-dimensional subspace of $\mathbb{R}^n$ has an orthonormal basis.*

Given an orthonormal basis, it is very simple to determine coordinates of a vector $\mathbf{v}$ relative to this basis.

Lemma 15.3.5. *If $\{\mathbf{v}_1, \ldots, \mathbf{v}_m\}$ form an orthonormal basis for the subspace V and if $\mathbf{v} \in V$ then*

$$(1) \qquad \mathbf{v} = \sum_{i=1}^{m} (\mathbf{v} \cdot \mathbf{v}_i) \mathbf{v}_i.$$

*The $\mathbf{v} \cdot \mathbf{v}_i$ are called the **Fourier coefficients** of $\mathbf{v}$ relative to $\mathbf{v}_1, \ldots, \mathbf{v}_m$.*

$$(2) \qquad |\mathbf{v}| = \sqrt{\sum_{i=1}^{m} |(\mathbf{v} \cdot \mathbf{v}_i)|^2}$$

Using an orthonormal basis for a subspace V of $\mathbb{R}^n$, we can solve the **closest vector problem**. That is, given a vector $\mathbf{w} \in \mathbb{R}^n$ find the vector $\mathbf{v} \in V$ that is closest to $\mathbf{w}$. The solution is given in the following theorem called the **Closest Vector Theorem**

Theorem 15.3.6 (Closest Vector Theorem). *Let V be a subspace of $\mathbb{R}^n$ and let $\mathbf{w}$ be a vector in $\mathbb{R}^n$. If $\{\mathbf{v}_1, \ldots, \mathbf{v}_m\}$ is an orthonormal basis for V then the unique vector $\mathbf{v} \in V$ closest to $\mathbf{w}$ is given by*

$$\mathbf{v} = \sum_{i=1}^{m} (\mathbf{w} \cdot \mathbf{v}_i) \mathbf{v}_i.$$

Notice that if $\mathbf{w}$ is actually in the subspace, then the expression in the theorem is just its expression in terms of the orthonormal basis. For a vector subspace the closet vector theorem actually provides an algorithm given a subspace V and a vector $\mathbf{w} \in \mathbb{R}^n$ to find the closest vector within the subspace. This works as follows. First determine an orthonormal basis for the subspace V and then use the formula provided in the Closest Vector Theorem.

If L is a lattice within V then there might be no orthogonal or orthonormal lattice basis of L. Hence, if we ask the closest vector question for the lattice L there is no general algorithm to solve this. We will look at this problem, which is the foundation for a cryptoprimitive in Section 15.4.

We have already introduced the use of the closest vector theorem in cryptography. Recall that in section 4.6.3 we described a secret sharing protocol developed by Chum, Fine, Rosenberger and Zhang [CFRZ] based on the use of orthonormal bases and the closest vector result. This was further enhanced and extended to a cryptosystem by Fine, Moldenhauer and Rosenberger [FMoR].

15.3.1 The Geometry of Numbers

The fundamental domain of a lattice L is a convex subset of $\mathbb{R}^n$ and hence a geometric object. If L is an integral lattice then L is a subset of $\mathbb{Z}^n$. The study of the geometric properties of the fundamental domains of integral lattices, or more generally of convex subsets of $\mathbb{Z}_n$, is a branch of mathematics called the **geometry of numbers**. This was initiated by Minkowski in 1889. Here we mention some results from this area relevant to cryptography.

Let L an n-dimensional integral lattice in $\mathbb{Z}^n$ with fundamental domain $\mathcal{F}$. We show first that the n-dimensional volume of $\mathcal{F}$ is an invariant of the lattice. An n-dimensional lattice in $\mathbb{R}^n$ is called a **complete lattice**. Hence for this section we will be considering complete integral lattices.

As an example, consider the lattice given by the Gaussian integers in real 2-space. Here $V = \mathbb{R}^2$, $\Gamma = \mathbb{Z} + \mathbb{Z}i = \mathbb{Z}[i]$ and the fundamental domain is

$$\phi = \{x + iy \mid 0 \leq x < 1, 0 \leq y < 1\}.$$

In $\mathbb{R}^n$ the cube spanned by the standard orthonormal basis $\mathbf{e}_1, \ldots, \mathbf{e}_n$ has volume 1 and more generally the parallelopiped

$$\phi = \{x_1\mathbf{v}_1 + \cdot + x_n\mathbf{v}_n \mid x_i \in \mathbb{R}, 0 \leq x_i < 1\}$$

spanned by the independent set of vectors $\mathbf{v}_1, \ldots, \mathbf{v}_n$ has a volume given by

$$\mathrm{vol}(\phi) = |\det(A)|$$

where $A = (a_{ij})$ is the transition matrix from the basis $\mathbf{e}_1, \ldots, \mathbf{e}_n$ to the basis $\mathbf{v}_1, \ldots, \mathbf{v}_n$, that is,

$$\mathbf{v}_i = \sum_{i=1}^{n} a_{ij}\mathbf{e}_j.$$

If we use the ordinary Euclidean inner product on $\mathbb{R}^n$ then

$$\text{vol}(\phi) = \lambda(\phi)$$

where λ is the Lebesgue measure. We call this the standard Euclidean volume.

Let L be the complete lattice in $\mathbb{R}^n$ with lattice basis $\mathbf{v}_1, \ldots, \mathbf{v}_n$ and $\mathcal{F}$ its fundamental domain relative to this basis. We let

$$V(\mathcal{F}) = \text{vol}(\mathcal{F})$$

be the standard Euclidean volume of $\mathcal{F}$. The basic questions in the geometry of numbers revolve about the measurement of $V(\mathcal{F})$ and by extension the length of the lattice basis vectors.

As we will see shortly, this volume is an invariant of the lattice, that is, it does not depend on the particular lattice basis. Notice that if $\mathbf{v}_1, \ldots, \mathbf{v}_n$ is an orthogonal basis then

$$V(\mathcal{F}) = |\mathbf{v}_1||\mathbf{v}_2| \cdots |\mathbf{v}_n|$$

and in general it can be proved that,

$$V(\mathcal{F}) \leq |\mathbf{v}_1||\mathbf{v}_2| \cdots |\mathbf{v}_n|.$$

Lemma 15.3.7 (Hadamard's Inequality). *Let L be a complete lattice in $\mathbb{Z}^n$ with integral basis $\mathbf{v}_1, \ldots, \mathbf{v}_n$ and fundamental domain $\mathcal{F}$. Then*

$$V(\mathcal{F}) \leq |\mathbf{v}_1||\mathbf{v}_2| \cdots |\mathbf{v}_n|,$$

with equality only if the $\mathbf{v}_1, \ldots, \mathbf{v}_n$ is an orthogonal basis.

For a given lattice basis $\mathbf{v}_1, \ldots, \mathbf{v}_n$ the **Hadamard ratio** $\mathcal{H}$ is the value

$$\mathcal{H} = \left(\frac{V(\mathcal{F})}{|\mathbf{v}_1| \cdots |\mathbf{v}_n|} \right)^{\frac{1}{n}}.$$

Thus the Hadamard ratio is a measure of how orthogonal a given basis is, being more orthogonal the closer it gets to 1. For cryptographic purposes lattice bases can be separated into good bases that are sufficiently orthogonal and bad bases that are not (some angles being very small).

Theorem 15.3.8. *Let L be a complete integral lattice with basis $\mathbf{v}_1, \ldots, \mathbf{v}_n$ and fundamental domain $\mathcal{F}$. Let M be the $n \times n$ integral matrix formed by letting the i-th row be the integer vector $\mathbf{v}_i$. Then we have $V(\mathcal{F}) = \det(M)$.*

As we mentioned, the transition matrix from one lattice basis to another is an integral matrix of determinant one. It follows then if $\mathbf{w}_1, \ldots, \mathbf{w}_n$ is another lattice basis for L with fundamental domain $\mathcal{F}'$ that $V(\mathcal{F}) = V(\mathcal{F}')$. That is, the volume of a fundamental domain depends only on the lattice and not on the lattice basis. It is hence an invariant of the lattice. Because of Theorem 15.3.2 the volume of a fundamental domain is called the **determinant of the lattice** denoted $\det(L)$.

Theorem 15.3.9. *If L is a complete integral lattice then the volume of a fundamental domain does not depend upon the particular basis, that is, it is an invariant of the lattice and not of the particular lattice basis.*

Along with the volume, an important problem is to determine the shortest non-zero vector in a lattice. The next result, due to Hermite, establishes an upper bound for the shortest vector.

Theorem 15.3.10 (Hermite's Theorem). *Let L be a complete n-dimensional integral lattice. Then there exists a vector* $\mathbf{v} \in L$ *satisfying*

$$|\mathbf{v}| \leq \sqrt{n} \, (\det(L))^{\frac{1}{n}}.$$

The proof of this result depends upon a result of Minkowski called **Minkowski's theorem**. This result was the beginning point for the study of the geometry of numbers.

Theorem 15.3.11 (Minkowski's Theorem). *Let L be a complete lattice in* $\mathbb{R}^n$ *(not necessarily integral). Let* $S \subset \mathbb{R}^n$ *be a symmetric convex subset whose Euclidean volume satisfies*

$$V(S) > 2^n \det(L).$$

Thus S must contain a non-zero vector in the lattice L. Further, if S is also closed as a subset of $\mathbb{R}^n$ *then the result is still true for* $V(S) \geq 2^n \det(L)$.

More information on Minkowski's theorem and its relation to Hermite's result can be found in [HPS]. A discussion placing Minkowski's theorem in the wider context of algebraic number theory can be found in the book by Fine and Rosenberger [FR].

Part of the geometry of numbers and of particular importance is the length of the shortest vector in a lattice. If L is a complete lattice then the **Gaussian expected shortest length** is

$$\sigma(L) = \sqrt{\frac{n}{2\pi e}} \, (\det(L))^{\frac{1}{n}}.$$

The **Gaussian heuristic** says that in a randomly chosen complete lattice we would have for the shortest vector $\mathbf{v}_s$ in L that

$$|\mathbf{v}_s| \approx \sigma(L).$$

The Gaussian heuristic is used in attempts to solve the shortest vector problem (see the next section).

15.4 Hard Lattice Problems

As we discussed in Sections 15.1 and 15.2 it is generally hard mathematical problems about a cryptographic platform that are used to construct cryptographic primitives.

For asymmetric or public key cryptography these problems must be hard to solve unless one has extra information to serve as the trapdoor.

For lattices the relevant hard mathematical problems are the **Shortest Vector Problem** or SVP and the **Closest Vector Problem** or CVP and variations of these two. In many situations they can be shown to be NP-hard and can actually provide security proofs for cryptosystems based on them.

The basic **shortest vector problem** for lattices, which we will denote by SVP, is the following:

Shortest Vector Problem: *Given a basis of a lattice L, find the shortest non-zero vector in the lattice.*

If $\lambda(L)$ denotes the length of the shortest non-zero vector in L the SVP can be further posed in the following three forms: the **computational SVP**, the **decision SVP** and the **search SVP**.

Computational Shortest Vector Problem: *Given a basis of a lattice L, find the shortest non-zero vector in the lattice.*

Decision Shortest Vector Problem: *Given a basis of a lattice L, and a real constant $d > 0$ distinguish between $\lambda(L) \leq d$ and $\lambda(L) > d$*

Search Shortest Vector Problem: *Given a basis of a lattice L, find a vector $\mathbf{v} \in L$ such that the length of $\mathbf{v}$ is $\lambda(L)$.*

The **closest vector problem** for lattices, which we will denote by CVP, is the following:

Closest Vector Problem: *Given a basis of a lattice L and a vector $\mathbf{w} \in \mathbb{R}^n$ not in the lattice, find the lattice vector $\mathbf{v}$ closest to $\mathbf{w}$, that is, the lattice vector with the minimum Euclidean distance to $\mathbf{w}$.*

The CVP can also be put in a computational, search and decision manner.

Computational Closest Vector Problem: *Given a basis of a lattice L, and a vector $\mathbf{w} \in \mathbb{R}^n$ find the shortest non-zero vector in the lattice and find the minimum distance from $\mathbf{w}$ to L.*

Decision Closest Vector Problem: *Given a basis of a lattice L, and a vector $\mathbf{w} \in \mathbb{R}^n$ and a real number $r > 0$ determine if the minimum distance is less than r or not.*

Search Closest Vector Problem: *Given a basis of a lattice L, and a vector $\mathbf{w} \in \mathbb{R}^n$ find a vector $\mathbf{v} \in L$ such that $|\mathbf{w} - \mathbf{v}| \leq |\mathbf{v} - \mathbf{u}|$ for all $\mathbf{u} \in L$.*

There are also approximate versions of these problems. For example:

Approximate Shortest Vector Problem: *Given a basis of a lattice L, and a constant γ depending on the lattice dimension, find a vector $\mathbf{v} \in L$ such that $0 < |\mathbf{v}| \leq \gamma\lambda(L)$.*

This is called the search approximate SVP denoted by SVP_γ. The problem with $\gamma = 1$ is the basic SVP. There are also approximate variants of the decision and the calculation SVP. There are further approximate versions of the CVP.

Notice that if we are in a vector subspace V in $\mathbb{R}^n$, these problems are easy to solve using an orthogonal basis for V. The closest vector theorem provides an easy solution. However, lattices do not necessarily have orthogonal lattice bases nor is there a general algorithm, like the Gram-Schmidt procedure, to find an orthogonal basis. Therefore within lattices these problems are normally hard to solve. In fact it can be proved that both the CVP and SVP in worst case hardness are NP-hard and thus provably hard in the complexity sense.

As we discussed in Chapter 11 on braid group cryptography, to provide a basis for a cryptographic security proof what is needed is not worst-case hardness but average-case hardness. Experimental evidence suggests that most NP-hard problems are probably only worst case hard. That is, their average case complexity is probably not NP-hard. However, the above lattice problems and variants of them have been conjectured or proven to be average-case hard, making them a potential class of problems to use to develop cryptographic schemes.

The closest vector problem is actually a generalization of the shortest vector problem (see [LPW]), and it has been proved that any hardness of SVP implies the same hardness for CVP.

Ajtai (see [Aj1]) in his foundational work, showed that the SVP problem is NP-hard and showed some connections between the worst-case complexity and average-case complexity of several lattice problems. However, there are so-called **good** bases for a lattice in which these problems can be solved. From these ideas Ajtai built a crypto-primitve and then Ajtai and Dwork created a public-key cryptosystem whose security could be proven using only the worst case hardness of a certain version of SVP. The Ajtai-Dwork method is easily implementable in many cases but not so in the cases where security could be proved. Subsequently their method was shown to be broken in many cases. We will discuss their technique in Section 15.5.

A complete discussion of these problems, relationships between them and their complexity can be found in the paper [LPW]. In that paper they also discuss the various solution techniques for the basic problems.

15.5 Lattice Reduction and Babai's Algorithm

Lattice basis reduction is a transformation of an integer lattice basis in order to find a basis with short, nearly orthogonal vectors. Such a basis is called a **good** basis for the lattice and can be defined formally depending on the dimension in terms of the Hadamard ratio. Recall this shows approximately how orthogonal a lattice is. A good basis for a lattice L is one in which this ratio has a value close to 1.

There are algorithms (see the Babai algorithm below) that make the above hard lattice problems easy to solve (in polynomial time) if we start with a good basis. Hence, if we can compute a good lattice basis, the CVP and SVP problems are easy to solve. A **lattice reduction algorithm** is an algorithm which, given a basis for a lattice, outputs a new basis consisting of relatively short, nearly orthogonal vectors. Combining a lattice reduction algorithm with solution algorithms using good bases provides a method to solve both the SVP and CVP and their variants.

The best known lattice reduction algorithm is the **Lenstra–Lenstra–Lovász** algorithm abbreviated **LLL algorithm**. This is an efficient algorithm which outputs an almost reduced lattice basis in polynomial time (see [LLL]). After lattice-based cryptosystems were introduced, the LLL algorithm and enhancements of it were used to break several lattice-based cryptographic protocols, making it a very important tool in cryptanalysis. The success of LLL on experimental data led to a belief that lattice reduction might be an easy problem in practice. However, this belief was challenged when new results, especially the result of Ajtai on the hardness of lattice problems, were obtained.

The LLL algorithm is a generalization of an older algorithm attributed to Gauss, and independently to Lagrange, that uses a modification of the Gram-Schmidt procedure to find a reduced lattice basis for a two-dimensional integral lattice. The LLL algorithm produces a relatively good basis in polynomial time, but does not solve the exact SVP problem. Another lattice reduction algorithm the **HKZ algorithm**, named for Hermite, Korkine and Zolotarev, solves the problem in time exponential in n where n is the dimension. Finally Schnorr presented an algorithm that interpolates between LLL and HKZ, called **Block Reduction**. What is important for cryptography is that while these algorithms do produce good bases the exact solutions are not easy.

Formal descriptions of the workings of these algorithms and the relationships between them are discussed in the paper [LPW] which is available on the internet. We leave these to the reader.

In 1986, Babai (see [Bab]) developed an approximation algorithm to the closest vector problem (CVP). This algorithm, now known as **Babai's Nearest Plane Algorithm**, determines a solution to the CVP given a good basis for a lattice.

Theorem 15.5.1 (Babai). *Let L be a lattice in $\mathbb{R}^n$ with a good lattice basis $\mathbf{v}_1, \ldots, \mathbf{v}_n$. Hence the basis vectors are assumed to be sufficiently orthogonal to each other. Let $\mathbf{w} \in \mathbb{R}^n$. Let*

$$\mathbf{w} = t_1 \mathbf{v}_1 + \cdots + t_n \mathbf{v}_n$$

be the standard expansion of $\mathbf{w}$ in terms of the basis vectors $\mathbf{v}_i$ of $\mathbb{R}^n$. Now let $a_i = [t_i]$, that is, the greatest integer less than t_i. Then the vector $\mathbf{v} \in L$ given by

$$\mathbf{v} = a_1 \mathbf{v}_1 + \cdots + a_n \mathbf{v}_n$$

solves the CVP.

Babai also shows that if the basis is not sufficiently orthogonal then this method will return a vector far away from the closest vector.

Given a general basis for a lattice L this provides an algorithm for potentially solving the CVP. Given a lattice basis $\mathbf{v}_1, \ldots, \mathbf{v}_n$ and a vector $\mathbf{w} \in \mathbb{R}^n$:

Step (1): Use a lattice reduction algorithm to get a good basis for L.
Step (2): Having obtained this good basis use Babai's algorithm to solve.

Since the lattice reduction algorithms that give the exact solution are in exponential time this provides a trapdoor. If we know a good basis then Babai's algorithm solves the CVP while if we start with a bad basis it does not.

15.6 Main Lattice Based Cryptosystems

Lattice problems are typically quite hard. The best known algorithms that exactly solve either the CVP or SVP run in exponential time. However, given a good lattice basis they can be solved in polynomial time. From this idea the field of lattice-based cryptography has been developed. There are two enticing features of lattice-based cryptography. First, the encryption can be quite efficient. The second, and perhaps most enticing aspect, is that there are currently no known quantum algorithms for solving lattice problems that perform significantly better than the best known non-quantum algorithms. On the other hand, RSA, Diffie-Hellman and elliptic curve methods are susceptible to quantum algorithm attacks. Attempts to solve lattice problems by quantum algorithms have been made since Shor's discovery of the quantum factoring algorithm but have so far been unsuccessful.

Worst-case hardness of lattice problems means that breaking the cryptographic construction (even with some small non-negligible probability) is provably at least as hard as solving several lattice problems (approximately, within polynomial factors) in the worst case. In other words, breaking the cryptographic construction based on a lattice problem would imply that there is an efficient algorithm to solve any instance of the lattice problem utilized. In most cases, a lattice-based cryptographic protocol uses as its underlying problem the approximate lattice problems such as SVP. This provides a strong security guarantee making lattice-based cryptography quite attractive.

The worst-case security guarantee assures us that attacks on the cryptographic construction are likely to be effective only for small choices of parameters and not asymptotically. This guarantees that there are no fundamental flaws in the design of our lattice-based cryptographic protocol.

In the next four subsections we describe Ajtai's hash function based on lattice problems and three public key cryptosystems constructed from these ideas. The basic idea is to encrypt within a lattice so that the decryption is based on solving some variant of either the CVP or SVP. The public key is then a bad basis for the lattice while the private key is a good basis. This provides a trapdoor, since by Babai's algorithm we can solve these problems if we have a good basis. However, the lattice reduction algorithms are too time consuming to solve the problems if we are given a bad basis.

15.6.1 Ajtai's Hash Function and Cryptosystem

In 1995, Ajtai described a problem that is hard on average, if some well-known lattice problems are hard to approximate in the worst case, and demonstrated how this problem can be used to construct one-way functions. Ajtai's method can also be used to construct families of almost collision-free hash functions.

The construction is quite straightforward. Given a parameter n, we choose a random $n \times m$ matrix M with entries from $\mathbb{Z}_q$ with q a large prime and where m and q are chosen so that $n \log q < m < \frac{q}{2n^4}$ and $q = \mathcal{O}(n^c)$ for some constant $c > 0$.

The hash function h_M maps strings of m bits into $\mathbb{Z}_q^n$, that is,

$$h_M : \{0, 1\}^m \to \mathbb{Z}_q^n.$$

If $s = s_1 s_2 \cdots s_m \in \{0, 1\}^m$ then h_M is defined by

$$h_M(s) \equiv Ms \pmod{q} \equiv \sum_i s_i M_i \pmod{q}$$

where M_i is the ith column of M.

Since the input to h_M has m bits, while its output has $n \log q$ bits and $m > n \log q$, there will be collisions in h_M. Ajtai showed However, that it is infeasible to find any of these collisions unless some well known lattice problems have good approximate solutions in the worst case. It follows that, although it is easy to find solutions for the equations $Ms \equiv 0 \pmod{q}$, it is hard to find binary solutions. This infeasibility makes this, although not collision-free, a workable hash function.

In his paper Ajtai described the following problem related to the hash function defined above.

For parameters n, m, q satisfying the conditions described above and a matrix $M \in \mathbb{Z}_q$ find a vector $x \in \mathbb{Z}_q^m$ with $|x| < n$ which satisfies $Mx \equiv 0 \pmod{q}$.

He then reduces the worst-case complexity of the shortest basis problem to the average case complexity of this problem. The shortest basis problem in a lattice is a variant of SVP which asks, given a lattice, for the shortest basis.

15.6.2 The Ajtai-Dwork Cryptosystem

Using the hash functions described in the last section, Ajtai made suggestions for one-way functions that could be used in a public key cryptosystem. Ajtai and Dwork then used this to develop a lattice-based probabilistic asymmetric cryptosystem. This cryptosystem was provably secure if the "unique" SVP is difficult in the worst-case. However, it has been shown by a detailed cryptanalysis and experiments that in order to be secure, implementations of the Ajtai-Dwork cryptosystem would require very large keys, making it impractical for real applications. In analyzing the Ajtai-Dwork cryptosystem it has been observed that lattice reduction algorithms behave surprisingly well and can provide much better approximations to SVP or CVP than expected.

Let h_M be the hash function Ajtai constructed with parameters n, m, q and matrix M. As a one-way trapdoor function Ajtai suggested the function

$$f(M, s) = (M, h_M(s))$$

where s is a string of bits. The trapdoor is based on the solvability of the SVP for good bases of a lattice.

The construction of the Ajtai-Dwork cryptosystem is as follows. First some notation: Let $\mathbf{v}_1, \dots, \mathbf{v}_n$ be a lattice basis. The width of the fundamental domain is the minimum over i of the Euclidean distance between $\mathbf{v}_i$ and the hyperplane spanned by the other $\mathbf{v}_j$'s. Reducing a vector $\mathbf{v}$ modulo the fundamental domain $\mathcal{F}$ means obtaining a vector $\mathbf{v}' \in \mathcal{F}$ such that $\mathbf{v}' - \mathbf{v}$ belongs to the lattice. This is denoted by $\mathbf{v}' \equiv \mathbf{v} \pmod{\mathcal{F}}$.

Given a security parameter n (which is also the precision of the binary expansion for real numbers), we let $m = n^3$ and $\rho_n = 2^{n \log n}$. We let B_n the n-dimensional cube of side-length ρ_n and S_n the n-dimensional ball of radius n^{-8}.

Given n, the private key is a uniformly chosen vector $\mathbf{u}$ in the n-dimensional unit ball. For such a private key, we denote by $H_{\mathbf{u}}$ the distribution on points in B_n that follows from the following construction:

(1) Pick a point a uniformly at random from $\{x \in B_n | x \cdot \mathbf{u} \in \mathbb{Z}\}$ with · the Euclidean inner product.
(2) Select $\delta_1, \dots, \delta_n$ uniformly at random from S_n.
(3) Output the point $a + \sum_i \delta_i$.

The public key is obtained by picking the vectors $\mathbf{w}_1, \dots, \mathbf{w}_n, \mathbf{v}_1, \dots, \mathbf{v}_m$ independently at random from the distribution $H_{\mathbf{u}}$, subject to the constraint that the width of the fundamental domain spanned by $\mathbf{w}_1, \dots, \mathbf{w}_n$ is at least $n^{-2}\rho_n$. We denote the fundamental domain of the lattice spanned by $\mathbf{w}_1, \dots, \mathbf{w}_n$ by $\mathcal{F}(\mathbf{w}_1, \dots, \mathbf{w}_n)$.

Encryption is done bit-by-bit. To encrypt a 0, uniformly select $s_1 \cdot s_m$ in $\{0, 1\}$, and reduce the vector $\sum_i s_i \mathbf{v}_i$ modulo the fundamental domain $\mathcal{F}(\mathbf{w}_1, \dots, \mathbf{w}_n)$. The vector obtained is the ciphertext. The ciphertext of 1 is just a randomly chosen vector in $\mathcal{F}(\mathbf{w}_1, \dots, \mathbf{w}_n)$.

To decrypt a ciphertext x with the private key $\mathbf{u}$, compute $\tau = x \cdot \mathbf{u}$. If $\tau \in \mathbb{Z} \pm n^{-1}$, then x is decrypted as 0, and otherwise as 1. Thus, an encryption of 0 will always be decrypted as 0, and an encryption of 1 has the probability of $\frac{2}{n}$ to be decrypted as 0. These decryption errors can be removed. The main result of [AD] states that a probabilistic algorithm distinguishing encryptions of a 0 from encryptions of a 1 with some polynomial advantage can be used to find the shortest non-zero vector in any n-dimensional lattice whenever the shortest vector $\mathbf{v}$ is unique, in the sense that any other vector whose length is at most $n^8 |\mathbf{v}|$ is parallel to $\mathbf{v}$.

15.6.3 The GGH Cryptosystem

The **Goldreich–Goldwasser–Halevi lattice-based cryptosystem** or **GGH cryptosystem** is another asymmetric cryptosystem based on lattices. There is also a GGH signature scheme.

The GGH cryptosystem was published in 1997 and uses as a trapdoor a one-way function that relies on the difficulty of lattice reduction and the fact that the closest vector problem is a hard problem and in fact NP-hard in its worst case form. The idea included in this trapdoor function is that, given any basis for a lattice, it is easy to generate a vector which is close to a lattice point, for example taking a lattice point and adding a small error vector. However, to return from this perturbed vector to the original lattice point a special basis is needed.

The private key in GGH is a good basis of a lattice, that is, a basis with short nearly orthogonal vectors and a unimodular matrix. The public key is another basis of the lattice but bad. Encryption and decryption works as follows.

Alice starts out with a good basis $\mathbf{v}_1, \ldots, \mathbf{v}_n$ for a complete lattice L. Hence the vectors $\mathbf{v}_1, \ldots, \mathbf{v}_n$ are reasonably short and orthogonal. She can check this in various ways, for example using the Hadamard ratio. These vectors are Alice's private key. Let V be the matrix whose rows are the $\mathbf{v}_i$.

Alice next chooses an $n \times n$ integral unimodular matrix M and computes $W = MV$. The row vectors in W, $\mathbf{w}_1, \ldots, \mathbf{w}_n$ are also a basis for L since M is integral and unimodular. These vectors are Alice's public key.

For Bob to send a message to Alice he chooses a small vector $\mathbf{m}$ as his plaintext message. Bob then chooses another small vector $\mathbf{r}$ which will serve as a session key. The ciphertext $\mathbf{p}$ for $\mathbf{m}$ is the vector

$$\mathbf{p} = \mathbf{m}W.$$

The encrypted vector $\mathbf{p}$ is not in L but is close to the lattice point $\mathbf{m}W$ since the session key vector $\mathbf{r}$ is small.

To decrypt, Alice uses Babai's algorithm with her private key $\mathbf{v}_1, \ldots, \mathbf{v}_n$ which is a good basis. She finds the vector in the lattice L close to $\mathbf{p}$. Since she has a good basis and $\mathbf{r}$ is small the vector she obtains is $\mathbf{m}W$. She then multiplies by W^{-1} to recover the plaintext message $\mathbf{m}$.

In 1999 Nguyen showed that the GGH encryption scheme has a flaw in the design of the schemes. He showed that every ciphertext reveals information about the plaintext and that the problem of decryption could be turned into a special closest vector problem much easier to solve than the general CVP.

15.6.4 NTRU Cryptosystem

The final lattice based cryptosystem that we will just mention is the **NTRU Cryptosystem** or NTRU family of cryptosystems.

The NTRU encryption algorithm is a lattice-based public key method based on the shortest vector problem in a lattice. Operations are based on objects in a truncated polynomial ring with convolution multiplication (see [HPS]) and all polynomials in the ring have integer coefficients and degree at most $N - 1$.

NTRU is actually a parameterised family of cryptosystems; each system is specified by four integer parameters (N, p, q, d) which represent the maximal degree for all polynomials in the truncated ring R, a small modulus and a large modulus, respectively, where it is assumed that N is prime, q is always larger than p, and p and q are coprime; and four sets of polynomials and (a polynomial part of the private key, a polynomial for generation of the public key, the message and a blinding value, respectively), all of degree at most $N - 1$.

It relies on the presumed difficulty of factoring certain polynomials in such rings into a quotient of two polynomials having very small coefficients. Breaking the cryptosystem is strongly related, though not equivalent, to the algorithmic problem of lattice reduction in certain lattices. Careful choice of parameters is necessary to thwart some published attacks.

Since both encryption and decryption use only simple polynomial multiplication, these operations are very fast compared to the other asymmetric encryption schemes that we have discussed, such as RSA, ElGamal and elliptic curve cryptography.

There is a related algorithm for digital signatures.

The NTRUEncrypt Public Key Cryptosystem, the first NTRU cryptosystem, was developed originally in 1996 by Jeffrey Hoffstein, Jill Pipher and Joseph H. Silverman. During the subsequent years people have been worked on improving this cryptosystem. Since the first presentation, changes were made to improve both the performance of the system and its security. Most performance improvements were focused on speeding up the process. In newer versions of NTRU, new parameters have been introduced that seem secure for all currently known attacks, including quantum attacks, and reasonable increases in computational power.

15.7 Security Proofs

Lattice-based cryptographic constructions hold a great promise for post-quantum cryptography. Many of them are quite efficient, and some even compete with the best known alternatives; they are typically quite simple to implement; and are all believed to be secure against attacks using conventional or quantum computers.

In terms of security, lattice-based cryptographic constructions can be divided into two types which includes practical proposals. The first type is typically very efficient, but lack a security proof. The second type admits strong provable security guarantees based on the worst-case hardness of lattice problems, but most of these are not sufficiently efficient to be used in practice.

The NTRU family of algorithms noted above are efficient but lack a strong security guarantee.

15.8 Exercises

15.1 Let $\{\mathbf{v}_1, \ldots, \mathbf{v}_m\}$ form an orthonormal basis for a subspace V of $\mathbb{R}^n$. Let $\mathbf{v} \in V$. Show that
$$\mathbf{v} = \sum_{i=1}^{m} (\mathbf{v} \cdot \mathbf{v}_i)\mathbf{v}_i.$$

15.2 Let V be a subspace of $\mathbb{R}^n$ and let $\mathbf{w}$ be a vector in $\mathbb{R}^n$. Let $\mathbf{v}_1, \ldots, \mathbf{v}_m$ be an orthonormal basis for V. Show that the unique vector $\mathbf{v} \in V$ closest to $\mathbf{w}$ is given by
$$\mathbf{v} = \sum_{i=1}^{m} (\mathbf{w} \cdot \mathbf{v}_i)\mathbf{v}_i.$$

15.3 Let W be the subspace of $\mathbb{R}^4$ spanned by $\mathbf{u}_1 = (0, 1, 0, 2)$, $\mathbf{u}_2 = (2, 1, 1, 0)$, and $\mathbf{u}_3 = (0, 2, 3, 0)$. Find the vector in this subspace closes to the vector $v = (0, 1, 0, 1)$.

15.4 Prove Babai's Theorem (Theorem 15.5.1): Let L be a lattice in $\mathbb{R}^n$ with a good lattice basis $\mathbf{v}_1, \ldots, \mathbf{v}_n$. Let $\mathbf{w} \in \mathbb{R}^n$ with $\mathbf{w} = t_1\mathbf{v}_1 + \cdots + t_n\mathbf{v}_n$. Let $a_i = [t_i]$, the greatest integer less than t_i. Show that the vector $\mathbf{v} \in L$ given by
$$\mathbf{v} = a_1\mathbf{v}_1 + \cdots + a_n\mathbf{v}_n$$

solves the closest vector problem.

15.5 Let L be complete integral lattice with basis $\mathbf{v}_1, \ldots, \mathbf{v}_n$ and fundamental domain $\mathcal{F}$. Let M be the $n \times n$ integral matrix formed by letting the i-th row be the integral vector $\mathbf{v}_i$. Show that $V(\mathcal{F}) = \det(M)$.

15.6 Let d be a square-free integer. Let $\omega = \frac{1+\sqrt{d}}{2}$ if $d \equiv 1 \pmod 4$ and $\omega = \sqrt{d}$ if $d \equiv 2 \pmod n$ or $d \equiv 3 \pmod 4$.
(a) Show that $\{1, \omega\}$ generates a lattice in $\mathbb{R}^2$.
(b) Determine a fundamental domain $\mathcal{F}$ for L and the volume $V = V(\mathcal{F})$.

15.7 Let S be a discrete subset of $\mathbb{R}^n$. Show that S has no accumulation points.

15.8 Let L be a lattice in $\mathbb{R}^n$. Show that the following properties hold.
(a) There exists a constant $c > 0$ such that for all $\mathbf{v} \in L$ there is no element $\mathbf{w} \in L$ with $|\mathbf{v} - \mathbf{w}| < c$.
(b) L is a closed set.
(c) If $S \subset \mathbb{R}^n$ is bounded than $L \cap S$ is finite.
(d) L is countable.

15.9 Let L be a complete lattice in $\mathbb{R}^n$. A subset $M \subset L$ that is also a complete lattice, is called a sublattice of L. Show:

(a) M is a subgroup of finite index, $|L : M|$, of M.

(b) If $\mathcal{F}_M$ and $\mathcal{F}_L$ are fundamental domains for L and M respectively, then

$$V(\mathcal{F}_M) = V(\mathcal{F}_K)|L : M|.$$

Bibliography

[AKr] P. Ackermann and M. Kreuzer, *Gröbner basis cryptosystems*, AAECC **17** (2006), 173-194.

[ARR] G. S. G. N. Anjaneyulu, P. Vasudeva Reddy, and U. M. Reddy, *Secured digital signature scheme using polynomials over non-commutative division semirings*, Int. J. Comput. Sci. and Network Security **8** (2008), 278–284.

[Aj1] M. Ajtai, *Generating hard instances of lattice problems*, in: Proc. 28th ACM Symposium on the Theory of Computing, Philadelphia, 1996, 99–108.

[Aj2] M. Ajtai, *The shortest vector problem in L_2 is NP-hard for randomized reductions*, in: Proc. 30th ACM Symposium on the Theory of Computing, Dallas, 1998, 10–19.

[AD] M. Ajtai and C. Dwork, *A public key cryptosystem with worst-case / average-case equivalence*, in: Proc. 29th ACM Symposium on the Theory of Computing, El Paso, 1997, 284–293.

[AKS] M. Agrawal, N. Kayal, and N. Saxena, *PRIMES is in P*, Annals of Math. **160** (2004), 781–793.

[AAG1] I. Anshel, M. Anshel, and D. Goldfeld, *An algebraic method for public key cryptography*, Math. Res. Lett. **6** (1999), 287–291.

[AAG2] I. Anshel, M. Anshel, and D. Goldfeld, *Key agreement, the algebraic eraser, and lightweight cryptography*, Contemp. Math. **418** (2006), 1–34.

[AKa] M. Anshel and D. Kahrobaei, *A non-commutative analog of the Cramer-Shoup Public Key Exchange Protocol*, preprint.

[Art] E. Artin, *Theory of braids*, Annals of Math. **48** (1941), 101–126.

[Arz] G. N. Arzhantseva, *Generic properties of finitely presented groups and Howson's Theorem*, Commun. in Alg. **26** (1998), 3783–3792.

[AO] G. N. Arzhantseva and A. Olshanskii, *Genericity of the class of groups in which subgroups with a lesser number of generators are free*, Mat. Zametki **59** (1996), 489–496.

[Atk] K. Atkinson, Numerical Analysis, Wiley, 2005.

[Bab] L. Babai, *On Lovász reduction and the nearest lattice point problem*, Combinatorica **6** (1986), 1–13.

[Bar] B. Barkee et al., *Why you cannot even hope to use Gröbner bases in public key cryptography*, J. Symb. Comput. **18** (1994), 497–501.

[BBDR] M. Batty, S. Braunstein, A. Duncan, and S. Rees, *Quantum algorithms in group theory*, Contemp. Math. **349** (2003), 1–62.

[Bau] G. Baumslag, Topics in Combinatorial Group Theory, Birkhäuser, Basel, 1993.

[BBFR] G. Baumslag, Y. Brjukhov, B. Fine, and G. Rosenberger, *Some cryptoprimitives for noncommutative algebraic cryptography*, in: Aspects of Infinite Groups, World Scientific Press, 2009, 26–44.

[BBFT] G. Baumslag, Y. Brjukhov, B. Fine, and D. Troeger, *Challenge response password security using combinatorial group Theory*, Groups – Complexity – Cryptology **2** (2010), 67–82.

[BFX1] G. Baumslag, B. Fine and X. Xu, *Cryptosystems using linear groups*, Appl. Alg. in Eng., Commun. and Comput. **17** (2006), 205–217.

[BFX2] G. Baumslag, B. Fine, and X. Xu, *A proposed public key cryptosystem using the modular group*, Contemp. Math. **421** (2007), 35–44.

[BCFRX] G. Baumslag, T. Camps, B. Fine, G. Rosenberger and X. Xu, *Designing key transport protocols using combinatorial group theory*, Contemp. Math. **418** (2006), 35–43.

[Ber] D. Bernstein, *Proving primality after Agrawal, Kayena and Saxena*, preprint, 2003, available at http://cr.yp.to/ntheory.html#aks.

[Big] S. Bigelow, *Braid groups are linear*, J. Amer. Math. Soc. **14** (2001), 471–486.

[BORS] J. C. Birget, A. Olshanskii, E. Rips, and M. V. Sapir, *Isoperimetric functions of groups and computational complexity of the word problem*, Annals of Math. **156** (2002), 467–518.

[Bir]   J. Birman, Braids, Links and Mapping Class Groups, Annals of Math. Studies Vol. **82**, Princeton Univ. Press, 1975.

[Bla]   G. Blakley, *Safeguarding cryptographic keys*, in: Proc. AFIPS National Computer Conference **48**, AFIPS Press, 1979, 313–317.

[BK]   H. Bluhm and M. Kreuzer, *Computation of syzygies over non-commutative rings*, Contemp. Math. **421** (2007), 187–200.

[Bog]   D. Bogdanov, *Foundations and properties of Shamir's secret sharing scheme*, preprint, Univ. of Tartu, 2007, available at research.cyber.ee/~peeter/teaching/seminar07k/bogdanov.pdf.

[Bor]   F. Bornemann, *PRIMES is in P: a breakthrough for everyman*, Notices of the AMS **50** (2003), 545–552.

[BMS]   A. Borovik, A. G. Myasnikov, and V. Shpilrain, *Measuring sets in infinite groups*, Contemp. Math. **298** (2002), 21–42.

[Buc]   J. Buchmann, Introduction to Cryptography, Springer, 2004.

[BD]   D. M. Burton and H. Dalkowski, Handbuch der elementaren Zahlentheorie, Heldermann Verlag, 2005.

[Came] P. J. Cameron, *Aspects of infinite permutation groups*, in: C. M. Campbell et al. (eds.), Groups St Andrews 2005, Vol. 1, London Math. Soc. Lect. Note Ser. **399**, Cambridge Univ. Press, 2007, 1–35.

[Camp] T. Camps, *Surface braid groups as platform groups and applications in cryptography*, dissertation, Universität Dortmund, 2009.

[CRR]   T. Camps, V. grosse Rebel, and G. Rosenberger, Einführung in die kombinatorische und die geometrische Gruppentheorie, Heldermann Verlag, 2008.

[CFR1] C. Carstensen, B. Fine, and G. Rosenberger, Abstract Algebra: with Applications to Algebraic Geometry and Cryptography, de Gruyter, Berlin, 2011.

[CFR2] C. Carstensen, B. Fine, and G. Rosenberger, *On asymptotic densities and generic properties in finitely generated groups*, Groups – Complexity – Cryptology **2** (2010), 113–121.

[CR]   CRAM-MD5, see http://en.wikipedia.org/wiki/CRAM-MD5

[Har]   B. Harris, RSA Key Exchange for the Secure Shell (SSH) Transport Layer Protocol, available at http://www.ietf.org/rfc/rfc4432.txt

[CJ]   J. Cheon, and B. Jun, *A polynomial time algorithm for the braid Diffie-Hellman conjugacy problem*, in: Proc. CRYPTO 2003, Springer Lect. Notes Comput. Sci. **2729**, 212–225.

[CFZ]   C. Chum, B. Fine, and X. Zhang, *Shamir's threshold scheme and its enhancements*, preprint 2011.

[CFRZ] C. Chum, B. Fine, G. Rosenberger, and X. Zhang, *A proposed alternative to the Shamir secret sharing scheme*, Contemp. Math. **582** (2012), 47–51.

[CP]   R. Crandall and C. Pomerance, Prime Numbers - A Computational Perspective, Springer Verlag, Berlin, 2001.

[Coh]   H. Cohn, A Course in Computational Algebraic Number Theory, Springer Verlag, Berlin, 1996.

[Cou]   N. T. Courtois, *How fast can be algebraic attacks on block ciphers*, in: E. Biham et al (eds.), Symmetric Cryptography (Dagstuhl 2007), Dagstuhl Sem. Proc. 07021

[DR]   J. Daemen and V. Rijmen, The Design of Rijndael, Springer Verlag, Berlin, 2002.

[Deh]   P. Dehornoy, *Braid-based cryptography*, Contemp. Math. **360** (2004), 5–34.

[DE]   A. Dickenstein and I. Z. Emiris, Solving Polynomial Equations, Springer Verlag, Berlin, 2005.

[Di1]   C. Diem, *On the discrete logarithm problem in class groups of curves*, Math. of Comput. **80** (2011), 443–475.

[Di2]   C. Diem, *On the discrete logarithm problem in elliptic curves II*, Alg. and Number Th. **7** (2013), 1281–1323.

[Die]   M. Dietzfelbinger, Primality testing in polynomial time. From randomized algorithms to "PRIMES is in P", Lect. Notes in Comp. Sci. **3000**, Springer Verlag, Berlin, 2004.

[DH]   W. Diffie and M. Hellman, *New directions in cryptography*, IEEE Trans. Inf. Theory **22** (1976), 644–654.

[EK]   B. Eick and D. Kahrobaei, *Polycyclic groups: a new platform for cryptology?*, preprint, available at http://arxiv.org/abs/math/0411077

[EO]   B. Eick and G. Ostheimer, *On the orbit-stabilizer problem for integral matrix actions of polycyclic groups*, Math. Comp. **72** (2003), 1511–1529.

[EM]   E. Elrifai and H. Morton, *Algorithms for positive braids*, Quart. J. Math. Oxford **45** (1994), 475–497.

[Eps]   D. B. A. Epstein, *Almost all subgroups of Lie groups are free*, J. Algebra **19** (1971), 261–262.

[Fau]   J.-C. Faugère, *Algebraic cryptanalysis of HFE using Gröbner bases*, INRIA Research Report **4738**, INRIA, Nancy, 2003.

[FGLM]   J.-C. Faugère, P. Gianni, D. Lazard, and T. Mora, *Efficient computation of zero-dimensional Gröbner bases by change of ordering*, J. Symb. Comput. **16** (1993), 329–344.

[FK]   M. R. Fellows and N. Koblitz, *Combinatorial cryptosystems galore!*, Contemp. Math. **168** (1996), 51–61.

[Fin]   B. Fine, The Algebraic Theory of the Bianchi Groups, Marcel-Dekker, 1989.

[FGR]   B. Fine, A. Gaglione, and G. Rosenberger, Abstract Algebra: From Groups, Rings and Numbers to Fields and Galois Theory, Johns Hopkins Press, 2014.

[FMoR]   B. Fine, A. I. S. Moldenhauer, and G. Rosenberger, *A secret sharing scheme based on the Closest Vector Theorem and a modification to a private key cryptosystem*, Groups – Complexity – Cryptology **5** (2013), 223–228.

[FMyR]   B. Fine, A. Myasnikov, and G. Rosenberger, *Generic subgroups of amalgams*, Groups – Complexity – Cryptology **1** (2009), 51–61.

[FR]   B. Fine and G. Rosenberger, Number Theory: An Introduction Using the Distribution of Primes, Birkhäuser, Basel, 2006.

[FP]   E. Formanek and C. Procesi, *The automorphism group of a free group is not linear*, J. Algebra **149** (1992), 494–499.

[Fre]   J. Freund, Mathematical Statistics, Miller and Miller, 2012.

[Gar]   D. Garber, Braid Group Cryptography, World Scientific Review Volume, 2008, available at http://arxiv.org/pdf/0711.3941.pdf

[Geb]   V. Gebhardt, *A new approach to the conjugacy problem in Garside groups*, J. Algebra **292** (2005), 282–303.

[GMO]   R. Gilman, A. G. Myasnikov, and D. Osin, *Exponentially generic subsets of groups*, Illinois J. Math. **54** (2010), 371–388.

[GGH]   O. Goldreich, S. Goldwasser and S. Halevi, *Collision-free hashing from lattice problems*, Electron. Colloq. Comput. Complex. **96**, 042 (1996)

[Gol]   R. Goldstein, *The density of small words in a free group is zero*, Contemp. Math. **360** (2004), 47–50.

[GMR]   S. Goldwasser, S. Micali, and R. Rivest, *A digitial signature scheme secure against chosen message attacks*, SIAM J. Comput. **17** (1988), 281–308.

[GP]   D. Grigoriev and I. Ponomarenko, *Homomorphic public-key cryptosystems over groups and rings*, Quaderni di Matematica **13** (2004), 305–325.

[HGS]   C. Hall, I. Goldberg, and B. Schneider, *Reaction attacks against several public key cryptosystems*, in: Proc. Information and Communications Security ICICS'99, Springer Verlag, Berlin, 1999, 2–12.

[HPS]   J. Hoffstein, J. Pipher, and J. Silverman, An Introduction to Mathematical Cryptography, Springer Verlag, Berlin, 2008.

[Hof]   P. Hoffman, Archimedes' Revenge, Fawcett Crest, 1988.

[HEO] D. Holt, B. Eick and E. O'Brien, Handbook of Computational Group Theory, Chapman and Hall/CRC Press, Boca Raton, 2005.

[HT] J. Hughes and A. Tannenbaum, *Length-based attacks for certain group based encryption rewriting systems*, preprint, 2003, available at http://arxiv.org/pdf/cs/0306032.pdf

[Jit] T. Jitsukawa, *Malnormal subgroups of free groups* , Contemp. Math. **298** (2002), 83–96.

[Joh] D. Johnson, Presentations of Groups, Cambridge Univ. Press, 1990.

[JK] P. Jovanovic and M. Kreuzer, *Algebraic attacks using SAT solvers*, Groups – Complexity – Cryptology **2** (2010), 247–259.

[KKh] D. Kahrobaei and B. Khan, *A non-commutative generalization of the ElGamal key exchange using polycyclic groups*, Proc. Conf. Globecom 2006, IEEE, NIS05-6.

[KMy] I. Kapovich and A. Myasnikov, *Stallings foldings and subgroups of free groups*, J. Algebra **248** (2003), 665–694.

[KKS] I. Kapovich, I. Kaimonovich, and P. Schupp, *The Subadditive Ergodic Theorem and generic stretching factors for free group automorphisms*, Israel J. Math. **157** (2007), 1–46.

[KCCL] K. H. Ko, D. Choi, M. Cho and J. Lee, *New signature scheme using conjugacy problem*, IACR Cryptology ePrint Archive **168** (2002), 1–13.

[KLCHKP] K. H. Ko, J. Lee, J. H. Cheon, J. W. Han, J. Kang, and C. Park, *New public-key cryptosystem using braid groups*, in: Advances in Cryptology, CRYPTO 2000, Lect. Notes Comp. Sci. **1880**, Springer Verlag, Berlin, 2000, 166–183.

[Kha] D. Kahn, The Codebreakers, Scribner, 1996

[KKi] C. Karpfinger and H. Kiechle, Kryptologie (in German), Vieweg and Teubner, 2010

[Kat] S. Katok, Fuchsian Groups, Univ. of Chicago Press, 1992.

[Ko1] N. Koblitz, A Course in Number Theory and Cryptography, Springer Verlag, Berlin, 1984.

[Ko2] N. Koblitz, Algebraic Aspects of Cryptography, Springer Verlag, Berlin, 1998.

[Ko3] N. Koblitz, *Elliptic curve cryptosystems*, Math. Comp. **48** (1987), 203–209.

[KMe] N. Koblitz and A. Menezes, *A survey of public key cryptosystems*, SIAM Review **46** (2004), 599–634.

[Kna] A. W. Knapp, Elliptic Curves, Princeton Univ. Press, 1992.

[Kra] D. Krammer, *Braid groups are linear*, Annals of Math. **151** (2002), 131–156.

[Kre] M. Kreuzer, *Algebraic attacks galore!*, Groups – Complexity – Cryptology **1** (2009), 231–259.

[KR1] M. Kreuzer and L. Robbiano, Computational Commutative Algebra 1, Springer Verlag, Berlin, 2000.

[KR2] M. Kreuzer and L. Robbiano, Computational Commutative Algebra 2, Springer Verlag, Berlin, 2005.

[LPW] T. Laarhoven, J. V. Pol, and B. Weger, *Solving hard lattice problems and the security of lattice-based cryptosystems*, preprint, 2012.

[Leh] J. Lehner, Discontinuous Groups and Automorphic Functions, Amer. Math. Soc., Providence, 1964.

[Lek] C. G. Lekkerkerker, Geometry of Numbers, North Holland, Amsterdam, 1969.

[Len] H. W. Lenstra, *Factoring integers with elliptic curves*, Ann. of Math. **126** (1987), 649–673.

[LLL] A. K. Lenstra, H. W. Lenstra, and L. Lovász, *Factoring polynomials with rational coefficients*, Math. Ann. **261** (1982), 515–534.

[LL1] A. K. Lenstra and H. W. Lenstra, *Algorithms in Number Theory*, in: Handbook of Theoretical Computer Science, Elsevier, 1990.

[LL2] A. K. Lenstra and H. W. Lenstra, The Development of the Number Field Sieve, Lect. Notes in Math. **1554**, Springer Verlag, Berlin, 1993.

[LS] R. Lyndon and P. Schupp, Combinatorial Group Theory, Springer Verlag, Berlin, 1978.

[LM] I. G. Lysenok, A. G. Myasnikov, *A polynomial bound on solutions of quadratic equations in free groups*, Proc. Steklov Inst. Math. **274** (2011), 136–173.

[MR] K. Madlener and B. Reinert, *Relating rewriting techniques on monoids and rings: Congruences on monoids and ideals in monoid rings*, Reports on Computeralgebra **14**, Universität Kaiserslautern, 1997.

[Mag] W. Magnus, *Rational representations of Fuchsian groups and non-parabolic subgroups of the modular group*, Nachr. Akad. Wiss, Göttingen II **9** (1973), 179–189.

[MKS] W. Magnus, A. Karrass, and D. Solitar, Combinatorial Group Theory, Wiley Interscience, New York, 1968.

[Mah] K. Mahlburg, *An overview of braid group cryptography*, preprint, 2004.

[Men] A. Menezes, Elliptic Curve Public Key Cryptosystems, CRC Press, 1997.

[MOV] A. Menezes, P. C. van Oorschot, and S. A. Vanstone, Handbook of Applied Cryptography, Kluwer Academic Publishers, 1993.

[Mil] C. F. Miller III, On Group-Theoretic Decision Problems and Their Classification, Princeton Univ. Press, 1971.

[MU] A. D. Myasnikov and A. Ushakov, *Length based attack and braid groups: Cryptanalysis of Anshel-Anshel-Goldfeld Key Exchange Protocol*, Proc. Conf. PKC 2007, Lect. Notes Comp. Sci. **4450**, Springer Verlag, 2007.

[MSU1] A. G. Myasnikov, V. Shpilrain, and A. Ushakov, Group-Based Cryptography, Adv. Courses in Math. – CRM Barcelona, Birkhäuser, Basel, 2008.

[MSU2] A. G. Myasnikov, V. Shpilrain, and A. Ushakov, *A practical attack on some braid group based cryptographic protocols*, Proc. Conf. CRYPTO 2005, Lect. Notes Comp. Sci. **3621** Springer Verlag, 2005, 86–96.

[New] M. Newman, Integral Matrics, Academic Pres, 1972.

[Pet] G. Petrides, *Cryptanalysis of the public key system based on the Grigorchuk groups*, Lect. Notes Comp. Sci. **2898**, Springer Verlag, 2003, 234–244.

[Pan] D. Panagopoulos, *A secret sharing scheme using groups*, available at http://arxiv.org/PScache/arxiv/pdf/1009/1009.0026v1.pdf.

[PP] The Prime Pages, see https://primes.utm.edu.

[Rai] T. Rai, *Infinite Gröbner bases and non-commutative Polly Cracker cryptosystems*, dissertation, Virginia Polytechnic Inst., Blacksburg, 2004.

[RSA] R. Rivest, A. Shamir, and L. Adelman, *A method for obtaining digital signatures and public-key cryptosystems*, Commun. ACM **21** (1978), 120–126.

[Rot] J. Rotman, An Introduction to the Theory of Groups, Springer Verlag, Berlin, 1999.

[Sc1] R. Schoof, *Elliptic curves over finite fields and the computation of square roots mod p*, Math. Comp. **44** (1985), 483–494.

[Sc2] R. Schoof, *Counting points on elliptic curves over finite fields*, J. Theor. Nombres Bordeaux **7** (1995), 219–254.

[Sch] B. Schoeneberg, Elliptic Modular Functions: An Introduction, Springer Verlag, Berlin, 1974.

[Ser] J. P. Serre, Trees, Springer Verlag, Berlin, 1980.

[Sha1] A. Shamir, *How to share a secret*, Commun. ACM **22** (1979), 612–613.

[Sha2] D. Shanks, *Class number, a theory of factorization and genera*, Proc. Symp. Pure Math. **20** (1971), 415–440.

[Sha3] C. E. Shannon, *Communication theory of secrecy systems*, Bell Sys. Tech. J. **28** (1949), 656–715.

[SZSI] J. Shikata, Y. Zheng, J. Suzuki, and H. Imai, *Realizing the Menezes-Okamoto-Vanstone (MOV) reduction efficiently for ordinary elliptic curves*, IEICE Trans. Fund., Vol. E **83-A4** (2000), 756–763.

[Sho] P. W. Shor, *Polynomial-time algorithms for prime factorization and discrete logarithms on a quantum computer*, SIAM J. Comput. **26** (1997), 1484–1509.

[SU]   V. Shpilrain and A. Ushakov, *The conjugacy search problem in public key cryptography: unnecessary and insufficient*, AAECC **17** (2006), 285–289.

[SZ]   V. Shpilrain and A. Zapata, *Using the subgroup membership problem in public key cryptography*, Contemp. Math. **418** (2006), 169–179.

[Sin]   S. Singh, The Code Book: The Science of Secrecy from Ancient Egypt to Quantum Cryptography, Harper Collins, 2000

[Sil]   J. H. Silverman, The Arithmetic of Elliptic Curves, Springer Verlag, Berlin, 1986.

[Ste]   R. Steinwandt, *Loopholes in two public key cryptosystems using the modular group*, in: Public Key Cryptography, Lect. Notes Comp. Sci. **1992** (2001), 180–189.

[Sti1]   D. R. Stinson, *An explication of secret sharing schemes*, Design, Codes and Cryptology **2** (1992), 357–390.

[Sti2]   D. R. Stinson, Cryptography: Theory and Practice, Chapman and Hall, 2002.

[Tho]   A. C. Thompson, Minkowski Geometry, Cambridge Univ. Press, Cambridge, 1996.

[vLy]   L. van Ly, *Gröbner Basen und das Kryptoverfahren Polly Two*, dissertation, University of Bochum, 2003.

[Wik]   *Challenge-Response Authentication*, Wikipedia: the Free Encyclopedia, see en.wikipedia.org/wiki/Challenge-response_authentication

[WM]   N. R. Wagner and M. R. Magyarik, *A public key cryptosystem based on the word problem*, Advances in Cryptology - CRYPTO 84, Lect. Notes Comp. Sci. **196**, Springer Verlag, Berlin, 1985, 19–36.

[Wag]   S. J. Wagstaff, Cryptanalysis of Number Theoretic Ciphers, Taylor and Francis, 2002.

[Was]   L. C. Washington, Elliptic Curves: Number Theory and Cryptography, Chapman and Hall, 2003.

[Wat]   W. C. Waterhouse, *Abelian varieties over finite fields*, Ann. Sci. Ec. Norm. Sup., Ser. 4 **2** (1969), 521–560.

[Xiu]   X. Xiu, *Non-commutative Gröbner bases and applications*, dissertation, Universität Passau, 2012.

[Xu]   X. Xu, *Cryptography and infinite group theory*, Ph.D. thesis, CUNY, 2006.

[Yam]   A. Yamamura, *Public key cryptosystems using the modular group*, Lect. Notes Comp. Sci. **1431**, Springer Verlag, Berlin, 1998, 203–216.

Index

79349367R00216

Made in the USA
Middletown, DE
09 July 2018